Dictionary of
SPACE

★

Malcolm Plant

Dictionary of
SPACE

★

Longman

About the author

Malcolm Plant, a physicist, is Principal
Lecturer at Trent Polytechnic, Nottingham.
His particular research interest is in space
physics and he is a Fellow of the British
Interplanetary Society. He has contributed to
various Schools' Council (School Curriculum
Development Committee) projects and to 'O'
and 'A' level examination syllabuses. His
publications include *Basic Electronics* (1978),
Instrumentation (1982), and *Microelectronics
A to Z* (1985).

Longman Group Limited,
*Longman House, Burnt Mill, Harlow,
Essex CM20 2JE, England
and Associated Companies throughout the world.*

First published 1986

British Library Cataloguing in Publication Data

Plant, M. (Malcolm)
  Longman dictionary of space.
  1. Outer space
  I. Title
  523   QB500

  ISBN 0-582-89296-1 Cased
  ISBN 0-582-89295-3 Pbk

Computer typeset in 9/10pt Linotron Times
by Computerset (MFK) Ltd, Ely, Cambs.
Printed in Great Britain
by Mackays Ltd, Chatham, Kent.

Illustrated by Jerry Collins

Contents

Illustrations

Abbreviations and symbols used in this dictionary

abbr	abbreviation or abbreviated form	m	milli-
AU	astronomical unit	M	mega-
b	born	M	million
C	celsius	mm	millimetre
esp	especially	ms^{-1}	metre per second
eV	electron volt	ms^{-2}	metre per second per second
fem	feminine	n	nano-
ft	foot	N	newton
in	inch	nm	nanometre
g	acceleration due to gravity	s	second
g	gram	specif	specifically
G	giga-	US	United States
Hz	hertz	USA	United States of America
J	joule	USSR	Union of Soviet Socialist Republics
k	kilo-	usu	usually
K	kelvin	V	volt
kg	kilogram	W	watt
kgm^{-3}	kilogram per cubic metre	Wm^{-2}	watt per square metre
km	kilometre		
kmh^{-1}	kilometre per hour	%	per cent
kms^{-1}	kilometre per second	°	degree
m	metre	′	minute

SMALL CAPITALS refer the reader to other entries

☐   a square box followed by 'see' and SMALL CAPITALS refers to other entries where the reader will find related information

[ ]   square brackets enclose the full forms of abbreviated terms entered in the dictionary

( )   round brackets enclose information about abbreviations, acronyms, and symbols for terms entered in the dictionary

⟨ ⟩   angle brackets enclose a phrase illustrating a typical use of the term in context. The term being illustrated is represented by a swung dash (∼).

Introduction

WHEN SPUTNIK CIRCLED THE EARTH nearly 30 years ago, its bleeps signalled the beginning of one of mankind's greatest adventures–the exploration of space.

Since 1957 an extraordinary variety of artificial satellites has followed Sputnik into space. Unimpeded by the clouds and dust of the Earth's atmosphere, the satellites have made dramatic discoveries. Their highly-sensitive instruments have detected the birthplaces of stars and the swiftly expanding remnants of gas and dust which mark their catastrophic end as supernovae. Powerful beams of X rays and radio waves signal the presence of rapidly rotating pulsars and space-warping black holes in distant parts of our own galaxy. And the glimmer of infrared radiation could mean Earth-like planets orbit other stars.

As well as finding out what goes on in the universe on the grand scale, very successful excursions have been made to most of the planets of our own solar system. The Soviet Venera spacecraft have reported that the climate of our sister planet, Venus, is far from suitable for life as we know it. The Red Planet, Mars, has a more promising climate, but the American Viking lander searched unsuccessfully for life there. And during its fifteen year 'planet hopping' interplanetary tour, the Voyager spacecraft has navigated past Jupiter, Saturn, and Uranus, and is soon to reach Neptune. Voyager has transmitted impressive pictures of Jupiter's Red Spot and Saturn's rings to Earth. In the next few years, the unmanned Galileo, Cassini, and Kepler missions to the planets will continue these voyages of discovery. By the end of the 1980s, all the planets except Pluto will have been visited by spacecraft.

Compared with visiting planets, it is much more difficult to rendezvous with fast-moving dusty and diffuse comets. Yet a fleet of spacecraft headed out to a meeting with Halley's comet as it swooped round the Sun. One of these spacecraft, Giotto, slipped through the tail to give us our first glimpse of this famous comet at close-quarters–what would Edmund Halley have made of that!

Learning more about what happens in the space around us is not the sole concern of spacefaring nations. From the start of the Space Age, it was realised that satellites offer a splendid bird's-eye view of the Earth's weather. And the potential of communications satellites in geostationary orbit for the transmission of TV and telephone calls was soon exploited. Indeed, communications satellites are now the main commercial spin-off from space exploration, and countries that have powerful rockets can launch satellites for nations that don't have the means to do so.

A number of satellites (eg Landsat) have been designed for the increasingly important job of searching for new mineral and energy resources and monitoring agricultural productivity and environmental pollution. It is to be hoped that any new wealth discovered in this way is not exploited wastefully but is managed to the long-term advantage of the people to whom it belongs.

Another way of capitalising on space exploration is to use the weightless conditions found in orbiting space stations to make products which are impossible to manufacture on Earth. These products include extremely pure crystals of semiconductors to enhance the performance of electronic equipment, new pharmaceutical products to improve the treatment of diseases, and new materials of remarkable strength.

Men and women from several nations have travelled into space in fuelled and provisioned craft much as the early mariners crossed the oceans to find new lands. Indeed many nautical terms are used to describe these journeys: space tugs, fuel tankers, and cargo vessels ferry supplies and equipment to orbiting space stations; and future interstellar spaceships may well hoist solar sails to catch solar wind!

Five hundred years after Columbus set sail with three small ships to the 'Indies', we are embarking on a similar voyage of discovery. But now we should be wiser in the use we make of new discoveries. Astronauts and cosmonauts say that 'spaceship' Earth looks good from space. Their pictures are a timely reminder that we live on a planet of finite resources that supports a remarkably complex and possibly unique assortment of lifeforms that have evolved over millions of years. To use the high ground of space to find more effective ways of delivering death blows to whole nations, or in the greedy exploitation of the Solar System, is to be avoided at all costs.

In thirty years we have learned far more about the universe than in the previous three thousand years of astronomical observations. What shall we know in another thirty years? Will our expanding horizons enable us to understand ourselves better and to solve some of the political and environmental problems on Earth? And how shall we behave if we meet extraterrestrial life in our travels?

This dictionary reflects the exciting developments and discoveries of the Space Age. It describes the latest findings about celestial bodies–stars, planets, galaxies, supernovae, black holes, etc–and the instruments sent aloft to study them. It explains the purpose of numerous satellites and spacecraft launched by increasingly powerful rockets which have carried those instruments into space. Relevant engineering principles and scientific concepts are also explained. And above all it includes the names of some of the men and women (and animals!) who have ventured into space, and their achievements. 'Dictionary of Space' has been written to keep anyone, casual reader or keen expert, in touch with all aspects of space exploration.

Malcolm Plant
November 1985

We shall not cease from exploration
And the end of all our exploring
Will be to arrive where we started
And know the place for the first time

T S Eliot
'Little Gidding', *Four Quartets*

A

AAM – see AIR-TO-AIR MISSILE

aberration 1 anything that prevents an optical system from producing a perfect image. For example, a properly designed telescope should produce a point image of a star (not including the Sun), but any of several aberrations causing blurring, coloured edges, etc may be present. □ see SPHERICAL ABERRATION, COMA 2, CHROMATIC ABBERATION **2** the apparent displacement of a star from its true position as a result of the motion of the Earth round the Sun. The maximum displacement is very small (20.5 arc seconds) and was first noticed and measured in 1729 by the English astronomer James Bradley who, during a sailing boat trip, realized that the light from a star and the Earth's speed was analogous to the combination of the wind and the boat's speed. □ see ARC SECOND

ablate to remove or be removed by cutting, erosion, melting, evaporation, or vaporization

ablating material a material covering part of the surface of an aerospace vehicle (eg the Space Shuttle) to help protect the vehicle from excessive heating due to hypersonic flight through a planetary atmosphere. The way an ablating material works is complex and involves a number of physical factors. The main process is the vaporization or melting of the material so that it takes part of the heat away with it. □ see THERMAL TILES, SPACE SHUTTLE, NOSE CONE

Able and Baker a 3kg rhesus monkey and a 300g squirrel monkey, both female, who were the first animals to survive a 480km high suborbital flight on 28 May 1959. (The Soviet dog, Laika, perished during her orbital flight in 1957). The monkeys were sent aloft in the nose cone of a Jupiter rocket and splashed down near the Caribbean island of Antigua. Though Miss Able died a few days later, Miss Baker survived and became known as America's 'first lady of Space'. She died at Auburn University, Alabama, on 29 November 1984 following a kidney failure. □ see LAIKA, ANITA AND ARABELLA, ABREK AND BION

ABM – see ANTIBALLISTIC MISSILE

abort to terminate preparations for the launch of a rocket, usu because of equipment failure □ see COUNTDOWN, MALFUNCTION

Abrek and Bion two (Soviet) monkeys launched into orbit on 14 December 1983 in Cosmos 1514 and returned to Earth six days later after they had been used to investigate adaption to weightlessness □ see COSMOS, ABLE AND BAKER, LAIKA

absolute 1 complete ⟨~ *vacuum*⟩ **2** of or being measurements related to a fundamental property ⟨~ *temperature*⟩

absolute temperature scale a scale of temperature that has the lowest possible temperature (absolute zero) at zero degrees kelvin (0K) and that has the melting point of ice at about 273 kelvin. Tem-

1

peratures on the absolute temperature scale are obtained from the celsius (formerly centigrade) scale by adding the number 273. Thus −273 celsius (−273C) is equal to 0K (absolute zero): 15C is equal to 288K, and so on. The absolute temperature scale is the natural scale for science and technology. □ see CRYOGENIC TEMPERATURE, MICROWAVE BACKGROUND RADIATION

absorption nebula – see DARK NEBULA

absorption spectrum a spectrum formed when electromagnetic radiation has been selectively absorbed by passage through an absorbing medium. The absorption spectrum of a star comprises a set of dark lines. Each line represents a narrow band of wavelengths which has been removed from the star's continuous spectrum by chemical elements and compounds in its cooler outer layers. Since the wavelengths of the dark lines are characteristic of the absorbing gas, astronomers can work out the elements present in the star's atmosphere – compare EMISSION SPECTRUM. □ SEE ELECTROMAGNETIC SPECTRUM, CONTINUOUS SPECTRUM, SPECTROMETER, FRAUNHOFER LINES

abundance – see COSMIC ABUNDANCE, NUCLEOSYNTHESIS, BIG BANG

accretion the growth of a larger body from smaller bodies □ see PROTOPLANET

accuracy the nearness of a measured value to its true value ⟨*the ~ of measuring the Gravitational Constant*⟩ – compare PRECISION

acquisition the process of tracking the path of a spacecraft or satellite using aerials and radio receivers, so that information can be gathered from it □ see TELEMETRY, DISH AERIAL, RADAR, PASS, GROUND-BASED ELECTRO-OPTICAL DEEP SPACE SURVEILLANCE

Active Magnetospheric Particle Explorer (abbr **AMPTE**) a three satellite programme launched in August 1984 and designed to study and explore the complex region of charged particles (plasma) and magnetic fields (the magnetosphere) that envelop the Earth out to distances beyond 100 000km. The US-built satellite AMPTE-1 (also known as Charge Composition Explorer) and the British-built AMPTE-3 (also known as United Kingdom Satellite) were equipped to detect lithium and barium tracer ions released at intervals by the German-built satellite AMPTE-2 (also known as Ion Release Module). On 18 July 1985, AMPTE-2 released a cloud of barium ions at an altitude of 110 000km above the Pacific Ocean. The ions formed an artificial comet as the solar wind drove the plasma away from the Sun. These experiments helped scientists to understand more fully how the plasma of ions in the magnetosphere interacts and moves under the influence of the solar wind. □ see ION, PLASMA, MAGNETOSPHERE

active satellite a satellite that transmits a radio signal (as most do) rather than merely reflecting signals beamed to it □ see SATELLITE LASER-RANGING

actuator – see SERVOMECHANISM

adapter skirt a ringlike extension round one stage of a launch vehi-

cle so that another stage can be fitted to it □ see STAGE, ROCKET
ENGINE

ADC – see ANALOGUE-TO-DIGITAL CONVERTER

Advanced X-ray Astrophysics Facility (abbr **AXAF**) a 9000kg
X-ray observatory which NASA plans to start building in 1988 and
launch into low-Earth orbit in 1993. The telescope will have an aperture of 1.2m and a focal length of 12m. AXAF will be deployed from
the Space Shuttle at an altitude of about 600km using the robot arm of
the Remote Manipulator System (RMS). A number of instruments at
the focus of the telescope will examine the energy and intensity of
X rays in the wavelength band 1.5 to 120 angstroms, emitted by black
holes, neutron stars, pulsars, supernovae remnants, and mysterious
X-ray 'bursters' from globular clusters. The AXAF will join the Hubble Space Telescope (HST) launched in 1986, and the Gamma-ray
Observatory (GRO) to be launched in 1988. These three orbiting
observatories and the Space Infrared Telescope Facility, which is
likely to be launched after AXAF, will help astrophysicists to under-
stand the complex processes inside stars and galaxies. □ see X-RAY
ASTRONOMY, GRAZING INCIDENCE X-RAY TELESCOPE, ANGSTROM,
X-RAY MULTI-MIRROR TELESCOPE, HUBBLE SPACE TELESCOPE,
GAMMA-RAY OBSERVATORY

aeon one billion (10^9) years ⟨*the Solar System is about 4.5 ~s old*⟩
□ see BIG BANG

aerial an arrangement of conductors usually placed in an elevated
position (eg on a mast), which transmits and/or receives radio waves

aeroballistics the study of how high-speed projectiles (eg ballistic
missiles) and aerospace vehicles (eg the Space Shuttle) behave in the
Earth's atmosphere or the atmosphere of another planet

aerocentric having Mars as a centre – compare GEOCENTRIC

aerodynamics the study of how bodies behave when moving
through gases (eg air) ⟨*the ~ of a Martian spaceship*⟩ □ see
AERONAUTICS, AEROBALLISTICS

aeroembolism *also* **decompression sickness, bends** the formation of
bubbles of gas (mostly nitrogen) in the blood of a pilot or astronaut
when there is a sudden drop in air pressure. Aeroembolism causes
pains in the chest and joints, cramps, swellings, and sometimes death.
□ see SPACE MEDICINE

aerofoil a body (eg the wing of the Space Shuttle) which is designed
to provide lift, drag, or some other aerodynamic force, when it moves
through air □ see SPOILER, CANARD, SPAN

aerography the study of the surface features of Mars

aeronautics the study of flight and of the aircraft and spacecraft
that operate in the Earth's atmosphere

aeronomy the study of the Earth's upper atmosphere, and the
atmospheres of other planets, where solar radiation causes gases to
ionize □ see IONIZATION, EARTH'S ATMOSPHERE

aerospace (of or relating to) the Earth's atmosphere and the space
beyond regarded as one environment ⟨*the Space Shuttle is an ~ vehi-*

cle⟩ □ see BRITISH AEROSPACE

aerospace vehicle a vehicle (eg the Space Shuttle) that is capable of operating both within the atmosphere and in outer space □ see SPACE SHUTTLE

AF – see AUDIO FREQUENCY

after body something that follows a satellite or spacecraft. The term includes any part of a launch vehicle or rocket that enters the atmosphere unprotected behind the protected front end of a vehicle. It also includes any unidentified object (eg a UFO) that may trail a manmade aerospace vehicle.

age of universe – see HUBBLE'S LAW, COSMOLOGY, BIG BANG, MAGIC NUMBER

AGORA [*A*steroid *G*ravity, *O*ptical and *R*adar *A*nalysis] – see EUROPEAN SPACE AGENCY

agravic of or being a condition in which there are no gravitational forces □ see ZERO-G, MICROGRAVITY

AI – see ARTIFICIAL INTELLIGENCE

airbreather a rocket engine in which the fuel is oxidized by air taken in from the atmosphere rather than by its own supply of oxygen. A cruise missile is an example of an air breather, while Ariane, for example, carries its own supply of oxygen so that it can operate independently of its surroundings. British Aerospace is developing a rocket called HOTOL, which would breathe air on its way up through the atmosphere. □ see BRITISH AEROSPACE, HOTOL

airflow a stream of air ⟨∼ *in a wind tunnel*⟩ □ see SUPERSONIC, HYPERSONIC, STREAMLINE FLOW

airframe the structure and aerodynamic surfaces of an aerospace vehicle, not including the engine and its various subsystems

airglow a faint light in the Earth's atmosphere which is most apparent at night. This background airglow can be a nuisance when making observations of faint celestial objects with Earth-based telescopes, and is one reason for putting telescopes in space above the atmosphere. Airglow is produced when atoms of oxygen and nitrogen, previously ionized by high energy radiation from the Sun, recombine to form neutral atoms. □ see HUBBLE SPACE TELESCOPE, IONIZATION, AURORA

airlock a (usu small) chamber in a spacecraft or space station which allows astronauts and equipment to pass between outer space and the spacecraft without depressurizing the spacecraft. An airlock is a standard module on board the Space Shuttle, Salyut, and other space stations, and permits astronauts and cosmonauts to spacewalk.

air shower – see COSMIC RAYS

airspace the atmosphere above a certain area of the Earth's surface, esp that controlled by a nation underneath ⟨*Russian* ∼⟩ □ see EARTH'S ATMOSPHERE

air-to-air missile (abbr **AAM**) a missile launched from an aircraft at a flying target

air-to-surface missile (abbr **ASM**) a missile launched from an aircraft at a target on the surface of the Earth

Airy disc a disc-shaped image of a star, or other point source of light, formed in the focal plane of a circular objective lens or mirror of an optical system (eg a telescope). The disc, rather than a point, is caused by the diffraction of light, and it is surrounded by fainter rings of light. About 87% of the total light entering the objective falls within the disc, the rest in the rings. The larger the diameter of the objective the smaller the Airy disc, and the better the resolving power of the optical system. Therefore, telescopes that have large diameters are not only able to collect more light, and therefore see fainter celestial objects, but they are also able to see more detail in the objects. □ see APERTURE, OBJECTIVE, RESOLVING POWER, HUBBLE SPACE TELESCOPE

albedo *also* **reflectance** the fraction of the incident light reflected from the surface of a nonluminous body such as a planet. Thus a surface having an albedo of 1.0 reflects all the light incident on it; an albedo of 0.5 reflects half the light incident on it, and so on. □ see CALLISTO

alpha (symbol α) the first letter of the Greek alphabet, used to indicate (**a**) the brightest star in a constellation of stars ⟨~ *Centauri is the brightest star in the constellation Centaurus and the second nearest star to our Sun* (**b**) the right ascension of a celestial object in the sky □ see RIGHT ASCENSION

Alpha Carina – see CANOPUS

ALSEP – see APOLLO LUNAR SURFACE EXPERIMENTS PACKAGE

altazimuth mounting a method of mounting a telescope so that it is free to move about a vertical axis (ie in azimuth), and a horizontal axis (ie in altitude). The altazimuth mounting is easier than an equatorial mounting to make; it enables a telescope to be moved quickly and is ideal for tracking artificial satellites. But since both altitude and azimuth must be adjusted to follow the path of a star across the sky, this type of mounting is difficult to operate by hand. The recent Multiple Mirror Telescope (MMT) and the Soviet 6m reflecting telescope have an altazimuth mounting, but they both make use of computer control mechanisms to accurately adjust the two axes simultaneously – compare EQUATORIAL MOUNTING □ SEE MULTIPLE MIRROR TELESCOPE, REFLECTING TELESCOPE

altimeter an instrument for measuring height usu above sea level ⟨*an ~ enables a spacecraft to make a controlled landing on a planet's surface*⟩ □ see RADAR

altitude 1 the angular distance, measured in radians or degrees, of a celestial object above or below the horizon – compare AZIMUTH **2** the height of an object above mean sea level ⟨*geostationary orbit is at an ~ of 35 900km*⟩ □ see CELESTIAL COORDINATES, RADIAN, GEOSTATIONARY ORBIT

aluminizing a process whereby a very thin and uniform layer of aluminium is deposited on a surface by evaporation. There are two main applications for this process: in astronomy, optical components, esp the main mirror of a telescope, are coated with a highly reflective aluminium film once the surface has been ground to the required shape; in the final stages of the preparation of an integrated circuit on a silicon chip, aluminium is deposited and then etched to produce the network of electrical connections between components on the chip. In both applications the aluminium is deposited on the surface in a vacuum chamber. □ see TELESCOPE

ambient surrounding on all sides ⟨*the ~ pressure on a spacecraft on the surface of Venus is about 90 times greater than that on the Earth's surface*⟩

American Society of Aerospace Pilots (abbr ASAP) a nonprofit organization based in Illinois which runs the first commercial spaceflight ground school for training future space pilots. Most of the trainee pilots are taken from the pool of airline pilots.

amplitude the range between the maximum and minimum values of something. In astronomy, amplitude is used to indicate the variation in brightness of a variable star. □ see MAGNITUDE

AMPTE – see ACTIVE MAGNETOSPHERIC PARTICLE EXPLORER

analogue 1 of or being information having a continuous range of values ⟨*atmospheric pressure is an ~ quantity*⟩ **2** of or being a device responding to or displaying information as a continuous range of values ⟨*an ~ watch*⟩ – compare DIGITAL

analogue computer a largely obsolete type of computer that responds to and produces continuously varying signals – compare DIGITAL COMPUTER

analogue display a readout of the value of something (eg the strength of a received radio signal) as a pointer moving over a continuous scale or as a band of light which lengthens or shortens – compare DIGITAL DISPLAY

analogue-to-digital converter (abbr ADC) *also* **digitizer** a device or circuit for changing an analogue signal into a coded digital signal. ADCs are commonly employed in instrumentation systems (eg weather satellites) to convert analogue information (eg temperature, pressure, and speed) into digital information for computer processing or for transmission in a communications system.

android a mechanical computer-controlled robot that looks human and walks and talks ⟨*R2-D2 was an ~ in the film* Star Wars⟩ □ see ROBOT

Andromeda galaxy a spiral galaxy just visible with the unaided eye as a fuzzy patch in the constellation of Andromeda. It is the largest galaxy in the Local Group and is about 2.2 million light-years from Earth. □ see GALAXY, HUBBLE, CLUSTER 3

anemometer an (electronic) instrument for measuring the speed or force of the wind. An anemometer is an essential part of an instrument package landed on another planet (eg by the Viking spacecraft on Mars, and Venera spacecraft on Venus). □ see TACHOMETER, INSTRUMENTATION

angels a colloquial aerospace term for images (radar echoes) on a radar screen caused by something unseen □ see RADAR, UFO

angle of incidence the angle at which radiation strikes a surface (eg a mirror). It is measured between the direction of the radiation and the perpendicular to the surface at the point of incidence. □ see SATELLITE LASER-RANGING

angle of reflection the angle at which reflected radiation leaves a surface (eg a mirror). It is measured between the direction of the reflected radiation and the perpendicular to the surface at the point of reflection.

angstrom *also* **angstrom unit** a distance equal to one ten thousand millionth of a metre (10^{-10}m), used for measuring the wavelength of visible light. Optical astronomy conducted from Earth-based telescopes receives electromagnetic radiation through a narrow optical window in the atmosphere. This window extends from about 3000 to 10 000 angstrom units. It should be no surprise to learn that the human eye has adapted to this range and has its maximum sensitivity at about 5500 angstrom units which is the colour of yellow-green light, the maximum emission wavelength of sunlight. The unit was named after A J Angstrom (1814–74), a Swedish physicist who spent his life teaching physics and astronomy and who studied the composition of light from the Sun and stars. □ see OPTICAL ASTRONOMY, ELECTROMAGNETIC SPECTRUM, HUBBLE SPACE TELESCOPE

angular diameter – see APPARENT DIAMETER

angular momentum (symbol L) the property of any body, or system of bodies, that is rotating. The value of the angular momentum depends on the distribution of the mass and speeds of the bodies about their centre of rotation. Angular momentum is defined as the product of the moment of inertia, I, of the body or bodies, and the angular velocity, ω, ie L = Iω. Angular momentum is one of the fundamental properties of elementary particles (eg electrons). □ see INERTIA, ANGULAR VELOCITY, SPIN

angular resolution – see RESOLVING POWER

angular velocity (symbol ω) the rate at which a body rotates about its axis, usu expressed in radians per second ⟨*a body rotating 3 times per second, has an* ~ *of* $3 \times 2\pi = 18.8$ *radians per second*⟩ □ see RADIAN

Anik any of a continuing series of satellites used to improve communications between Canada's widely distributed population – Canada's 24.3 million people are unevenly scattered over the second largest geographical area in the world. The first Anik satellite, Anik A–1, was launched by Delta rocket on 17 November 1972 and began commercial operations on 11 January 1973 from geostationary orbit. Anik A–2 and Anik A–3 were launched in 1973 and 1975. These were three identical cylindrical and spin-stabilized satellites, 1.8m in diameter and 3.4m high, and 87% built by Hughes Aircraft Company in California. They each had a despun oval dish antenna, and were powered by solar cells which completely covered their bodies. Their 12 channels operated uplink at 6GHz, downlink at 4GHz. This series

was followed by Hermes (1976) and Anik B (1978) which were largely
used to test the sociocultural use of communications satellites, as well
as their technical aspects. In this decade, the Anik C and D satellites
have been launched by the Space Shuttle. Anik C–3 was launched in
November 1982, before Anik C–1 and C–2, on mission STS–5. Anik
C–2 (also known as Telesat 7, after the Canadian organization which
administers the satellite system) was launched on STS–7 in June 1983,
and Anik C–1 in April 1985 on Mission 51D. These Anik C satellites
have a design life of 10 years, are 6.4m high, 2.13m in diameter and
have a mass of 1080kg at launch. The second of the slightly heavier
and larger Anik D satellites, Anik D–2, was launched on Mission 51A
in November 1984, Anik D–1 having been launched by Delta rocket in
August 1982. Anik D–2 was 'stored' in orbit for up to 2 years until
needed. As well as distributing data, and voice and television signals
to all parts of Canada, including remote communities in its northern
territories, the Anik satellites are used to distribute cable TV to
almost every city in the country via Earth stations. The name 'Anik'
was chosen in a nationwide contest: it is the Eskimo word for brother.

Anita and Arabella two spiders taken aboard the US space station, Skylab, in May 1973 so that their skill at spinning webs in zero
gravity could be observed. School children often select spiders when
they are invited to suggest experiments to be carried out during flights
of the Space Shuttle. □ see ABLE AND BAKER, LAIKA

annihilation the reaction between an elementary particle (eg an
electron) and its antiparticle (a positron) to produce electromagnetic
energy. For example, annihilation between an electron and positron
produces two gamma-ray quanta each having an energy of 0.51 million
electron-volts – compare PAIR PRODUCTION □ SEE COSMIC RAYS,
ANTIMATTER, POSITRON, PROTON, MASS-ENERGY EQUIVALENCE,
ELECTRON VOLT, QUANTUM THEORY

anode the terminal of a device (eg a cathode-ray tube) towards
which electrons flow – compare CATHODE

antenna (plural **antennae**) a relatively complex form of aerial used
in specialized radio communications ⟨*a satellite tracking* ∼⟩ □ see
AERIAL

anthropic principle a principle stating that the existence of life on
Earth determines the structure of the universe. This principle does not
mean that we actively shape the universe to our taste; rather that time
must pass before intelligent life can develop and equip itself with the
technology needed to observe the universe in its present form. For
example, life on Earth is dependent on the element carbon. Oxygen
and nitrogen are also vital chemical building blocks of life. These elements were not present in the primaeval universe, but accumulate
inside stars by nucleosynthesis. And if the star is rather massive, at the
end of its life it will explode violently as a supernova and spread carbon and the other elements of life throughout interstellar space.
Eventually these elements become the stuff from which solar systems
are made; and from the ashes of these long-dead stars, life evolves to

see the universe as it is now. Thus during the evolution of the
universe, from the fiery furnace of the Big Bang to a widely dispersed
collection of burnt-out galaxies, life (as we know it) can only exist for
a limited time span. There are two main objections to this argument
that our temporal location in the universe is constrained by the fact
that the universe evolves. First, it is possible for intelligent life to be
based on processes not involving carbon and, second, human beings
may be able to develop technologies capable of sustaining life well
beyond the time when all the stars have burnt out. □ see CARBON,
NUCLEOSYNTHESIS, SUPERNOVA, MAGIC NUMBER, UNIVERSE,
ASTROBIOLOGY

antiballistic missile (abbr ABM) a missile for intercepting and
destroying a ballistic missile in flight □ see BALLISTIC MISSILE, INTER-
CONTINENTAL BALLISTIC MISSILE

antielectron – see POSITRON

antimatter matter in which the ordinary particles are replaced by
their antiparticles. For example, ordinary hydrogen gas is composed of
protons and electrons. If the hydrogen were composed of antimatter,
the electrons would be replaced by positrons and the protons by
antiprotons. If matter and antimatter were to come together, they
would mutually annihilate each other with the release of energy.
Although antiparticles have been found for most of the particles in our
universe (many are produced in cosmic ray showers and particle
accelerators), the search for antimatter in the universe goes on
through ground-based and space-based telescopes. □ see ANNIHILA-
TION, POSITRON, PROTON, COSMIC RAYS, BIG BANG

antimatter propulsion the (proposed) use of the matter/antimat-
ter reaction to produce a high-speed rocket engine suitable for
journeys to the stars. One study of this idea suggests that the charged
pions, produced by the mutual annihilation of protons (matter) and
antiprotons (antimatter), could be directed by magnetic fields into a
high-speed beam providing the necessary forward thrust. However,
rather than using equal quantities of protons and antiprotons, a
smaller quantity of antiprotons could be used to heat and accelerate a
much larger amount of propellant (eg liquid hydrogen which contains
protons). Such a rocket engine could propel a 1-tonne spacecraft to
the nearest star, Alpha Centauri, in about 22 years. The rocket would
use 20 tonnes of liquid hydrogen, and 180kg of antihydrogen
(hydrogen atoms having antiproton nuclei), to accelerate the
spacecraft to about two-tenths of the speed of light in less than one
year. A rocket flight using conventional propellants would take 100
thousand years to reach Alpha Centauri. The idea of an antimatter
rocket is not so far-fetched. The European Centre for Nuclear
Research (CERN) includes a 7km diameter synchrotron straddling the
Swiss-French border near Geneva. Beams of antiprotons are readily
produced by this machine and stored in a magnetic storage ring. However, the cooling and storage of antihydrogen on board a 'starship'
presents considerable engineering problems, for any ordinary matter it

came into contact with would result in a violent explosion. But who would have thought, less than 50 years ago, that we would now be using, as a matter of course, highly reactive liquid hydrogen and oxygen in our rockets? □ see ANTIMATTER, ANNIHILATION, ELECTRIC PROPULSION, ELEMENTARY PARTICLES, HYPERSPACE

antineutrino the antimatter equivalent of the neutrino □ see NEUTRINO, QUARK, ANTIMATTER

antineutron the antiparticle equivalent of the neutron □ see NEUTRON, ANTIMATTER

antiparticle an uncommon opposite form of the more common elementary particles like electrons, protons, and neutrons. For example, the positron has the same mass as an electron, but it has an equal and opposite (ie positive) electrical charge. □ see ANTIMATTER, ANNIHILATION, PAIR PRODUCTION, PROTON QUARK

antiproton the antiparticle equivalent of the proton □ see PROTON, ANTIMATTER

antislosh baffle an important mechanical device (eg a vane) built into liquid hydrogen and liquid oxygen propellant tanks to prevent the liquids sloshing around □ see EXTERNAL TANK

apastron the point in the orbit of a body about a star that is furthest from the star. The closest distance is called the periastron. The term is mainly used in connection with the orbit of one member of a binary star system. □ see BINARY STAR, PERIHELION

apereon that point in an orbit about Mars at which a satellite or spacecraft is at its greatest distance from the planet □ see MARS

aperture the diameter of the lens, mirror, or dish aerial of a telescope. The apertures of optical and radio telescopes have increased over the years so as to collect more light or radio waves from faint celestial objects. Increasing the aperture of a telescope has two advantages: it increases the telescope's sensitivity by increasing its radiation-collecting area – the sensitivity is proportional to the area of the lens, mirror, or dish, (eg doubling the aperture quadruples the telescope's sensitivity); and it increases its resolving power so enabling more detail to be seen in these objects, (eg doubling the aperture doubles the amount of detail that can be seen). Since it is impractical to build large single mirrors and dish aerials, the technique of combining several small collectors to produce, in effect, a larger collector is becoming popular. In optical astronomy, several small computer-controlled mirror segments can be combined to produce what is known as a multimirror telescope. In radio astronomy several separated dish aerials can be used to increase sensitivity and resolving power in a technique known as aperture synthesis. This technique was used to receive data from Voyager 2's flyby of Uranus in January 1986. The Australian Parkes 64m radio telescope was arrayed with the Australian DSN 64m dish aerial 300km distant. Combining radio telescopes in this way will be all the more important when Voyager 2 reaches Neptune in August 1989, a distance twice that of Uranus from Earth, which reduces the signal strength of the data received by a factor of four. □ see OPTICAL ASTRONOMY, RADIO ASTRONOMY, RESOLVING POWER, AIRY DISC, MULTIPLE MIRROR TELESCOPE, APERTURE SYNTHESIS, VERY LARGE ARRAY

aperture ratio the ratio of the useful diameter of a lens to its focal length. Two lenses that have the same aperture ratio provide the same exposure of film.

aperture synthesis a technique used in radio astronomy for obtaining detailed radio maps of cosmic radio sources by combining the signals from two or more separated radio antennae. The amplitudes and phases of these signals are fed to a computer which produces interference patterns from which the radio map is produced. The Five Kilometre radio telescope at Cambridge, England, uses four movable and four fixed dish aerials, mounted on an east-west railway line, which gives 16 interferometer pairs, as they are called. The dishes are programmed to track a radio source (eg a quasar) for 12 hours during which time the baseline for each interferometer pair rotates by half a revolution. By moving the four movable antennae each half day, a dish aerial of much larger aperture, and hence resolution, can be synthesized. Since the resolving power obtained by aperture synthesis increases with increasing distance between the antennae (it is then called very long baseline interferometry), it seems natural to produce an interferometer pair by placing one aerial in space aboard a satellite to give an extended baseline with an aerial on Earth. NASA and ESA have therefore proposed a joint project (called QUASAT, from quasar satellite) in which a 15m diameter telescope would be placed in Earth orbit to study quasars. QUASAT would send data to Canadian, Australian, American, and European radio astronomers so that radio-emitting sources could be studied with a resolution five times greater than that obtainable with Earth-based radio telescopes. □ see ARRAY, RESOLVING POWER, RADIO ASTRONOMY, QUASAR, QUASAT, VERY LARGE ARRAY

aphelion the point in the orbit of an artificial satellite, planet, or comet in solar orbit when it is farthest from the Sun ⟨*the Earth is at ~ on about July 3 each year*⟩ – compare PERIHELION □ SEE APSIDES, KEPLER'S LAWS

apoapsis – see APSIDES

apogee the point in the orbit of the Moon or an artificial satellite when it is at its farthest point from the Earth, and at which its speed is least – compare PERIGEE □ SEE GEOCENTRIC

apogee motor *also* **apogee kick motor, apogee rocket** a rocket motor with which a spacecraft or satellite can be boosted from a parking orbit into a higher (eg geostationary) orbit. The apogee motor is fired when the satellite is at apogee, the furthest position from the Earth in the parking orbit. Sometimes an apogee motor provides a spacecraft with its escape velocity. The Payload Assist Module (PAM) is used as an apogee motor. □ see PARKING ORBIT, PAYLOAD ASSIST MODULE

Apollo The American space programme of the 1960s and early 1970s for landing astronauts on the Moon and returning them safely. Follow-ing the first manned spaceflight by Yuri Gagarin in 1961, President J F Kennedy set America the challenge of landing a man on the Moon by the end of the decade. The project was successful: on 20 July 1969, Neil Armstrong and Edwin Aldrin, on the Apollo 11 mission, landed their lunar module, *Eagle*, in Mare Tranquillitatis (Sea of Tran-quillity). Six further landings followed, ending with Apollo 17 in December 1972 which explored the Taurus-Littrow region. The three astronauts aboard Apollo 13 in April 1970 did not make a landing and were never to do so. Because of an explosion in a fuel cell in their command and service module (CSM) the mission had to be aborted, and the astronauts, cold and short of oxygen and water, simply looped round the Moon for a return to Earth. The method chosen for the Moon landings was lunar orbit rendezvous. Three astronauts travelled to and from the Moon in the command module (CM) attached to which was the service module (SM). The SM housed the main rocket engine and the lunar module (LM). Once in orbit round the Moon, one astronaut remained in the CM and the other two entered the LM for descent to the Moon. After exploring the Moon's surface, the ascent of the LM brought the two astronauts back to the orbiting CM. The LM was released to crash into the Moon and the three astronauts returned to Earth in the CM. The astronauts brought back Moon rocks for study on Earth, and placed five automatic scientific stations on the Moon which returned data until October 1977 when they were switched off. The first six Apollo flights tested the feasibility of the Saturn 5 rocket launch vehicle and other aspects of the complex system developed for Apollo. In January 1967, three astronauts, Gus Grissom, Ed White, and Roger Chaffee were burned to death in a fire which swept through their module just before they were due to make the first Apollo test flight. There was an 18-month delay while the command module was completely redesigned and all sorts of faults which were worrying the astronauts were put right. Apart from this fire on the ground, not one life was subsequently lost during this most ambitious and hazardous of all manned space journeys to date. □ see WHITE, ARMSTRONG, SATURN 5, LUNAR ROVER, SURVEYOR, SPACE STATION, MOON BASE, LUNAR ORBITER PROBES, LUNIK PROBES, LUNOKHOD, RANGER

Apollo Lunar Surface Experiments Package (abbr **ALSEP**) any of the several scientific instrument packages that were set up by astronauts on the Moon during the Apollo missions of 1969 to 1972, and which automatically transmitted data to the Earth. The packages became more sophisticated as the missions proceeded. On the last mission, Apollo 17, a thermoelectric power generator operated experi-ments which included tests for a residual lunar atmosphere, moonquakes, and impacts of meteorites. □ see APOLLO, RADIOISOTOPE THERMOELECTRIC GENERATOR

Apollo-Soyuz Test Project (abbr **ASTP**) the historic docking at an altitude of 225km of the Soviet Soyuz 19 spacecraft with the Apollo

18 spacecraft on 19 July 1975. For nearly 48 hours the three US astronauts and the two Soviet cosmonauts exchanged gifts, shared meals, and spoke with the Soviet and US Presidents while the 'show' was televised for all the world to see. This rendezvous was the first Soviet-American 'space handshake' and a similar joint effort is being planned between the two nations in the 1980s, though such an event would involve major design modifications to both the Soyuz and (now) the Space Shuttle. □ see APOLLO, DOCKING, SOYUZ

apolune a point in the orbit of an artificial satellite or spacecraft about the Moon that is furthest from the Moon – compare PERILUNE □ SEE ORBIT

apparent diameter *also* **angular diameter** the observed diameter (not the actual diameter) of a celestial object, that is given in degrees, minutes, and seconds of arc □ see RESOLVING POWER

apparition the act of becoming visible. Since comets only become visible as they near the Sun and develop a tail, the event is called an apparition ⟨*the next ~ of Halley's comet is in the year 2061AD*⟩. □ see RECOVER, HALLEY'S COMET

approach 1 the final part of the flight of an aerospace vehicle before landing ⟨*"Challenger is now on its ~ to the Edwards Air Force Base"*⟩ **2** any movement of a (manned) spacecraft towards another spacecraft for the purpose of docking with or capturing the spacecraft ⟨*the Shuttle's ~ to a space station*⟩

apsides (singular **apsis** or **apse**) the two points of an orbit that are closest to (periapsis) and furthest from (apoapsis) the centre of gravitational attraction. For an elliptical orbit, the line joining these two points is the major axis of the ellipse. The perigee is the periapsis of the Moon's orbit about the Earth; the apogee its apoapsis. □ see ORBIT, ELLIPSE

APU – see AUXILIARY POWER UNIT

Arabella – see ANITA AND ARABELLA

arc 1 part of a curved line ⟨*an ~ of a circle*⟩ **2** a luminous discharge of electricity ⟨*the ~ of an electric welder*⟩

arc jet engine – see ELECTRIC PROPULSION

arc second (abbr **arc sec**) an angle equal to 1/360 of a degree, used to give the apparent diameter of objects, the resolving power of telescopes, and the pointing accuracy of artificial satellites ⟨*the Hubble Space Telescope will have a pointing accuracy of 0.1 ~s*⟩ □ see HUBBLE SPACE TELESCOPE, BINARY STAR, RESOLVING POWER

Arecibo radio telescope the largest semi-steerable radio telescope on Earth comprising a 305m reflecting dish laid down in a natural bowl-shaped valley in the remote hinterland of Puerto Rico. Cornell University operates the telescope for the National Science Foundation. Mounted on cables high above the dish is a feed arm which receives radio signals reflected from the surface of the dish. Alternatively, when the telescope is being used as a transmitter, the feed arm sends signals to the dish for transmission into space. The Arecibo telescope has been used both to search for intelligent signals from possible civilizations in space, and, just once, to transmit a message to the globular cluster, M13, in the constellation of Hercules. The signal contained 1679 bits of information, which is the product of two prime numbers, 73×23, suggesting (to intelligent life forms) that the bits can be arranged in a 73×23 array. This small amount of data yields a surprisingly large amount of information presented as an array of blocks in rows: the first row establishes a binary number system; the second row states the atomic numbers of the elements hydrogen, carbon, nitrogen, oxygen, and phosphorus, of which human beings are made; and another row represents part of the structure of DNA. Other blocks show a rough picture of a human being, and the number of such creatures on the third planet of a solar system (shown as a set of blocks associated with a larger block – the Sun). Another set of blocks shows the radio telescope that transmitted the message at a wavelength of 0.126m. The message will take 25 000 years to reach the globular cluster – a reply can be expected in 50 000 years time! □ see SEARCH FOR EXTRATERRESTRIAL INTELLIGENCE, CLUSTER 2, ASTROBIOLOGY, RADIO ASTRONOMY

Ariane a three-stage rocket which launches satellites for the European Space Agency (ESA). The current version of this rocket is Ariane 3 which is capable of putting a 2240kg satellite into Earth orbit. Ariane 3 is 49m high and has a mass of 233 tonnes at lift-off. 90% of the mass is propellant, 9% is structure, and 1% is payload. Its first stage engine comprises four Viking-V engines and the second stage a single Viking-IV engine, which burns a hypergolic fuel; its third stage uses liquid oxygen/liquid hydrogen as fuel, the first cryogenic engine to be developed in Europe. The Viking-V engines can be swivelled in pairs to control the direction of the rocket, and the Viking-IV can also be swivelled to control pitch and yaw. In addition, Ariane 3 has two strap-on booster rockets which burn a solid fuel. Ariane 4, which will serve ESA between 1986 and 1995, has four strap-on boosters to enable it to lift payloads between 1.9 and 4.2 tonnes into orbit. After 1995, Ariane 5 is planned and should be able to take payloads of 15 tonnes into geostationary orbit. Ariane 5 will be propelled by a cryogenic engine designated HM60, which will run on liquid hydrogen and oxygen. The Ariane rockets mostly launch communications satellites (eg Meteosat and Intelsat), though an earlier version, Ariane 1, launched the Giotto spacecraft to Halley's comet in July 1985. Among the European countries which finance Ariane through ESA, France has most influence through its establishment of a company called Arianespace. This is a consortium of banks and high technology firms, two-thirds controlled by CNES (Centre National d'Etudes Spatiales). The company is responsible for planning and selling all future Ariane launches. During the launch of two communications satellites, Spacenet-F3 and ECS-3, on 13 September 1985, an Ariane 3 was destroyed in flight when a faulty hydrogen valve caused its first-stage rocket to misfire. □ see EUROPEAN SPACE AGENCY, BRITISH AEROSPACE, CENTRE NATIONAL D'ETUDES SPATIALES,

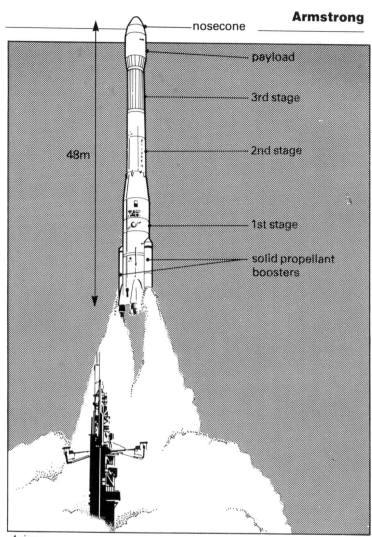

nosecone

payload

3rd stage

2nd stage

48m

1st stage

solid propellant
boosters

Ariane

HERMES, HOTOL, COLUMBUS, INTELSAT

Armstrong, Neil (*b*1930) an American pilot who became an
astronaut in 1965 aboard the Earth-orbiting Gemini spacecraft and
who, in 1969, made history by being the first human being to set foot
on the Moon. On that occasion he said "That's one small step for a
man, one giant leap for mankind". He was commander of the Apollo
11 spacecraft which carried the lunar module that he and Edwin
Aldrin took down to the surface of the Moon. The greatest tangible
honour given to Armstrong is the naming of the Armstrong Air and

Space Museum in his home town of Wapokoneta, Ohio, which has on show many of the objects connected with the astronaut's career. These include an aircraft he has flown, the Gemini 8 spacecraft he shared with Dave Scott in 1966, and his Gemini and Apollo space suits. There are at least six schools named after him, and one Boeing 747 'The Neil A Armstrong' belonging to Alitalia. □ see APOLLO

array a (usu geometric) pattern of devices which form part of an imaging system. By using an array of mirrors or dish aerials it is possible to greatly increase the sensitivity and resolving power of optical and radio telescopes, respectively. The technique enables weak cosmic sources of radio and optical energy to be detected and separated from other sources close to them. For example, the Multiple Mirror Telescope (MMT) on Mount Hopkins in Arizona uses six 1.8m diameter mirrors to create an equivalent 4.3m mirror. The Very Large Array (VLA) is a complex of radio aerials in New Mexico which comprises 27 movable 25m diameter radio telescopes to synthesize an aerial which has great sensitivity and resolving power. □ see APERTURE, RESOLVING POWER, DEEP SPACE NETWORK, OPTICAL ASTRONOMY, RADIO ASTRONOMY, X-RAY ASTRONOMY, X-RAY MULTIMIRROR TELESCOPE, VERY LARGE ARRAY

articulated having two or more parts flexibly connected and intended to operate as a unit ⟨an ~ robot arm⟩ □ see REMOTE MANIPULATOR SYSTEM

artificial gravity simulated gravity. It is not possible to make gravity for it is a basic property of all matter. But it is possible to give astronauts a feeling of weight even though they are far removed from a planet. A simulated gravity can be produced in a spacecraft by rotating the spacecraft or part of it, about an axis. An astronaut standing in this rotating section feels a centrifugal force pressing him or her on the rotating surface. □ see SPACE STATION, SPACE SETTLEMENT, CENTRIFUGAL FORCE, CENTRIFUGE

artificial intelligence (abbr **AI**) a branch of computer science which aims to construct computer programs that display adaptive behaviour in novel situations and that can discern relationships between facts. The resulting behaviour of a machine controlled by AI would be considered intelligent if done by humans. AI is goal-oriented: ie the machine is trying to accomplish a task set by its builder, and in a way that is not as rigidly set out as with most of the present-day (von Neumann) computers that execute instructions one after the other. From time-to-time, computers on board the Voyager spacecraft are fed (by radio) with a sequence of operating instructions which direct Voyager's cameras and instruments to carry out a programme of scientific investigation. The programs were developed at the Jet Propulsion Laboratory (JPL), California, and are an adaptation of existing AI programs. Other space missions are expected to benefit from AI, including the Space Shuttle and a proposed Martian Rover. NASA's Johnson Space Centre has an Artificial Intelligence and Information Science Office to support the development of the US

Space Station. □ see FLIGHT DATA SUBSYSTEM, FIFTH GENERATION COMPUTER, EXPERT SYSTEM, SUPERCOMPUTER, VON NEUMANN MACHINE

artificial satellite a (usu unmanned) manmade object that is sent into orbit round the Earth, Moon, or another celestial body. Artificial satellites are expensive to build and launch into orbit and, once there, they cannot (usu) be repaired. They contain sophisticated electronic equipment which may be damaged by cosmic rays, micrometeorites, and electrical and magnetic fields. They have a wide range of functions: astronomy, communications, remote sensing, weather forecasting, navigation, surveillance, etc. □ see SPACECRAFT, SPUTNIK 1, SPACE INSURANCE, COMMUNICATIONS SATELLITE, LANDSAT, INMARSAT, NATIONAL OCEANIC AND ATMOSPHERIC ADMINISTRATION, BIG BIRD, INTELSAT, MOLNIYA, VIKING, [1]GALILEO, [1]GIOTTO, GAMMA-RAY ASTRONOMY, X-RAY ASTRONOMY, OPTICAL ASTRONOMY, RADIO ASTRONOMY, HUBBLE SPACE TELESCOPE

ASAP – see AMERICAN SOCIETY OF AEROSPACE PILOTS

ASAT [*anti-sat*ellite] any system for destroying hostile Earth-orbiting satellites. ASAT technology is becoming increasingly sophisticated and now involves research into the use of laser beam weapons to disable enemy satellites. For example, one design for a ground-based ASAT proposes the use of a 200kW laser which could be pointed with an accuracy better than 1 microradian (equivalent to an accuracy of 0.8m at a range of 800km). A further development is laser-defence satellites (DSATs) which would be in orbital patrol to destroy enemy ASATs before they have a chance to attack. The first successful test of an F-15-launched ASAT took place on 14 September 1985. The F-15 fighter launches a highly sophisticated anti-satellite missile, called the Miniature Vehicle (MV), while it is in level subsonic flight, or in a supersonic climb if the target satellite is in a higher orbit. The MV is sped into orbit by a two-stage 1200kg 5.4m long solid-fuel rocket at the base of which are fins to control the rocket during its ascent into orbit. The MV in the upper stage is a small object (about 47×51mm) and contains a cryogenically-cooled infrared sensor which receives information about the target by means of eight small telescopes. Just before release from the upper stage, the MV is made to spin at 20 revolutions per minute to stabilize it and to help it lock onto the enemy satellite. It destroys its target by direct collision, travelling at a speed of about 22 000kms⁻¹. □ see STAR WARS, LASER

ascending node *also* **northbound node** the point at which a celestial body crosses from the south to the north of the celestial equator – compare DESCENDING NODE □ SEE CELESTIAL EQUATOR

ASM – see AIR-TO-SURFACE MISSILE

assemble to fit together a number of parts ⟨*to ~ a radio antenna*⟩

assembly a collection of parts (eg the modules making up a space station) joined together to provide something useful. Each of the individual modules might be called a subassembly. □ see SYSTEM

asteroid *also* **minor planet, planetoid** any of many small irregular-shaped solid bodies which orbit the Sun, most of them between the orbits of Mars and Jupiter (the asteroid belt). The largest asteroid is Ceres which is 1003km in diameter; Pallas, Juno, and Vesta are next in order of size. Ceres, unlike Vesta, does not reflect much light and cannot be seen by the naked eye. Most asteroids, like the planets, have circular orbits, though a few have highly elliptical orbits. The World Space Foundation's Asteroid Project, using Earth-based tele-scopes, is continuing to discover new asteroids. Some asteroids (known as the Amor class of asteroids) have orbits which bring them close to the Earth on occasion, while some actually cross the Earth's orbit (the Apollo class). However, the chance of the Earth colliding with an asteroid is negligible. By the 1990s, asteroids will be the only major class of objects in the Solar System still awaiting a visit by a spacecraft. The spacecraft Galileo will examine the asteroid Amphitrite during its trip to Jupiter in 1986 to 1988. Some scientists are not keen on the idea, though, since a collision with an asteroid would certainly wreck the Galileo spacecraft and wipe out NASA's major planetary mission of the decade. However, the European Space Agency (ESA) is studying the feasibility of sending a spacecraft to the asteroids in the 1990s. This spacecraft, called AGORA (Asteroid Gravity, Optical and Radar Analysis) would survey several targets using an array of instruments. □ see EUROPEAN SPACE AGENCY, [1]GALILEO, LANGRANGIAN POINTS, TROJAN GROUP

asteroid belt – see ASTEROID

Asteroid Gravity, Optical and Radar Analysis (abbr AGORA) – see EUROPEAN SPACE AGENCY

ASTP – see APOLLO-SOYUZ TEST PROJECT

astral (consisting) of or relating to the stars □ see STELLAR

astroballistics the study of the effects of bodies moving through gases at hypersonic speeds (eg the use of ablation materials on the Space Shuttle and the disintegration of meteroids on entering the Earth's atmosphere) □ see HYPERSONIC

astrobiology *also* **exobiology** the study of living organisms that may exist on other planets and satellites in the Solar System, or on plane-tary systems of other stars. Despite frequent claims of UFO sightings and close encounters with alien (usu humanoid) beings, there is, to date, no direct evidence for extraterrestrial life. However, exobiology is a fascinating field of study, and there are a number of ways of look-ing for life on other worlds. The Apollo programme (1969–1972) brought back soil and rocks from the Moon for analysis, and Viking, which landed on Mars, produced ambiguous results when its labora-tory carried out an analysis of Martian soil in 1976. Even the search for intelligent messages in the radiation from the stars (eg Tau Ceti and Epsilon Eridani) have so far been fruitless. Of course, intelligent life in our Galaxy may already share a 'telephone system' and we just do not have the right equipment to hook up to it. Furthermore, there is no reason why the life we are looking for is humanoid in appear-ance: we may not have the 'eyes' to see it and the 'ears' to hear it.

☐ see SEARCH FOR EXTRATERRESTRIAL INTELLIGENCE, ENCOUNTER 2, UFO, APOLLO, VOYAGER, PIONEERS 10 AND 11, PANSPERMIA, SAGAN

astrodynamics – see ASTRONAUTICS

astrometry a branch of astronomy concerned with measuring the precise positions and movements of celestial objects ☐ see HIPPARCOS, ABERRATION 2, GROUND-BASED ELECTRO-OPTICAL DEEP SPACE SURVEILLANCE

astronaut somebody from a non-communist country who travels beyond the Earth. On 20 July 1969 Neil Armstrong of the United States became the first astronaut to land on the Moon – compare COSMONAUT ☐ SEE ARMSTRONG, SPACEWALK

astronautics *also* **astrodynamics** the science and technology of spaceflight. Astronautics includes the fields of celestial mechanics, ballistics, propulsion theory, chemistry, physics, engineering, and aerodynamics ☐ see ASTROPHYSICS

astronomer someone who takes an interest in the behaviour, properties, and composition of celestial bodies ☐ see ASTRONOMY, ASTROPHYSICS

astronomical 1 of astronomy **2** enormously or inconceivably large ⟨~ *distances between the galaxies*⟩ ☐ see INFINITE, ASTRONOMICAL UNIT, PARSEC

astronomical unit (symbol **AU**) a unit for measuring distances within the Solar System, defined as the mean distance between the centre of the Earth and the centre of the Sun, and closely equal to 150 million kilometres. Pluto at the edge of the Solar System is about 25 AU from the Sun. ☐ see LIGHT-YEAR, PARSEC

astronomy a branch of science which studies celestial bodies, the space between them, and the universe as a whole. Visual astronomy is one of the oldest sciences but modern-day astronomers have access to the latest advances in instrumentation (eg the charge-coupled device and other microelectronics devices), space technology (eg artificial satellites), and data processing (eg the computer). These developments contribute to two more recent branches of astronomy, astrophysics and astronautics. The new technology available to the astronomers has extended observations of the universe in the gamma-ray, X-ray, ultraviolet, infrared, and radio regions of the electromagnetic spectrum. ☐ see OPTICAL ASTRONOMY, RADIO ASTRONOMY, X-RAY ASTRONOMY, GAMMA-RAY ASTRONOMY, ULTRAVIOLET ASTRONOMY, INFRARED ASTRONOMY, ASTROPHYSICS, CELESTIAL MECHANICS, CHARGE-COUPLED DEVICE

astrophysics the study of the physical and chemical processes and behaviour of celestial bodies and the regions of space which separate them. Astrophysics attempts to explain how stars, galaxies, pulsars, quasars, and other celestial bodies produce and expend their energy. It is also concerned with understanding the structure and processes of the universe as a whole. It draws on knowledge in the fields of nuclear physics, thermodynamics, spectroscopy, and relativity, and is supported by ground-based and space-based astronomy. ☐ see OPTICAL

ASTRONOMY, RADIO ASTRONOMY, GAMMA-RAY ASTRONOMY, ULTRAVIOLET ASTRONOMY, INFRARED ASTRONOMY, BIG BANG, BLACK HOLE, PULSAR, QUASAR, STEADY STATE, RELATIVITY

AT – see ATOMIC TIME

Atlas a family of launch vehicles used in a number of programmes, including Mercury and Ranger, since the first launch in 1957. Atlas boosters have a cluster of three liquid propellant engines and two small vernier engines. The Atlas rocket is unique in that it gains most of its structural strength from its inflated thin stainless steel fuel tanks. If the pressure is lost, the tanks sag like punctured balloons. The Atlas launcher has recently been redesigned to take the Centaur upper stage engine, and the Atlas-Centaur combination is capable of putting a 2300kg payload into geostationary orbit. □ see CENTAUR, REMOTE SENSING

atmosphere the gases, liquids, and solids which surround a planet or satellite. The molecules of a gas are in constant and rapid motion and they escepe if their speeds exceed the escape velocity of the planet. Since the speed of gas molecules increases with their temperature and decreases with their molecular mass, light gases such as hydrogen, helium, and methane are more easily lost from small hot bodies (eg the planet Mercury) than from larger cooler bodies (eg Jupiter). Heavier gases such as nitrogen, carbon dioxide, and oxygen are more easily retained by the smaller bodies (eg the Earth and Mars), though Pluto (a small planet) is so cold it probably has an atmosphere locked up as frozen gases on its surface. A study of planetary atmospheres is important since it can provide clues as to whether the planet could support life. The atmospheres of three planets (Venus, Mars, and Jupiter), other than the Earth, have been examined at close range by spacecraft. Jupiter and the other Jovian planets have deep atmospheres composed mainly of hydrogen, helium, methane, and ammonia which were probably captured from the nebula from which the Solar System is thought to have been formed. But the smaller terrestrial planets (Earth, Mars, Venus, and Mercury) were too small to capture these gases: their atmospheres are thought to have come from volcanoes after being liberated from their surfaces by radioactive heating of the rocks. The weather systems of Venus, Earth, Mars, Jupiter, and Saturn have also been studied by spacecraft. It is probably internal energy sources which control the circulation of the atmospheres of Jupiter and Saturn, whilst the weather on Earth, Mars, and Venus is driven by energy coming from the Sun. □ see EARTH'S ATMOSPHERE, VENUS, MARS, JUPITER, SATURN, URANUS, NEPTUNE, VIKING, VOYAGER, VENERA, [1]GALILEO, PIONEER VENUS, VEGA, SAGAN

atmospheric drag the small force experienced by a satellite or spacecraft in passing through the upper atmosphere. Spacecraft whose perigee is lower than a few hundred kilometres lose energy in this way and eventually burn up dramatically in the atmosphere. To compensate for the effects of atmospheric drag, the manned Soviet Salyut space station has to be lifted periodically into a higher orbit using the engine of the visiting Progress 'cargoship'. □ see SALYUT

atmospheric pressure the force per unit area at any point in an atmosphere due solely to the weight of the atmospheric gases above it. Atmospheric pressure is measured in units of newtons per square metre (Nm^{-2}), millibars (mb), or atmospheres. The standard atmospheric pressure at the Earth's surface is 101 325 Nm^{-2} or 1013 mb. The pressure on the surface of Venus is about 90 000mb (90 atmospheres); and on the surface of Mars, 7.5mb. □ see ATMOSPHERE, EARTH'S ATMOSPHERE

atom the smallest particle of a chemical element that can exist alone or in combination with other atoms without being destroyed. Nearly all the mass of an atom is concentrated in its nucleus which is made up of protons and neutrons. In a neutral atom, a cloud of electrons occupies the rest of the volume of an atom. Life in all its varied forms is largely the result of the chemical behaviour of atoms which rearranges the energy levels of electrons in atoms. But the much more energetic nuclear reactions, the source of energy in stars, supernovae, quasars, and pulsars, is determined by changes to the nucleus of atoms. □ see PROTON, NEUTRON, ELECTRON, ISOTOPE, NEUTRON STAR, IONIZATION

atomic bomb a bomb which produces its violent explosive power by the sudden release of energy when heavy atoms (eg plutonium and uranium) are split by neutrons in a chain reaction □ see NUCLEAR FISSION, THERMONUCLEAR

atomic clock – see ATOMIC TIME

atomic cloud – see RADIOACTIVE CLOUD

atomic number (symbol **Z**) the number of protons in the nucleus of an atom. Each chemical element is identified by its atomic number (eg 1 for hydrogen, 103 for lawrencium). □ see MASS NUMBER, ATOM, ISOTOPE

atomic time (abbr **AT**) the precise measurement of time using vibrations of molecules or atomic nuclei as the basic timing mechanism. Electronic circuits are used to sense and control these vibrations to produce 'ticks' accurate to one second in thirty million years. Atomic clocks are widely used in scientific and military systems. □ see QUASAT, GLOBAL POSITIONING SYSTEM

attenuation the reduction in the strength of electromagnetic radiation (eg light), or of particles (eg protons), by absorption and scattering in the medium between their source and destination. Gas and dust in the plane of the Galaxy attenuates visible light quite strongly and makes it difficult to see the nucleus of the Galaxy. However, information about this region has been obtained at radio and infrared wavelengths which are absorbed less strongly than light by this dust and gas. □ see RADIO ASTRONOMY, INFRARED ASTRONOMY, MILKY WAY

attitude the position of a spacecraft, artificial satellite, or aircraft, at rest or in motion, relative to a particular point of reference (eg the horizon) or a fixed system of reference axes □ see CELESTIAL COORDINATES, STABILITY, NAVIGATION

attitude control automatic adjustment of the position of a spacecraft or artificial satellite using on-board control systems. The control system may be internal (eg based on a gyroscope), or external using Sun-sensors, Earth-sensors, or star trackers. Usually a combination of these methods is used. The Hubble Space Telescope (HST) provides a good example of the precision with which attitude control systems aboard satellites can point instruments at their targets – in this case, stars, galaxies, and other celestial objects. The main structure of the HST is the Support Systems Module (SSM). The SSM contains the communications, data handling, thermal control, power, and attitude control systems. The attitude control system makes use of a computer which receives error signals from optical and inertial sensors, and computes the required torque moments that command reaction wheels to keep the telescope pointing at its target. Coarse control is achieved using stars and the Sun, and fine pointing is done with signals from the fine guidance system looking through the telescope. This technique enables the telescope to point with an accuracy of 0.007 arc seconds – equivalent to identifying a 1p coin in London from Edinburgh! □ see HUBBLE SPACE TELESCOPE, SPIN STABILIZED, GYROSCOPE, THRUSTERS

AU – see ASTRONOMICAL UNIT

¹audio of or being oscillations (eg sound waves, mechanical vibrations, or electrical oscillations) having frequencies in the range of human hearing ⟨an ~ amplifier⟩ □ see AUDIO FREQUENCY

²audio the transmission, reception, and reproduction of sound by electronic means

audio frequency (abbr **AF**) (of or being waves having) a frequency in the range of human hearing. Audio frequency is usu taken to refer to the range 20Hz to 20kHz. □ see FREQUENCY

aurora (plural **aurorae, auroras**) a display of streamers and arches of coloured light which is seen in the Earth's atmosphere at high altitudes. The lights are produced at altitudes of approximately 100km when charged particles from the solar wind become trapped in the Earth's magnetic field. The particles spiral in the lines of magnetic field and interact with gases in the upper atmosphere to produce varying colours from red to green. An artificial aurora has been 'painted' during one of the missions of the Space Shuttle by injecting a stream of electrons into the upper atmosphere. The aurora borealis occurs in northern latitudes, while the aurora australis occurs in southern latitudes. □ see SOLAR WIND, IONOSPHERE, MAGNETOSPHERE, EARTH'S ATMOSPHERE

Aussat [*Au*stralian *Sat*ellite] an Australian communications satellite launched by the Space Shuttle during Mission 51I, in August 1985. An Australian space industry is financially feasible, but Australia does not have an active involvement in space, and it has deferred the opportunity to train a payload specialist to accompany the Aussat mission. However, a report published by the Australian Academy of

Technological Sciences in June 1985 recommended that Australia should set up a NASA-style independent space agency with a major interest in the field of remote sensing in which Australia has particular expertise. □ see SPACE SHUTTLE, MISSION 51I

Australian National Radio Astronomy Observatory an observatory near Parkes, New South Wales, the main instrument of which is the fully steerable 64m dish aerial □ see RADIO ASTRONOMY, APERTURE, DEEP SPACE NETWORK

autoignition the spontaneous combustion of a propellant in the presence of an oxidizer □ see HYPERGOLIC

automatic controlled by built-in devices ⟨~ pilot⟩

automatic pilot also **autopilot** (electronic) equipment which automatically stabilizes the attitude of an aerospace vehicle about its pitch, roll, and yaw axes □ see PITCH, ROLL, YAW

autopilot – see AUTOMATIC PILOT

auxiliary power unit (abbr **APU**) any electrical power supply which can be used in addition or as a backup to a main power supply in the event of overload or failure of the main source of power on a spacecraft or space station

avionics [aviation and electronics] the use of electronic devices and systems (eg automatic pilots) in aeronautics and astronautics □ see RADAR

AXAF – see ADVANCED X-RAY ASTROPHYSICS FACILITY

axis an imaginary line about which a body (eg a satellite) rotates ⟨the ~ of rotation of a satellite passes through its centre of gravity⟩

azimuth the clockwise angle, measured horizontally in radians or degrees, of a celestial object from a north or south point – compare ALTITUDE 1

background radiation atomic or electrical radiation which occurs in the natural environment □ see COSMIC RAYS, RADIOACTIVITY, STATIC, COSMIC BACKGROUND RADIATION

backup duplicating the function of something in case of emergencies ⟨*a ~ computer on Voyager 2*⟩ □ see FLIGHT DATA SUBSYSTEM, EXPEDITION TO MARS

BAe – see BRITISH AEROSPACE

baffle – see ANTISLOSH BAFFLE

Baikonur – see COSMODROME

Baker – see ABLE AND BAKER

ballistic missile a missile that is guided and propelled (unlike a cruise missile) only during the initial phase of its flight. During the nonpowered phase of its flight, its trajectory, like that of an artillery shell, is largely determined by gravity and atmospheric drag. □ see ANTIBALLISTIC MISSILE, CRUISE MISSILE, GUIDANCE, ASTROBALLISTICS

ballistics the study of the motion, behaviour, and effects of projectiles such as missiles and bullets □ see BALLISTIC MISSILE

band a range of frequencies. A number of bands (eg short wave band) are used for radio communications. These are: very low frequency (VLF), below a frequency of 30kHz; low frequency (LF), 30 to 300kHz; medium frequency (MF), 300 to 3000kHz; high frequency (HF), 3000 to 30 000kHz; very high frequency (VHF), 30 to 300MHz; ultra-high frequency (UHF), 300 to 3000MHz; super-high frequency (SHF), 3 to 30GHz; extremely high frequency (EHF), above 30GHz. During World War 2, radio frequency bands were designated by letters (eg K band, L band, S band, and X band). □ see FREQUENCY, SPECTRUM, MEDIUM WAVES, SHORT WAVES, LONG WAVES, MICROWAVES, X-BAND, C-BAND

bandwidth the frequency range of the messages handled by an information processing system. Bandwidth is important when considering the volume of information transmitted along a communications channel. As a general rule in a digital system each bit (ie 0 or 1) that is transmitted per second requires a bandwidth of 2 hertz (Hz). Thus a bandwidth of 600Hz is needed to transmit 300 bits per second; normal speech along a telephone line needs a minimum bandwidth of 3000Hz; and a colour TV picture requires a bandwidth of about 8MHz to be transmitted. The bandwidth of a communications system or an element of it (eg a communications satellite) is often expressed in terms of the number of telephone conversations it can carry. Thus the largest coaxial cable has a bandwidth of 500MHz and can therefore carry about 167 telephone conversations. And a hair-thin glass fibre can carry about 10 000 telephone conversations. Bandwidth is also used to describe the frequency range over which a sensor (eg an antenna) or an instrument (eg a radiometer) responds to signals. □ see DATA RATE, OPTICAL COMMUNICATIONS, TELEMETRY, CABLE TV, DIRECT BROADCAST SATELLITE, BIG COMMUNICATOR, COAXIAL CABLE

barbecue mode the slow roll of a space station, esp the Space Shuttle, that helps to maintain an even temperature over its surface. This technique is necessary because the Sun heats one side of the space station intensely while the other side cools in shadow. □ see THERMAL CONTROL SYSTEM

Barnard's star a red dwarf star about 6 light years away, in the constellation Ophiuchus, and the fourth nearest star to the Sun. Barnard's star was discovered by E E Barnard in 1916 by its large back and forth movement against the background of more distant stars. This movement also has a wobble superimposed on it which suggests that the star may have planets orbiting it. Barnard's star would be the mission objective of the starship, *Daedalus*. □ see DAEDALUS, STARSHIP

barycentre – see CENTRE OF MASS

baryon the general name for the two elementary particles, the proton and the neutron, which bind together in the nuclei of atoms. Baryons are a type of hadron and are made up of triplets of quarks. □ see ELEMENTARY PARTICLE, HADRON, LEPTON, QUARK

beacon a device that emits radio or light signals for guidance or warning □ see GUIDANCE, COSPAS-SARSAT

beam a narrow well-defined stream of particles (eg electrons) or electromagnetic radiation (eg radio waves from a pulsar) down which energy passes □ see LASER

beam-rider guidance – see GUIDANCE

bends – see AEROEMBOLISM

berth to bring a satellite alongside a space station or into the payload bay of the Space Shuttle, using a mechanical arm or equivalent device □ see REMOTE MANIPULATOR SYSTEM, CAPTURE

beta (symbol β) the second letter of the Greek alphabet, used to identify the second brightest star in a constellation ⟨~ *Gemini*⟩ □ see BETA PICTORIS

Beta Pictoris a faint fourth magnitude star in the constellation Pictor (the Painter), 50 light years away, and visible from the Southern Hemisphere. In 1983, the Infrared Astronomy Satellite (IRAS) noted that this star emits strong infrared radiation which indicates that cool planet type material may exist round it. In 1984, two astronomers, Dr Richard Terrile and Dr Bradford Smith, turned the 2.5m telescope of the Las Campanas Observatory in Chile, on Beta Pictoris and detected a flattened disc of dust extending some 60 thousand million kilometres out from the star. The observation was made possible by the use of a charge-coupled device at the focus of the telescope, and a coronograph to block out the overpowering light from the star. The ring of dust does not tell us that Beta Pictoris has a planetary system, but no doubt this star will be one of the first to be studied by the Hubble Space Telescope. □ see INFRARED ASTRONOMY SATELLITE,

Big Bang

CHARGE-COUPLED DEVICE, HUBBLE SPACE TELESCOPE, DAEDALUS

Big Bang the explosive event thought to have begun our universe. Many cosmologists believe that sometime between 10 and 20 thousand million years ago all the matter and energy now in the universe was concentrated in a very small, extremely dense sort of 'cosmic egg'. Then a cataclysmic fireball took place – the Big Bang – and the universe began an expansion which continues to this day. There is evidence for this expansion: first, all the observable galaxies are rushing away from each other, as measured by the red shift in their spectral lines; second, the microwave background radiation is regarded as the considerably cooler remnant of the intense radiation content of the early universe; and third, the present cosmic abundance of helium agrees with calculations of its formation in the Big Bang. The model of the Big Bang theory suggests that events occurred very rapidly in the first few minutes after the explosion. When time begins, the universe exists as a soup of electrons, positrons, neutrinos, antineutrinos, and photons (radiation) at a temperature of perhaps 100 thousand million degrees, and with a density of 4 thousand million times that of water. One second later its temperature has dropped to 10 thousand million degrees, protons and neutrons have begun to form, and the density is about 400 thousand times that of water. Fourteen seconds later, the temperature is about 3000 million degrees, and positrons and electrons are liberating energy through their mutual annihilation which slows down the rate of cooling of the universe. Various stable nuclei such as helium now form. Three minutes later the temperature is 900 million degrees and low enough for deuterium nuclei to form. Thirty minutes later, the temperature has dropped to about 300 million degrees, and electrons and protons have almost been annihilated by positrons and antiprotons, their antimatter counterparts. The density of the universe is now about a tenth that of water. The nuclear particles are now for the most part bound into helium and hydrogen, but the universe is still too hot for stable neutral atoms to form. The universe goes on expanding for about 700 thousand million years when its temperature is low enough for neutrons and protons to form stable nuclei. Matter, initially mostly helium and hydrogen, begins to form into galaxies and stars. The galaxies continue rushing apart, and about 10 thousand million years after the Big Bang some of the stars have planets on which intelligent life is trying to make sense of what their eyes and instruments reveal of the universe about them. Many cosmologists agree with the Big Bang (called the standard model) of the universe's beginning, but an alternative theory called the Steady State model is philosophically more attractive: there was no beginning, no genesis; the universe has always been here, and we don't have to account for our beginnings in the variety of inventive ways that characterize religious beliefs in many cultures. □ see RED SHIFT, HUBBLE'S LAW, MICROWAVE BACKGROUND RADIATION, HELIUM, HYDROGEN, COSMIC ABUNDANCE, STEADY STATE, NUCLEOSYNTHESIS, BIG CRUNCH, HOYLE, ANTIMATTER, PROTON

Big Bird a class of US spy satellites which take very detailed photographs of the Earth from the relatively low orbital altitude of between 150km to 250km. These 10-tonne satellites are fitted into an upper-stage Agena rocket and launched several times a year by the huge Titan 3D rockets. Since they are at low altitude, their orbits decay quickly so every few months they are given a boost to a higher orbit by firing the Agena rocket. Film is automatically processed on board, placed in a re-entry capsule, and sent back to Earth. High-flying aircraft are used to scoop the capsule out of the air as it comes down by parachute, or ships are used to collect it from the sea. Though obtaining film in this way gives high quality pictures, increasingly photographs, having been processed on board Big Bird, are scanned by a laser, converted into a stream of electronic data, and transmitted to US Air Force bases round the world. These pictures show aircraft, military vehicles, and men on the ground, and since Big Bird sweeps round the Earth every 90 minutes, it can monitor the progress of military developments. □ see TITAN 2, COSMOS, STAR WARS

Big Communicator a class of powerful communications satellites that British Aerospace is developing to replace the current series of Olympus-class satellites. They will be launched by Ariane rocket, the Space Shuttle, or by the new launcher HOTOL, in the 1990s. Each Big Communicator would be capable of handling millions of two-way telephone calls, compared with the present-day limit of about 25 000 two-way calls (or their equivalent in video or data). The Big Communicators will serve an ever-expanding population of telecommunications users, and their increased power will reduce the size of indoor dish aerials to about 0.3m. Commercial users would be using 1m diameter aerials capable of handling 100 Mbits per second of mixed voice, data, and video. Once in orbit, the Big Communicators will unfurl solar panels capable of delivering up to a hundred kilowatts of electrical energy to their electronic circuits to provide this increased communications traffic. About two-thirds of this power is dissipated as waste heat by the internal electronic circuits, and this has to be got rid of by radiation to space. Since conventional spin-stabilized and three-axis stabilized communications satellites cannot dissipate more than about 8kW of waste heat, the Big Communicators will dissipate heat through the part of their body which is in the shade of their solar panels. They will lock onto the Sun or the Earth so that they will not have any rotating parts as in spin-stabilized satellites. □ see BRITISH AEROSPACE, DIRECT BROADCAST SATELLITE, SPIN STABILIZED, HOTOL

Big Crunch the possible finale to a contracting universe when all the galaxies smash together. At the moment, the universe is expanding; the galaxies are moving away from each other. But the gravity of all the galaxies and other cosmic material (eg intergalactic dust) is restraining the outward motion. Eventually, the restraint may be strong enough to reduce the expansion to zero. The universe would begin to contract, slowly at first but over many billions of years at an ever increasing rate until all the galaxies come together in a Big Crunch resulting in the end of the universe – a Big Bang in reverse! Thus the universe as we know it only exists for a finite time. But might the Big Crunch be followed by another period of expansion and go on for ever as a series of expansions and contractions? □ see BIG BANG, COSMOLOGY, UNIVERSE, NEUTRINO, PULSATING UNIVERSE

billion 1 a thousand millions (10^9). This is the value used in this book. ⟨*the age of the universe is about 18 ~ (18 × 10^9) years*⟩ **2** *in British use* a million millions (10^{12}) □ see SI UNITS

binary – see BINARY STAR

binary digit *also* **bit** either of the two digits 0 and 1 used to express a number (eg 10110) in the binary system of numbers □ see DATA RATE

binary star *also* **binary** a pair of stars orbiting a common centre of mass. Although the Sun does not have a close companion star (life on Earth would be interesting if it had!), more than 50% of stars in the Galaxy are binaries. It is possible to estimate the mass of the stars from the motion of the stars about each other. There are several types of binary star systems: visual binaries can be resolved by an optical telescope; spectroscopic binaries can be detected using the Doppler shift of their spectral lines; and eclipsing binaries are revealed by the regular variation in their brightness as they rotate about each other. □ see PULSAR, X-RAY ASTRONOMY, BARNARD'S STAR, POLE STAR

binding energy the minimum energy required to separate neutrons and protons in the nucleus of an atom. The binding energy of atomic nuclei is much larger than the energy required to remove electrons from their orbital positions in atoms. □ see ELECTRON VOLT, EINSTEIN 2

biological shield a shield placed round a nuclear reactor or radioactive material to reduce atomic radiation to a level safe for human beings and other animals □ see ELECTRIC PROPULSION

Bion – see ABREK AND BION

bionic [*bi*ological electr*onic*] of or being an electronic device which has the characteristics of living organisms □ see ANDROID

biopak a container for keeping living organisms alive and well while their activities are recorded during flights into space □ see LIFE SCIENCES, LAIKA, SHUTTLE PALLET SATELLITE

biosensor a sensor (eg an electrode) used to obtain information about a life process (eg heart rate) □ see BIOTELEMETRY

biosphere the narrow zone between the lower and upper part of the Earth's surface, within which varied life forms can live – compare ECOSPHERE □ SEE EARTH'S ATMOSPHERE

biotechnology the application of engineering knowhow to the life sciences □ see SPACE MEDICINE, CONTINUOUS-FLOW ELECTROPHORESIS SYSTEM, LIFE SCIENCES

biotelemetry the remote measurement of the life functions (eg heart beat and respiration) of animals using radio, and often involving the use of communications satellites as relay stations ⟨*~ of astronauts* *aboard the Shuttle*⟩ □ see TELEMETRY, SPACE MEDICINE

bird a colloquial name for a rocket, spacecraft or satellite □ see BIG BIRD

BIS – see BRITISH INTERPLANETARY SOCIETY

bit – see BINARY DIGIT

bit rate – see DATA RATE

Black Arrow – see BLACK NIGHT

black body a body that absorbs all the radiation falling on it. A black body is a theoretical concept, but it is useful for working out the surface temperature of stars by analysing their spectra. As the temperature of a black body increases, the wavelength at which it emits most energy decreases, ie blue stars are hotter than red stars. The product of the wavelength and (absolute) temperature for a black body is a constant (this statement is known as Wien's displacement law). The sun has a black body temperature of about 6000K corresponding to the emission of most radiation as yellow light. Thus the black body temperature of a distant star can be worked out merely by measuring the wavelength at which the star emits most radiation.

black box any electronic component or equipment which is looked at from the point of view of what it does rather than how it works. An experiment package on a satellite might be regarded as a black box.

black hole the invisible remnants of a star which has died catastrophically. When most, but not all, stars cease to produce energy, they contract in about one millionth of a second and retreat into black holes of their own creation. Being black, a black hole is hard to spot: it effectively disappears from the visible universe since its intense gravitational field stops all light from leaving it. There are probably countless numbers of black holes in the universe, but their existence can only be implied by the awesome activity in some regions of space: the enormous energy emitted by a quasar is probably caused by a black hole at its centre. Observations by telescopes in orbit round the Earth are helping astronomers to identify events which may be caused by black holes. A better understanding of the behaviour of black holes will help astrophysicists to predict the ultimate fate of the universe and to validate Einstein's theory of relativity which predicts their existence. □ see WHITE DWARF, PULSAR, QUASAR, BIG BANG, RELATIVITY, X-RAY ASTRONOMY, SCHWARZCHILD RADIUS

Black Night a British high-altitude research rocket first launched from Woomera, in Australia, on 7 September 1958. A series of 22 of these rockets were equipped with nose cones to study how different metals behaved at high reentry speeds of up to 19 thousand kmh^{-1} when the rockets returned from altitudes of 960km. The information gathered has been very useful up to this day, and particularly so in the subsequent development of Black Arrow. This rocket had two liquid propellant stages and a solid booster rocket to inject a satellite into orbit. Following a successful development programme starting in 1968, Black Arrow placed *Prospero* in orbit on 28 October 1971 – it is still functioning on command today. These British programmes were very

blackout

cost-effective but the political scene of the 1970s demanded that European countries discontinue their individual programmes and combine their talents to form the European Space Agency (ESA). □ see EUROPEAN SPACE AGENCY, WOOMERA TEST RANGE

blackout 1 a communications failure between a spacecraft and an Earth station, esp that caused by the high-temperature shockwave which surrounds a spacecraft as it rapidly enters the Earth's atmosphere. The term also applies to the loss of radio signals caused by a spacecraft passing behind a celestial object (eg the Moon). □ see SHOCKWAVE, PLASMA **2** a temporary loss of vision caused by reduced blood pressure in the head and a consequent lack of oxygen. An astronaut may suffer a blackout if a spacecraft undergoes a rapid change of speed, esp when a rocket takes off from the launch pad. An astronaut undergoes rigorous training in order to combat blackouts. □ see CENTRIFUGE, SPACE MEDICINE, AEROEMBOLISM

blast-off – see LIFT-OFF

bleed to remove or draw off fluid from a system ⟨to ~ a cryogenic propellant⟩ □ see BOILOFF

blip also **pip** a small image on an oscilloscope or radar screen caused by a distinct signal (eg reflection from a satellite) □ see RADAR

blockhouse a reinforced concrete building, esp one at a rocket launch site, which protects electronic and other control instruments from the effects of blast, heat, and explosions □ see REVETMENT

bloomed lens also **coated lens** a lens (eg of a telescope) which is coated with a thin film of transparent material to improve the brightness of the image it produces. The thickness of the film is chosen so that destructive interference occurs between light reflected from the lens-coating surface, and that reflected from the air-lens surface. The result is more light passing through the lens and a brighter image. Bloomed lenses often have a blue-grey colour. □ see REFRACTING TELESCOPE

blue shift a displacement towards shorter wavelengths of lines in the spectrum of a celestial object. If the light is in the visible spectrum, the lines are shifted towards the blue end of the spectrum. The blue shift is due to the Doppler Effect on electromagnetic waves from a light source and indicates that it is moving towards us. In our own Milky Way galaxy, stars show both red shift (they are moving away from us) and blue shift. But all the distant galaxies are moving away from our own galaxy and show a red shift in their spectrum. This is taken as evidence that the universe is expanding. □ see DOPPLER EFFECT, RED SHIFT, BIG BANG, HUBBLE'S LAW, ELECTROMAGNETIC SPECTRUM, SPECTROMETER

Bode's law a law stating that there is a mathematical progression in the distances of the planets from the Sun. This law (for which there is no scientific justification) was formulated by Johann Titius (1729–96) of Wittenberg in 1766, and subsequently in 1772 presented as a virtual law by Johann Bode (1747–1826) director of the Berlin Observatory. The Titius progression assigned numbers to the position of the planets. Mercury was 0; Venus 3; Earth 6; Mars 12; and so on. When 4 is added to each of these numbers and the result divided by 10, the progression 0.4, 0.7, 1.0, 1.6, 2.8, 5.2, ... is found to be in good agreement with the approximate distances of the planets from the Sun in astronomical units (AU). For example, for Mercury the formula gives 0.4AU, though the actual mean value is 0.387AU. For Earth, the distance is correct. For Mars 1.6AU, (actually 1.524), and for Jupiter 5.2AU (actually 5.203). Bode's law fails to predict the positions of Neptune and Pluto, but its validity seemed to be confirmed when Uranus was predicted and discovered at 19.2AU in 1781. The law predicts that a body should be orbiting between Jupiter and Mars. Nothing of planetary size was discovered though astronomers searched the skies for decades. Then in 1801, an Italian astronomer, Guiseppe Piazzi (1746–1826) discovered a small planetoid at 2.8AU. He named it Ceres after the patron goddess of Sicily. It turned out to be the largest of the minor planets (asteroids). By 1890 over 300 asteroids had been charted in the region where Bode's law indicated there should be a planet. □ see SOLAR SYSTEM, ASTRONOMICAL UNIT, LAGRANGIAN POINTS

boiloff the vaporization of liquid hydrogen or liquid oxygen from a propellant tank while a rocket is being readied for launch □ see BLEED

boost to provide additional power, pressure, or force ⟨to ~ a radio signal⟩

booster – see BOOSTER ROCKET

booster engine – see BOOSTER ROCKET

booster rocket *also* **booster engine, booster, rocket booster** a rocket engine which assists the normal propulsive system during some stage of a mission, esp during its launch phase ⟨*Shuttle* ~s⟩ □ see SPACE SHUTTLE, SOLID ROCKET BOOSTER

bowshock – see MAGNETOSPHERE

Braun, Wernher Magnus Maximillian von (1912–77) German-American rocket engineer (son of a baron) who helped in the development of the V-2 missile in Germany during World War 2, gave himself up to the Americans in 1945, led the team which put the first US artificial satellite, Explorer 1, in orbit in 1958, and was the driving force behind the development of Apollo which took American astronauts to the Moon in 1969. □ see V-2, EXPLORER 1, APOLLO, ARMSTRONG, ROCKET ENGINE, GODDARD, EHRICKE

British Aerospace (abbr BAe) a British company with over 20 years of experience in space technology, the European leader in space communications, and the supplier of European communications satellites. BAe designed, built, and launched (1967) the first all-British satellite, Ariel 3, and, in 1968, BAe launched Esro 2 (a satellite designed to collect scientific data) for the European Space Agency (ESA). BAe is currently designing the Space Platform which is to work in conjunction with the NASA space station in the 1990s. The 11 tonne Space Platform would be taken into orbit aboard the Space

Shuttle for manned assembly in situ: solar panels would provide it with 35kW of electrical power to operate scientific and communication payloads serviced by personnel from the space station. In partnership with other European countries and Canada, BAe is now building Olympus 1 for ESA. The Olympus satellites will be the most advanced communications satellites in the world. Olympus 1 will generate 3.5kW of electrical energy from its solar panels to provide a variety of broadcasting and telecommunications services. These satellites produce waste heat due to the inefficiencies of the electronic equipment carried inside their bodies. This waste energy is dissipated through the 'north' and 'south' faces of the satellite which do not face the Sun. The three-axis stabilization of these satellites is adequate for input powers from the solar panels of up to 8kW. Above this figure, three-axis stabilization is inadequate for handling the waste heat and the ever-increasing communications traffic they will be called upon to handle. BAe is looking into the design of the Big Communicator where the electronic equipment is contained in a rotating drum which is kept in the shade of the large solar panels and radiates excess heat from the payload through its structure. BAe is also looking further ahead into the design of a Horizontal Take-off and Landing (HOTOL) launcher which will provide a more efficient and more economic launching than today's vertical take-off rockets. □ see NASA, EUROPEAN SPACE AGENCY, HOTOL, DIRECT BROADCAST SATELLITE, COLUMBUS, US SPACE STATION, HUBBLE SPACE TELESCOPE

British Interplanetary Society (abbr **BIS**) an organization formed in 1933 to promote the exploration and use of space. Through a range of literature, lectures, and conferences BIS constantly stimulates new ventures. Its monthly magazine *Spaceflight* keeps its 3500 members updated in space matters, and technical data is published monthly in *JBIS* (*Journal of* the *British Interplanetary Society*). In the 1930s BIS prepared detailed designs for a manned Moon spaceship – three decades before the real thing – and in the late 1970s it produced a blueprint for an unmanned starship, *Daedalus*, to cross the vast gulf of interstellar space. □ see STARSHIP, NASA, DAEDALUS

broadcast to transmit the same message to a widely scattered audience □ see DIRECT BROADCAST SATELLITE, COSPAS-SARSAT

buffeting the shaking or beating of part or all of an aircraft or aerospace vehicle that results from turbulent airflow over its surfaces □ see POGO, RUMBLE

bug anything which prevents a computer program, device, or circuit from working properly ⟨*a software* ~ *in a spacecraft's computer*⟩

buggy *also* **rover** a remote-controlled wheeled or tracked vehicle for exploring the surface of a planet or satellite. A buggy is usu controlled by an on-board computer which can be overridden by commands direct from Earth, or via a parent spacecraft in orbit round the planet □ see LUNAR ROVER

bulkhead a strong partition separating compartments in a space station which are at different pressures □ see AIRLOCK

burn a firing of a rocket engine ⟨*the third ~ placed the spacecraft on a trajectory to Mars*⟩ □ see BOOSTER ROCKET, LIFT-OFF, APOGEE MOTOR

burnout velocity the speed of a rocket after its engine has ceased firing

burn rate the rate (eg in kgs⁻¹) at which a solid propellant is used up in a rocket engine □ see SPECIFIC IMPULSE

burnup 1 the disintegration and vaporization of a satellite or other spacecraft when it enters a planetary atmosphere at high-speed □ see SHOOTING STAR, ABLATE **2** the percentage of nuclear fuel which has undergone fission in a nuclear reactor □ see NUCLEAR FISSION

burst any sudden short-lived pulse of radiation or particles ⟨*radio ~s from Jupiter*⟩ □ see JUPITER, X-RAY ASTRONOMY, GAMMA-RAY OBSERVATORY, PULSAR, VEGA

c the symbol used for the speed of light □ see SPEED OF LIGHT, ELECTROMAGNETIC RADIATION, SPECIAL RELATIVITY
C the symbol used for the degree celsius (formerly centigrade) □ see ABSOLUTE TEMPERATURE SCALE, KELVIN
cable a group of electrical conductors or optical fibres running side-by-side in a protective sheath □ see CABLE TV, OPTICAL COMMUNICATIONS
cable TV (abbr **CATV**) the distribution of TV programmes by means of underground cables. Though well established in the USA, cable TV is still under review in the UK. As well as offering a very wide choice of programmes, cable TV can be used for shopping and banking from home, accessing library resources, voting on local affairs, entering competitions, and other interactive purposes. □ see DIRECT BROADCAST SATELLITE, COAXIAL CABLE
Callisto the fourth in increasing distance and the second largest (4900km diameter – about the size of Mercury) of Jupiter's Galilean moons. Flying above the north pole of Callisto on 6 March 1979, at a range of about 120 000km, the Voyager 1 spacecraft photographed a surface covered with small 20 to 30km diameter craters packed shoulder-to-shoulder. Callisto has far more craters than even Ganymede, which suggests that it is the oldest satellite so far discovered in the Jovian planets, as its surface has probably remained unchanged since the final stages of planet formation 4000 million years ago. The Voyager photographs revealed one very prominent feature, unique in the Solar System, on Callisto's surface: a huge 2500km-diameter bull's-eye formation which has a light coloured central basin, about 300km in diameter, surrounded by at least eight concentric evenly-spaced mountainous rings. Unlike some of our Moon's giant craters, this feature does not have any ejected material radiating from its centre, though it is clearly caused by the impact of a large object. Perhaps this object broke through Callisto's surface and was completely swallowed up in its interior. The concentric circles are then the frozen waves which rippled out from the centre. Callisto is the darkest of the Galilean satellites having an albedo about 0.17 which is about half as reflective as its sister moon Ganymede. This suggests that Callisto has a surface composed of a frozen mixture of ice and rock. □ see JUPITER, IO, EUROPA, GANYMEDE, ALBEDO, VOYAGER
camera – see SHUTTLE CAMERAS, VIDEO CAMERA, CHARGE-COUPLED DEVICE
canals linear markings on the surface of Mars which were mapped in detailed drawings by the Italian astronomer, Giovanni Schiaparelli, in 1877. He called them 'canali', meaning 'channels', but the Italian word was mistranslated into the English word 'canals'. This word, combined with the suspicious straightness of the lines, set the public's

imagination alight, helped by the American astronomer, Percival
Lowell (1855–1916), who championed the theory that there was
intelligent life on Mars. Using his own telescope specially built for the
task, he saw more than Schiaparelli ever did; his sketches showed
'oases' linking canals together, and seasonal changes which he thought
marked the ebb and flow of agriculture. The suggestion was that the
lines were irrigation ditches dug by intelligent Martians to conserve
and distribute water across an arid planet. Martian fever lasted well
into the 20th century but modern telescopes, and in particular the
observations made by the Mariner flyby and Viking lander spacecraft,
have shown conclusively that Martian canals do not exist. However,
close-up pictures of Mars do show clear evidence of giant canyons and
riverbeds which have the marks characteristic of flowing water. If, as
suggested, water is frozen under the Martian soil, what might trigger
its melting to form large rivers flowing across the surface? And did the
most recent melting of the ice coincide with the period when
Schiaparelli and Lowell focussed their telescopes on Mars? □ see
MARS, VIKING, MARINER, VOYAGER, PHOBOS AND DEIMOS

canard an aerodynamic vehicle that has horizontal control surfaces
mounted in front of the main lifting surfaces

Canopus *also* **Alpha Carina** (Alpha Carina) the brightest star in the
constellation, Carina, and the second brightest star (after Sirius) in the
sky. Canopus is often used as a guide star by spacecraft (eg Voyager)
navigating the Solar System. To Earth-bound observers, Canopus is
visible only in the Southern hemisphere.

Cape Canaveral part of the Florida coastline, USA, and the site of
the Kennedy Space Centre which comprises a complex of launching
pads from which many rockets, including Space Shuttle, have been
launched. Cape Canaveral also has an industrial complex, offices,
astronaut quarters, and spacecraft assembly areas where cargoes for
the Space Shuttle and Spacelab are readied for launch. □ see VAN-
DENBERG AIR FORCE BASE, LAUNCH PAD, SPACE SHUTTLE, APOLLO,
JOHNSON SPACE CENTRE, COSMODROME

captive firing – see HOT TEST

captive test – see HOT TEST

capture to bring under control 〈*to ~ a satellite using the Remote
Manipulator System*〉 □ see RMS

captured rotation – see SYNCHRONOUS ROTATION

carbon (symbol **C**) an element of atomic number 8 and, in its most
abundant form, of mass number 12, which forms the basis for life on
Earth. Carbon is made inside stars by the fusion of three helium
nuclei: first the coming together of two helium nuclei forms the unstable beryllium nucleus, and then the beryllium captures a third helium
nucleus. Astrophysicists have noted that the capture of the third
helium nucleus is a rather fortuitous event (for life on Earth!): the
thermal energy inside stars is just right for the chance matching of the
(nuclear) resonances of the helium and beryllium nuclei, and hence
their mutual capture. Furthermore, this newly synthesized carbon sur-

vives the subsequent nuclear activity inside the star. As the star ages, it progressively burns these lighter nuclei to produce heavier nuclei, ie oxygen is formed by the addition of yet another helium nucleus to the carbon nucleus. But this time the resonances of helium and carbon do not coincide, and a large proportion of the carbon survives to be flung into space by the subsequent explosion of the star as a supernova. As a constituent of the cosmic dust which condensed to form the Solar System, carbon subsequently became the basis of the molecules of life on Earth. It is tempting to suggest, as Hoyle the astrophysicist has, that the delicate positioning of the nuclear resonances for carbon and oxygen is the work of a superintellect, who (or which) has tampered with physics and chemistry to ensure the subsequent development of terrestrial life. □ see NUCLEOSYNTHESIS, SUPERNOVA, HOYLE, ANTHROPIC PRINCIPLE

cargo any equipment (eg experimental packages provided by a user) that is carried by the Space Shuttle and which is not part of the basic payload as are, for example, space suits carried for astronauts □ see PROGRESS

cargo bay – see PAYLOAD BAY

carrier – see CARRIER WAVE

carrier wave (abbr **CW**) *also* **carrier** a relatively high frequency wave (eg a radio wave) by which a message is sent along a communications channel at a lower speed □ see MODULATE

Cass A – see CASSIOPEIA A

Cassegrain telescope a reflecting telescope which has a pierced primary mirror through which light passes to the final image position after reflection from a hyperboloid secondary mirror placed on the telescope's axis. The secondary mirror increases the magnification of the telescope without undue increase in its length. The Cassegrain telescope was designed in 1672 by the Frenchman, Guillaume Cassegrain; it is compact and easily mounted on its support, and it was the design chosen for the orbiting Hubble Space Telescope. □ see REFLECTING TELESCOPE, HUBBLE SPACE TELESCOPE, HYPERBOLOID

Cassini an interplanetary spacecraft which may be launched in 1993–4 and go into orbit around Saturn in the year 2000. The Cassini project is a joint ESA-NASA proposal and would be designed to carry out a more sophisticated and extended study of this interesting planet than was possible with the Pioneer and Voyager flyby spacecraft. Cassini would carry a probe for injection into Titan's atmosphere. □ see ¹GALILEO, VOYAGER, SATURN, TITAN

Cassiopeia A (abbr **Cass A**) an intense source of radio waves situated in the constellation Cassiopeia about 7 kiloparsecs distant. It is the remnant of a supernova which probably took place in the late 17th century though, surprisingly, it was never recorded. It also emits X rays and its radiations are frequently observed by observatories in space. □ see RADIO ASTRONOMY, X-RAY ASTRONOMY, VERY LARGE ARRAY, SUPERNOVA

cathode the terminal of a device (eg an electric welder) away from

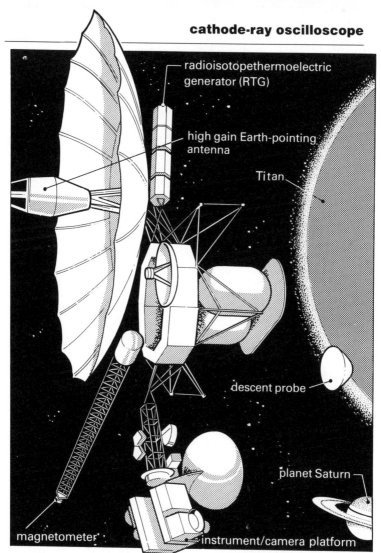

radioisotopethermoelectric generator (RTG)

high gain Earth-pointing antenna

Titan

descent probe

planet Saturn

magnetometer

instrument/camera platform

Cassini

which electrons flow in an external circuit – compare ANODE

cathode-ray oscilloscope (abbr **CRO**) *also* **oscilloscope** a test and measurement instrument for showing the patterns of electrical waveforms and for measuring their frequency and other characteristics. The cathode-ray oscilloscope is an indispensible aid to circuit designers and test engineers, esp those involved in the design and maintenance of computer and telecommunications systems. Although most oscilloscopes are based on the cathode-ray tube, recent designs make use of a large-screen liquid crystal display for increased portability and lower power consumption. □ see VDU

cathode-ray tube (abbr **CRT**) a glass tube (eg a TV tube) in which a beam of electrons moves in a vacuum to 'draw' information on a screen. The electron beam is produced by an electron gun and it is focussed and deflected by coils and/or electrodes. □ see VDU

CATV – see CABLE TV

C-band a range of radio waves having frequencies in the range 3.9GHz to 5.2GHz. The C-band is one of the bands used by the Intelsat 5A communications satellites operating at 6GHz uplink and 4GHz downlink. □ see BAND, X-BAND, K-BAND, INTELSAT

CCD – see CHARGE-COUPLED DEVICE

CCE [*Charge Composition Explorer*] – see ACTIVE MAGNETOSPHERIC PARTICLE EXPLORER

celestial of or in the heavens ⟨*a comet is a ~ object*⟩

celestial coordinates coordinates which locate the position of a star on the celestial sphere. There are four main coordinate systems: the equatorial, horizontal, ecliptic, and galactic coordinate systems. In the equatorial system, right ascension locates a star in the east-west direction, and declination locates a star's position north or south of the celestial equator. □ see DECLINATION, RIGHT ASCENSION

celestial equator the great circle in the celestial sphere which is an extension of the Earth's equatorial plane. It is the reference plane for the equatorial coordinates, right ascension, and declination. □ see CELESTIAL SPHERE, CELESTIAL COORDINATES

celestial guidance – see GUIDANCE

celestial mechanics the study of how planets, artificial satellites, spacecraft, and other celestial bodies move through space under the influence of gravitational fields □ see NEWTON'S LAWS OF MOTION

celestial sphere an imaginary sphere of immense radius centred on the Earth, on which the stars appear to be fixed. The principal reference circles on the celestial sphere are the celestial equator, the ecliptic, and an observer's horizon. □ see CELESTIAL COORDINATES

cellular radio a reliable and fast computer-controlled communications system for people on the move, especially in-car telephone subscribers. Areas of a city, town, or other region are divided into radio transmission zones known as cells. Each cell has its own transmitter/receiver, usu on a high building. A cluster of cells is controlled by a central computer so that calls made by a car driver, cyclist, or rambler are routed to the public telephone system or to another group of cells. A scanning system keeps track of the signal level. Should it fade in strength, the computer automatically transfers the call to another transmitter/receiver in another cell without the caller knowing. And, of course, as with 'normal' telephone calls, international calls are connected using existing communications satellites. Cellular radio can easily handle 50 000 calls per hour compared with 15 000 calls using conventional shortwave radio, and it is generally regarded as being the most important innovation in telephony since its invention in the 1870s. □ see COMMUNICATIONS SATELLITE

CELSS [Controlled Ecological Life Support System] – see SPACE GREENHOUSE

Centaur a liquid hydrogen-fuelled rocket engine which is lifted into Earth orbit by the Space Shuttle and then used to launch satellites and spacecraft. The payload bays of some of the current Space Shuttles have been modified to take the new Centaur. This Shuttle/Centaur combination is scheduled to send the Galileo spacecraft on its way to Jupiter. Once the craft is in orbit, the cradle holding the Centaur in the Shuttle payload bay is raised by 45° and the crew fires an explosive cord to release the Centaur and spacecraft; springs then push it out into space. Once the Centaur is free, the Shuttle moves off and turns to protect its windows from the rocket exhaust. The Shuttle/Centaur development is a joint effort between NASA and the US Air Force (USAF) and two versions of the Centaur have been developed: the G-prime type for NASA is 8.9m long and 4.33m in diameter, and the 4.95m long (same diameter) version for USAF – shorter to enable 4800kg, 12m long satellites to be launched into geostationary orbit.
□ see SPACE SHUTTLE, CRYOGENIC ENGINE, [1]GALILEO, INERTIAL UPPER STAGE, [2]TITAN, ATLAS

Centre National d'Etudes Spatiales (abbr **CNES**) the French space agency formed in 1962. CNES launched its first satellite by the Diament A launch vehicle in 1965 which made France the world's third space-faring nation. CNES proposed the 3-stage Ariane rocket as the European launch vehicle. CNES cooperates with NASA and ESA in its gamma-ray astronomy programme, and gives strong support to the Soviet Academy of Sciences in its exploration of Venus. French scientists have produced experiments for the Venera 17 and 18 Venus flyby and lander spacecraft, and for the Vega missions to Venus and Halley's comet. □ see ARIANE, VEGA, VENERA

centre of inertia – see CENTRE OF MASS

centre of mass *also* **centre of inertia, barycentre** the point in a system (eg two stars orbiting each other) at which the total mass of the system can be considered as being concentrated. The centre of mass of the Earth-Moon system lies somewhere between the Earth and the Moon.

centrifugal force the force that appears to act outwards on bodies rotated in a circular path. The centrifugal force is directed along the radius of the path and is equal to $mr\omega^2$, where m is the mass of the body, r is the distance from the centre of rotation, and ω is the angular velocity (in radians per second) of the body. Future space stations will make use of centrifugal force to provide artificial gravity. A cylinder of radius 10 metres, rotating once every 6 seconds would simulate Earth's gravity on someone standing on its inside surface.
□ see CENTRIFUGE, SPACE STATION

centrifuge a large motor-driven device which has a long arm at one end in which human and animal subjects can be revolved and rotated at high speed. Riding in a centrifuge is an essential part of an astronaut's training for it equips him or her to tolerate rapid accelera-

tions experienced at launch and at other phases of a rocket flight.
□ see SPACE MEDICINE

centripetal force a force that causes a body to follow a curved
path, the force acting towards the centre of curvature of the body's
motion. The positive electrical charge on the nucleus of an atom
causes a centripetal force to act on negatively charged electrons making them follow a curved path round the nucleus. Gravitation
produces centripetal forces on orbiting Earth satellites – compare
CENTRIFUGAL FORCE

Cepheid variable one of a class of very luminous yellow or orange
giant stars whose brightness, or stellar magnitude, pulsates in a regular
way. The prototype Cepheid variable is Delta Cepheid, discovered in
1784, which varies in brightness over a period of 5.36 days. The period
of Cepheid variables ranges from a few days to several weeks and is
caused by regular changes of their temperature and radius as pulsa-
tions occur in their outer layers. Over 700 Cepheid variables are
known in our own Milky Way galaxy, and several thousand in the
Local Group. From 1908 to 1912 the astronomer, Henrietta Leavitt,
discovered a relationship between the brightness (absolute magnitude)
and the period of Cepheid variables: the longer the period, the brighter the star. This discovery led to a period-luminosity law which has
become an interstellar scale for working out distances to galaxies
which contain Cepheid variables. The scale is first calibrated by
measuring the absolute magnitude of Cepheid variables at known distances. Then, once the period of a Cepheid variable in a distant galaxy
is known, its absolute magnitude can be obtained from the period-
luminosity law. But, its apparent magnitude is less than its absolute
magnitude due to the dimming of its light before it reaches us, so if we
assume that, in clear space, the dimming of the light is caused by the
inverse square law (ie doubling the distance reduces the apparent
brightness by a factor of 4) the distance of the Cepheid variable (and
hence the galaxy in which it 'lives') can be determined. □ see HUBBLE,
MAGNITUDE, CLUSTER 3, GIANT STAR, T TAURI STARS

cermet [*cer*amic-*met*al] material comprising ceramic particles
bonded with a metal. Cermet is used as a resistive material in
electronic components and in aerospace applications because of its
great strength and resistance to high temperatures.

CFES – see CONTINUOUS-FLOW ELECTROPHORESIS SYSTEM

chain reaction a self-sustaining chemical or nuclear reaction which
produces energy, or products that cause further reactions of the same
kind □ see NUCLEAR FISSION

charge a basic property of matter that occurs in separate natural
units and is considered as negative or positive ⟨*the negative ~ of an
electron*⟩ □ see COULOMB, PROTON, ELECTRON

Charge Composition Explorer (abbr CCE) – see ACTIVE MAG-
NETOSPHERIC PARTICLE EXPLORER

charge-coupled device (abbr CCD) a data storage device consisting of an array of metal-oxide semiconductor cells on a silicon chip.

Each cell stores a small electric charge representing a bit of information which can be moved through the memory by means of electrodes formed on the surface of the chip. CCDs are being increasingly used in television cameras for storing images, especially in the field of robotics, since a CCD can recognize not just black and white but also a range of grey tones. The CCD is about 30 times more sensitive than a photographic plate so it is used in astronomy for taking pictures of faint celestial objects in minutes rather than hours. In May 1983, some of the best images of Neptune ever recorded were obtained using a CCD positioned at the focus of the 2.5m reflector of Las Campanas Observatory in Chile. □ see PLANET-A, TELESCOPE, NEPTUNE

Charon – see PLUTO

CHASE [*Coronal Helium Abundance Spacelab Experiment*] – see HELIUM

chase pilot the pilot of an aircraft that follows and advises the pilot of another aircraft undergoing trials (eg unpowered flight tests of the Space Shuttle)

checkout any sequence of steps taken to check the readiness of a system for action, or to familiarize a person with the operation of something ⟨*pre-launch* ~⟩

chemical energy energy changes (eg the burning of liquid hydrogen in liquid oxygen) which produce new substances (eg water) but leave the atoms unchanged. In chemical reactions, energy changes liberate heat (are exothermic) or absorb heat (are endothermic) and involve changes only to the outermost electrons of the atoms leaving the nuclei intact – compare NUCLEAR ENERGY □ SEE ATOM, IONIZATION

chemical fuel a (rocket) fuel (eg liquid hydrogen) that reacts with an oxidizer (eg liquid oxygen) to produce hot gases that provide thrust – compare NUCLEAR ENERGY □ SEE ROCKET ENGINE

chilldown the cooling of all parts of a cryogenic engine prior to ignition by circulating cryogenic liquid (eg liquid hydrogen) through the system □ see CRYOGENIC ROCKET

chromatic aberration a fault of lenses (but not of mirrors) that gives images false coloured edges. It arises because the lens is wedge-shaped (like two prisms with their bases in contact) and this causes white light to disperse into its constituent colours. □ see ABERRATION, REFLECTING TELESCOPE, REFRACTING TELESCOPE, NEWTON

chromosphere – see SUN

civilian not supported by or existing for the armed forced ⟨*INMARSAT is for* ~ *use*⟩

Clarke, Arthur C (*b*1917) science fiction writer and early member of the British Interplanetary Society which was founded in 1933. In the October 1945 issue of the journal *Wireless World* (before the days of artificial satellites) he was the first to propose the use of satellites orbiting in geostationary orbits for global television transmissions. He cooperated with Stanley Kubrick in the MGM production of his novel *2001: A Space Odyssey*, and recently in its sequel *2010: Odyssey Two*. Clarke's writings combine strict technological accuracy with a belief in a mystical significance for human destiny. For example, in *Child-hood's End,* devil-like superior aliens come to Earth to assist mankind in achieving his cosmic destiny. And in *Rendezvous with Rama* he describes the close transit and exploration by a giant alien spacecraft. □ see GEOSTATIONARY ORBIT, SEARCH FOR EXTRATERRESTRIAL INTELLIGENCE

Clarke orbit – see GEOSTATIONARY ORBIT

closed-loop *of a control system (eg a servosystem)* having a means of controlling the output by feeding a proportion of the output signal to the input so as to oppose the input – compare OPEN-LOOP

closed-loop system a system in which the output is used to control the input □ see SERVOSYSTEM

closing rate the speed at which two spacecraft or satellites approach each other for an (orbital) rendezvous □ see RENDEZVOUS

cluster 1 two or more rocket motors joined together to operate as a single propulsion unit **2** a group of stars in the Galaxy whose members are close enough together to be influenced by each other's gravitational attraction. The stars of open clusters are loosely bound by gravitational attraction, and contain up to several hundred stars (eg Pleiades). Globular clusters are compact spherical groups of several thousand to a million stars (eg the Great Cluster in Hercules). Open clusters are generally young, ie a few tens of million years old, and star formation is still taking place in them. The stars are so loosely bound together that open clusters may only survive a few revolutions of the Galaxy. Globular clusters are distributed in and out of the galactic plane. The stars in them are older than the Sun and were probably formed early in the life of the Galaxy, perhaps 12 thousand million years ago – compare CLUSTER 3 □ SEE STAR **3** a group of galaxies that may contain a few thousand members (eg the Perseus cluster) but generally contain only two or three galaxies. Our own Galaxy is a member of the Local Group, an irregular assembly of about 30 known members that includes the Andromeda galaxy and the Large Magellanic cloud – compare CLUSTER 2 □ SEE GALAXY, MILKY WAY, VIRGO A, LOCAL GROUP

CNES – see CENTRE NATIONAL D'ETUDES SPATIALES

coated lens – see BLOOMED LENS

coaxial cable an electrical conductor (eg a TV aerial) for carrying high frequency electrical signals. A coaxial cable has a central wire surrounded by a layer of insulation, and then by a sheath of copper mesh (braid) which reduces electrical interference between neighbouring cables. □ see FEEDER, CABLE TV, OPTICAL COMMUNICATIONS

coherent light any light, esp the light from a laser, which comprises many waves all in step (in phase) with each other □ see LASER

collimate 1 to make parallel ⟨*to ~ a beam of light*⟩ **2** to adjust optical components in an optical instrument so that they are in proper alignment to each other

colonize to make a place for people to live and work as a community ⟨*to ~ Mars*⟩ □ see MOON BASE, SPACE SETTLEMENT

Columbus a long-term programme being undertaken by Germany and Italy for the European Space Agency (ESA), to build on the technologies of Spacelab and Eureca in the scientific and commercial use of space. There are three main parts to the programme: a pressurized module (PM), a payload carrier (PC), and a resources module (RM). The PM would be a further development of Spacelab and be manned or man-tended depending on its payload. The PC would be developed from Eureca and be designed to carry experiments, materials, processing equipment, etc. The RM would provide power communications, data management, and other housekeeping facilities for the PM and PC. Demonstration missions, launched by the Space Shuttle or Ariane, are planned for 1993. The PM would initially be attached to and serviced from the US space station. But by the end of the century, there is a possibility that *Columbus* would separate from the US Space Station programme and become an independent European system serviced by European launches. *Columbus* is funded by the German BMFT (the Ministry of Research and Technology), and the Italian MRST (the Ministry of Scientific and Technological Research). Christopher Columbus sailed westwards from Spain on 3 August 1492 for what he thought were the Indies. In the 1990s, the first flights of *Columbus* will be taking place about 500 years after his voyage of exploration. □ see SPACELAB, EURECA, US SPACE STATION, EUROPEAN SPACE AGENCY

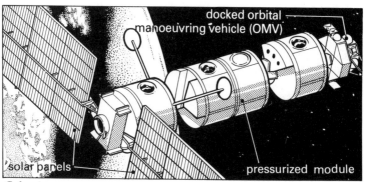

Columbus

coma 1 the gaseous envelope that is seen to develop round the nucleus of a comet as it approaches the Sun □ see COMET **2** an aberration of an optical system (eg a telescope) in which the image of a point source of light is formed as a comet-shaped blur rather than as a point of light □ see ABERRATION, IMAGE

combustion chamber that part of a rocket engine in which fuel burns at high pressure □ see NOZZLE, ROCKET ENGINE

comet a (loose) assembly of frozen gases and dust orbiting the Sun, some of which vaporizes on nearing the Sun to produce a long veil-like

command

luminous tail. Since comets usu have very elliptical orbits, they spend most of their time out of sight in the far-reaches of the Solar System. They are of particular interest to astrophysicists since the material from which they are made is thought to be samples of the primordial material from which the planets were formed 4 thousand million years ago. Space technology has given us the means to examine comets at close-range and missions have sent spacecraft to rendezvous with Halley's comet in 1986 as it nears the Sun. This comet is the most spectacular of the long-period comets (it returns to our region of the Solar System every 76 years). Of particular interest is the composition of the comet's nucleus which is believed to consist of frozen gases (eg methane and ammonia) and dust. On being heated by the Sun's rays as it approaches perihelion, the nucleus of a comet forms a coma which extends away from the Sun to form a 100 thousand kilometre long tail. □ see [1]GIOTTO, SOLAR WIND, PERIHELION, HALLEY'S COMET

command 1 a signal transmitted to a spacecraft which starts something (eg switches on cameras) in the spacecraft **2** a computer programming instruction telling a computer what to do

command destruct a control system that is activated by radio to destroy an airborne rocket which becomes a safety hazard

commander the crew member of a manned spacecraft who has overall responsibility for the safety of the astronauts aboard, and who has the authority to make changes to the prearranged flight plan in the event of possible danger to crew or spacecraft ⟨*Robert Crippen was ~ of Shuttle Mission 7 launched on 18 June 1983*⟩. □ see MISSION SPECIALIST, PAYLOAD SPECIALIST, PILOT

command guidance – see GUIDANCE

command module – see APOLLO

commercial supported by or existing for profit ⟨*Intelsat is a ~ organization*⟩ – compare MILITARY

communicate to send information from one place to another

communications satellite (abbr **comsat**) any of many artificial satellites put into orbit round the Earth to relay data, radio, and TV programmes between distant Earth stations across the world. Most communications satellites carry solar panels to power their complex circuits and control equipment. They are usu placed in geostationary orbit so that Earth stations can keep them 'in sight' day and night. □ see CLARKE, GEOSTATIONARY ORBIT, SOLAR CELL, DIRECT BROADCAST SATELLITE, BIG COMMUNICATOR, MOLNIYA, BRITISH AEROSPACE, INTELSAT

communications system an arrangement of components, circuits, and equipment which enables messages to be sent from one place to another ⟨*Intelsat is part of a global ~*⟩ □ see COMMUNICATIONS SATELLITE

companion body a last stage rocket, nose cone, or other body that accompanies an artificial Earth satellite – compare AFTER BODY

component a part (eg a bolt) or a combination of parts (eg a rocket motor) having a particular purpose □ see MODULE, SYSTEM

composite materials structural materials of metal, ceramic, or plastic which are strengthened by built-in filaments, foils, powders, or threads of a different material

compression chamber a strongly built chamber in which air can be pressurized to several times normal atmospheric pressure

computer science the study of how computers work, the mathematics of their operation, and the languages with which we communicate with them □ see FIFTH GENERATION COMPUTER

computer system one or more computers and associated equipment used for solving problems □ see SUPERCOMPUTER

computing 1 the process of working with numbers to solve a problem ⟨*navigational ~ used to plot the course of a spacecraft*⟩ **2** the science, activity, or profession of using computers ⟨*a course on ~*⟩ □ see COMPUTER SCIENCE

comsat – see COMMUNICATIONS SATELLITE

concave curving inwards ⟨*a ~ mirror*⟩ □ see PARABOLOID, REFLECTING TELESCOPE

condensation shock wave a (shortlived) trail of fog formed over the wing and other surfaces of a rapidly moving aircraft or aerospace vehicle. The fog is condensed water vapour in air which has been rapidly cooled by a sudden fall in pressure as it flows over the aircraft's surface – compare CONDENSATION TRAIL

condensation trail *also* **contrail, vapour trail** a trail of condensed water vapour or ice particles formed behind an aircraft or aerospace vehicle. The water vapour is generally produced as aviation or rocket fuel is burnt, especially when liquid hydrogen burns in liquid oxygen – compare CONDENSATION SHOCK WAVE

configuration the way parts of equipment or systems are arranged ⟨*the ~ of a space station*⟩

conic section a family of curves obtained by cutting through sections of a right circular cone at different angles. Thus a horizontal section gives a circle; an inclined one an ellipse; one parallel to the edge of the cone a parabola; and one with an even greater inclination a hyperbola. Conic sections are important in astronomy and space technology for they represent the paths of bodies (eg planets and artificial satellites) that move in a gravitational field about a parent body. □ see KEPLER'S LAWS, ELLIPSE, HYPERBOLA

conjunction the lining up of two celestial bodies so that they have the same longitude as seen from Earth. Most often, conjunction refers to that between a planet and the Sun. Inferior conjunction occurs when Mercury or Venus lies between the Sun and the Earth; superior conjunction when the planet is on the far side of the Sun. Planets (eg Mars and Jupiter) whose orbits lie outside the Earth's, cannot be in inferior conjunction. Two planets, the Moon and the Sun, or the Moon and a planet can be in conjunction. Conjunction is also the moment at which the two bodies line up. □ see ELONGATION, OPPOSITION, ECLIPSE

console an array of controls and indicators for monitoring the working of something ⟨*an instrument* ~ *in Spacelab*⟩

constellation any of the 88 areas into which the celestial sphere is divided. The constellations contain groups of stars, many of which were given mythological names (eg Perseus) by early civilizations. In 1930, the International Astronomical Union unambiguously divided the celestial sphere into the present areas. Stars within a constellation are identified by a Latin name preceded by a letter of the Greek alphabet (eg Alpha Ursae Minoris – the brightest star in the constellation 'Little Bear', and more familiarly known as the Pole Star). □ see MAGNITUDE

continuous-flow electrophoresis system (abbr **CFES**) a method for separating the components of biological materials using an electrical field. Particles of different charges and masses travel at different speeds through a solution towards an oppositely charged electrode. The electric field is applied across the flow, and the different particles are collected at separate outlets. The CFES experiment was first carried aboard the Space Shuttle (STS-6) in January 1983, and is part of a continuing series of experiments which aims for commercial success. In zero gravity, the CFES produces improved yield and purity of materials than on Earth. □ see MATERIALS PROCESSING, MONODISPERSE LATEX REACTION

continuous spectrum a spectrum consisting of an unbroken band of wavelengths. The spectrum of the Sun's photosphere is a continuous spectrum ranging from ultraviolet, through visible, to infrared. □ see ELECTROMAGNETIC SPECTRUM, EMISSION SPECTRUM, ABSORPTION SPECTRUM

contrail – see CONDENSATION TRAIL

¹control to direct or regulate something ⟨*the Voyager camera scan platform can be* ~led *from Earth*⟩

²control a device for directing or regulating something ⟨*a thrust* ~⟩

Controlled Ecological Life Support System (abbr **CELSS**) – see SPACE GREENHOUSE

controller a device that converts an input signal from a controlled quantity (eg fuel flow rate to an engine) into a signal which operates a servomechanism (eg a valve) to control that quantity □ see SERVOMECHANISM, THERMAL CONTROL SYSTEM

control station a usu complex console or building which masterminds the operation of something □ see DEEP SPACE NETWORK, EARTH STATION, MISSION CONTROL

convex curving outwards ⟨*a* ~ *mirror*⟩ □ see REFLECTING TELESCOPE

coolant a liquid or gas circulated through pipes to remove heat from something □ see SPACESUIT

¹Copernicus – see ORBITING ASTRONOMICAL OBSERVATORY

²Copernicus, Nicolas (1473–1543) Polish astronomer who became interested in astronomy while studying medicine and canon law in Italy. At that time it was generally believed that the Earth was at the centre of the universe with the Sun, stars, and planets revolving round it, a view put forward by the ancient Greek, Ptolemy, and firmly believed by the Catholic Church in Copernicus's time. Copernicus realized, and went on to prove, that calculating the positions of the planets would be much easier if the planets were moving through space about the Sun. In 1543, his revolutionary views about the universe were published in his book *On the Revolutions of the Heavenly Spheres* in which he wrote: "In the centre of everything is the Sun. Nor could anyone have placed this luminary at any other, better point in this beautiful temple, than that from which it can illuminate everything uniformly". The book began to make converts immediately. The scientific revolution began with Copernicus, but the idea of placing the Earth in orbit round the Sun was hard for the Church to accept: it was not until 1835 that Copernicus's book was removed from the list of those banned by the Catholic Church. □ see ²KEPLER, ²GALILEO, NEWTON

core 1 the central portion of a nuclear reactor where fission of nuclear fuel produces heat □ see NUCLEAR FISSION **2** the central region of a star where thermonuclear reactions produce energy □ see SUN, STAR, NUCLEAR FUSION **3** the centre of the Earth or other planet or satellite

Coronal Helium Abundance Spacelab Experiment (abbr CHASE) – see HELIUM

cosmic of the universe ⟨~ *dust*⟩ □ see COSMIC RAYS

cosmic abundance the relative proportions of all the elements in the universe. The proportions are estimated from observations and measurements made within the Solar System. These include measuring the relative strengths of the lines in the spectrum of the Sun and the composition of meteorites, and geological surveys of the Earth, Moon, and Mars using spacecraft. The cosmic abundance, given as relative masses of elements, turns out to be: hydrogen 73%, helium 25%, oxygen 0.8%; carbon 0.3%; and less than 0.1% of each of neon, nitrogen, silicon, magnesium, and sulphur. All the other elements are present in minute proportions. □ see SUPERNOVA, NUCLEOSYNTHESIS, STAR, BIG BANG, ANTHROPIC PRINCIPLE, CARBON

cosmic background radiation electromagnetic radiation from space which has a more or less continuous spread of wavelengths. It is caused by innumerable individual sources of X rays, radio, infrared, and other wavelengths. One important component of this radiation is the microwave background radiation which has a maximum intensity at a wavelength of 2.5mm and is considered to be the radiation remaining from the Big Bang from which the universe began. □ see MICROWAVE BACKGROUND RADIATION, BIG BANG

cosmic dust (small) particles of rock found throughout the Galaxy. Cosmic dust has collected in the craters on the Moon, forms massive interstellar clouds which obscure the centre of the galaxy, and is the raw material from which stars are born. □ see MICROMETEOROID, COSMIC RAYS, ZODIACAL LIGHT

cosmic radiation – see COSMIC RAYS

cosmic ray hit a mysterious malfunction of some kind on a satellite or spacecraft (eg loss of attitude control), which is put down to damage caused by cosmic rays striking some sensitive electronic component

cosmic rays *also* **cosmic radiation** very penetrating atomic particles which bombard Earth from outer space. Nuclei of all atoms up to those with an atomic number of 56 (iron) are present, but these heavy nuclei are very rare in cosmic rays. Most of the particles are protons (nuclei of hydrogen), but a few electrons, positrons, neutrinos, and gamma ray photons are also present. Primary cosmic rays are those in space before meeting the Earth. On passing into the Earth's atmosphere most of the primary cosmic rays produce an 'air shower' of secondary cosmic rays, and most of these reach the Earth's surface. A Geiger tube shows that we are all exposed to this secondary radiation unless we happen to be in a deep mine which gives some protection. When a heavy nucleus in primary cosmic rays strikes an air molecule, most of the initial particles produced are charged and neutral pi mesons (pions). But neutral pions almost immediately decay into gamma rays which, in turn, give rise to cascades of positrons and electrons by pair production. The charged pions decay into mu mesons (muons) which subsequently decay into electrons and muon-neutrinos. In this way, a primary cosmic ray nucleus with an energy of, say 10^{15}eV produces over a million secondary particles – the 'air shower'. Many spacecraft have been fitted with detectors to measure the nature of cosmic rays and the direction from which they reach us from space. Most of the lighter nuclei are thought to be produced when heavier nuclei collide with interstellar dust in space. Some particles are created in supernova explosions, such as those which gave rise to the relatively near Crab nebula and the Vela pulsar. Solar flares on the Sun produce low energy cosmic rays. Most cosmic rays are generated within the Galaxy and confined within it by the galactic magnetic field which acts on the positive charge carried by the atomic nuclei in cosmic rays. Cosmic rays with the highest energy, ie greater than 10^{15}eV, probably have an extragalactic source. □ see NUCLEUS, ATOMIC NUMBER, POSITRON, NEUTRINO, GAMMA RAYS, SUPERNOVA, NUCLEOSYNTHESIS, CRAB NEBULA, MESON, PAIR PRODUCTION, INTERSTELLAR MEDIUM, ELECTRON VOLT, EXPLORER 1

cosmodrome a place from which Soviet space vehicles (eg Soyuz) are launched. The main cosmodrome in Siberia was once called Tyuratam (the nearest town), but is now called Baikonur after a town 370km away! □ see CAPE CANAVERAL

cosmogony – see COSMOLOGY

cosmological model a 'picture' of the universe in simple terms. Most models of the universe (eg the Big Bang theory) are based on advanced mathematics, especially Einstein's general theory of relativity, which is often applied to an expanding universe in which gravitation supplies the only force between galaxies. □ see BIG BANG, STEADY STATE, HUBBLE'S LAW, RELATIVITY

cosmology *also (but rarely)* **cosmogony** the study of the overall structure and origin of the universe. The variety and complexity of the stars, galaxies, and other objects that adorn the universe are so baffling that it would seem hopeless to try and find laws to describe what we see. However, the laws of physics, particularly the quantum theory, have such a tremendous predictive power that they can explain phenomena as diverse as the formation of a snowflake and the explosion of a supernova. The cornerstone of modern cosmology is Edwin Hubble's observation in the 1920s that the space between the galaxies is increasing, which implies that the universe is expanding. By working backwards from the known rate of expansion, it is possible to predict that about 18 thousand million years ago the universe was infinitely compressed and expanding infinitely rapidly. Much of modern cosmology rests on this idea: that the universe began with a cataclysmic convulsion, generally known as the Big Bang. □ see UNIVERSE, BIG BANG, HUBBLE'S LAW, STEADY STATE, QUANTUM THEORY

cosmonaut somebody from a communist country who travels beyond the Earth. On 12 April 1961 Yuri Gagarain of the Soviet Union became the first human being to orbit the Earth in a spacecraft – compare ASTRONAUT □ SEE LAIKA, SPACEWALK

Cosmos a continuing series of Soviet satellites which have a variety of missions, including photographic reconnaissance, and studying the composition of the upper atmosphere, electron densities in the ionosphere, and the interaction of the solar wind with the magnetosphere. Others have been used for military communications, biological studies, and testing rocket propulsion units. Since the launch of Cosmos 1 on 16 March 1962, over 1700 Cosmos satellites have been launched. For example, Cosmos 1443 docked automatically with the Salyut 7 space station in March 1983, doubling the station's capacity and adding 40m^2 of solar array, or 3kW of electrical power. Salyut 7 was boarded by the Soyuz T-9 cosmonauts who docked with the hybrid space station on 28 June 1983. This event marked the first manned use of a space station assembled automatically in space from modules. Cosmos 1609, launched on 14 November 1984 from Plesetsk, was a military photo-reconnaissance satellite and was recovered after 145 days to retrieve its film capsule. Cosmos 1614, launched on 19 December 1984 from Kapustin Yar, was a delta-winged vehicle about 3.5m long with a 2.5m span. Evidently this was a test and recovery flight of a spaceplane, and it landed in the Black Sea after a 110-minute orbit. □ see MOLNIYA, SALYUT, SOYUZ, GROUND-BASED ELECTRO-OPTICAL DEEP SPACE SURVEILLANCE, GLOBAL POSITIONING SYSTEM

cosmos the observed universe regarded as an orderly collection of galaxies, stars, planets, comets, etc. In this rather old-fashioned and restricted sense, 'cosmos' has a different meaning from 'universe' which embraces all things discovered and, as yet, undiscovered. □ see COSMIC, COSMOLOGY

COSPAS-SARSAT an international rescue satellite relay system presently using three Soviet and one American satellite. COSPAS is an acronym based on the Russian words for 'search and rescue': SAR-SAT is an acronym for 'search and rescue satellite'. The Soviet satellite (COSPAS-1) was launched as Cosmos 1383 on 30 June 1982 and is equipped to pick up emergency signals from the radio beacons of downed aircraft and ships in distress, and relay their location to earth stations. COSPAS works along with NASA's SARSAT which was launched on 28 March 1983 and the weather satellite, NOAA 9. COSPAS-SARSAT picked up a distress signal from a floating life-jacket after the Air India Boeing 747 crashed into the sea off the coast of Ireland in June 1985, but, tragically, all lives were lost. □ see INMARSAT, NOAA 9

coulomb the SI unit of electrical charge equal to 1 ampere second (As). The (negative) charge carried by an electron is 1.6×10^{-19} coulombs. □ see SI UNITS, ELECTRON, PROTON

countdown 1 a continuous step-by-step process leading up to the start of an event, esp the launch of a rocket or other spacecraft ⟨*a Shuttle launch* ~⟩ **2** the continuous count backwards to zero of the time remaining before an event □ see LIFT-OFF

course correction (usu small) changes made to the speed of a spacecraft to ensure that it reaches its destination. For example, on 14 November 1984, the trajectory of the spacecraft Voyager-2 was adjusted slightly to make sure it was on target for the flyby of Uranus in January 1986. This adjustment was made by turning on Voyager's engine (the first time for three years!) for about 10 minutes to give it a speed change of about 1.54ms^{-1}. Later, two smaller changes of speed were made, one five days before its closest distance to Uranus on 24 January. After each manoeuvre, Voyager-2 was carefully tracked by antennae at the Deep Space Network to find out whether further course correction was required. In this way it was possible to deliver Voyager-2 within 100km of the path required by the Voyager Navigation Team.
□ see GUIDANCE, ATTITUDE CONTROL, THRUSTERS, VOYAGER

Crab nebula the crablike filaments of glowing gas and dust in the constellation Taurus which are the remnants of a star which blew itself up in the year 1054. It was a supernova, and the event was recorded by Chinese and Japanese astronomers. They said it was so bright it was visible in daylight, and you could read by it at night. In 1967, an optical pulsar was discovered at the heart of the Crab nebula (about 5000 light-years away from us and in our Galaxy), which has the shor-test period of all known pulsars. It radiates energy at radio, infrared, ultraviolet, and X-ray wavelengths, as well as in the visible spectrum. The ultraviolet radiation ionizes the expanding filaments of gas and dust causing the atoms to 'glow in the dark'. □ see PULSAR, SUPER-NOVA, SYNCHROTRON RADIATION

craft any machine designed to fly through space ⟨*a space*~⟩

crater a depression in the solid surface of something caused by the impact of a high-speed object ⟨*a lunar* ~⟩ □ see MICROMETEOROID, MOON

Crippen, Bob (*b*1937) veteran US astronaut, who accompanied John Young on the first flight of the Space Shuttle (STS-1) on 12 April 1981. He was also commander of STS-7, STS-11 (Mission 41C), and STS-13 (mission 41G). Crippen, a US Navy Captain, joined NASA in 1969, and in Spring 1985 he was appointed deputy director of flight crew operations at the Johnson Space Centre, though he continued on active flight duty. □ see SPACE SHUTTLE, YOUNG, JOHNSON SPACE CENTRE

CRO – see CATHODE-RAY OSCILLOSCOPE

crosstalk unwanted signals on a communications channel (eg a satellite link) which come from a nearby channel □ see INTERFERENCE, OPTICAL COMMUNICATIONS

CRT – see CATHODE-RAY TUBE

cruise missile a self-propelled low-flying atomic bomb which is programmed to reach a target automatically. A cruise missile carries an on-board computer which compares the view of the ground over which it flies with an image of the target area stored in its memory.

crust the outer region of a body such as a planet, satellite, or pulsar □ see EARTH, PULSAR

cryogenic of, involving, or being the use of very low temperatures ⟨*a ~ propellant*⟩ □ see ABSOLUTE TEMPERATURE SCALE, INFRARED ASTRONOMY SATELLITE

cryogenic engine a rocket engine that uses liquid hydrogen or kerosene as fuel and liquid oxygen as oxidizer □ see CRYOGENICS, ROCKET ENGINE, SATURN 5, ²TITAN, ARIANE

cryogenic propellant a rocket fuel, esp liquid oxygen and hydrogen, which is used in its liquid state at very low temperatures

cryogenics the physics, production, and use of very low temperatures. Cryogenics has enabled the gases, hydrogen and oxygen, to be liquefied for use as a cryogenic propellant in rocket engines. Liquid hydrogen (LH_2) and liquid oxygen (LO_2) become liquid at 20K and 90K, respectively. Like Saturn 5, which launched the Apollo spacecraft to the Moon, the main engines of the Space Shuttle are fuelled by LH_2 and LO_2. Helium liquefies at about 4K and it is extremely useful for laboratory use and space research since it is chemically inert. The Infrared Astronomy Satellite (IRAS) carried infrared sensors cooled by liquid helium so that they could measure very weak sources of infrared radiation from new-born stars. □ see SPACE SHUTTLE, CENTAUR, INFRARED ASTRONOMY SATELLITE, ARIANE

cryogenic temperature temperatures below the boiling point of nitrogen (-195.6C or 77.4K), esp temperatures near absolute zero (about -273C) □ see ABSOLUTE TEMPERATURE SCALE

CW – see CARRIER WAVE

cybernetics the study of how communications and control systems can be combined to simulate human functions. Cybernetics increasingly makes use of the latest achievements of microelectronics, particularly computing, pattern recognition, and voice recognition. □ see ARTIFICIAL INTELLIGENCE, ROBOTICS, ANDROID

cyborg [*cyb*ernetic *org*anism] a living organism which uses mechanical and/or electrical parts. A person with a heart pacemaker is a cyborg. □ see ANDROID

cycle 1 a complete sequence of events ⟨*the solar ~*⟩ □ see SUNSPOTS **2** one complete wave ⟨*50 ~s per second mains current*⟩ □ see FREQUENCY

Cyclops – see SEARCH FOR EXTRATERRESTRIAL INTELLIGENCE

cyclotron radiation electromagnetic radiation produced by high speed electrons and other charged particles spiralling round lines of force in a magnetic field. If the electrons are moving fast enough, the radiation is beamed forwards in the direction of motion of the electrons – this radiation is properly known as synchrotron radiation. The radio emissions from extragalactic radio sources and supernovae remnants are thought to be due to synchrotron radiation. This type of radiation is nonthermal as opposed to the production of X rays by heat as in X-ray binary stars. □ see BLACK HOLE, X-RAY ASTRONOMY

Cygnus A an intense source of radio waves that is about 170 megaparsecs (555 million light-years) away and located in the constellation Cygnus. Cygnus A has the typical structure of a binary extragalactic radio source: two bright outer lobes situated about 50 kiloparsecs from a central galaxy which also emits low-energy X-rays. □ see CASSIOPEIA A, SUN, RADIO ASTRONOMY, QUASAR, RADIO GALAXY

Cygnus X-1 an intense source of X rays in the constellation Cygnus located in the Milky Way. This was the first galactic X-ray source to be discovered by X-ray detectors aboard high-flying rockets, and it is being extensively studied by X-ray astronomy satellites. The X rays are generated in a binary star system – a luminous supergiant star and a dark companion star – which has a period of about 5.6 days. The dark companion is thought to be a black hole since it is too massive for a neutron star. Gas from the supergiant star spirals in on the black hole, becomes heated to about ten million degrees, and emits the X rays detected at Earth. □ see BLACK HOLE, X-RAY ASTRONOMY

D

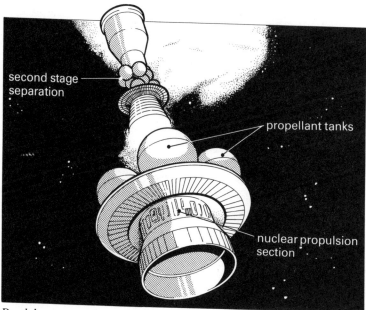

second stage separation

propellant tanks

nuclear propulsion section

Daedalus

Daedalus an unmanned starship designed by the British Interplanetary Society (BIS) for a 50-year mission to Barnard's star. This star is the fourth nearest star to the Sun and may have planets in orbit about it. *Daedalus* would be propelled by a nuclear reactor used to produce the heat and electricity for generating and controlling the flow of high-speed plasma from its engine. Propellant tanks would be jettisoned in pairs en route. Instructions from Earth up to six years from the encounter with Barnard's star would be handled by 'wardens' which would be artificially-intelligent robots responsible for preparing experiments, making decisions, and releasing smaller star probes in flight. It is unlikely that *Daedalus* could be built before the middle of the next century. □ see BRITISH INTERPLANETARY SOCIETY, BARNARD'S STAR, STARSHIP, PLASMA, ELECTRIC PROPULSION, ANTIMATTER PROPULSION, HYPERSPACE

damped the progressive decrease in amplitude of vibrations resulting from loss of energy by friction

dark nebula *also* **absorption nebula** a cloud of interstellar dust and gas that partially or completely obscures light from objects behind it. Dark nebulae occur in the spiral arms of the Galaxy (eg the Coalsack

and Horsehead nebula) and show up as dark regions in the Milky
Way. □ see MILKY WAY, SPIRAL GALAXY
data any information which can be processed or communicated.
Data can be divided into two types: that acquired by remote sensing
(eg pictures of the Earth's surface from the Landsat satellite), and that
obtained by measurement of the environment of the spacecraft (eg
magnetic field strengths and the characteristics of particles such as cosmic rays). □ see DATA RATE, DATA TRANSMISSION, DATA
PROCESSING, DATA ACQUISITION
data acquisition *also* **data capture** any method for collecting data
and converting it into a form which is meaningful □ see DATA,
REMOTE SENSING
data capture – see DATA ACQUISITION
data processing *also* **information processing** the acquisition,
recording, and manipulation of data by electronic, esp computer-
based means ⟨*~ from the Voyager spacecraft*⟩ □ see DATA ACQUISI-
TION, SUPERCOMPUTER
data rate the rate at which data is sent from one place to another.
Data from satellites and interplanetary spacecraft is received by Earth
stations as streams of binary data, ie groups of the binary digits (bits)
0 and 1, that represent the measurements made by on-board instruments. The maximum rate at which data can be transmitted depends
mainly on the size of the spacecraft's dish-aerial, the distance of the
spacecraft from Earth, and the frequency of the radio transmissions.
Thus the highest data rate from the Voyager at Jupiter was 115.2
kilobits per second (kbs^{-1}), and will be 29.9kbs^{-1} at Uranus. This rate
should be compared with the Earth-orbiting Infrared Astronomy
Satellite (IRAS) which played back data from its tape recorders at
1000kbs^{-1}. □ see TELEMETRY, VENUS RADAR MAPPER, VOYAGER,
INFRARED ASTRONOMY SATELLITE
data sink the end of a communications channel at which data is
received – compare DATA SOURCE
data source the end of a communications channel from which data
is transmitted – compare DATA SINK
data transmission the sending of data from its point of acquisition to a display, storage, or data processing device. It is usual to
transmit the data by electromagnetic waves (esp radio), although
actual samples (eg Moon rocks) can be brought back to Earth. □ see
DATA ACQUISITION, LASER
DBS – see DIRECT BROADCAST SATELLITE
debug to correct any errors in a computer program or in the circuits
of the computer itself
decay to decrease gradually in quantity, activity, or force ⟨*the
satellite burnt up as its orbit ~ed*⟩
decay product the new substance formed when a radioactive
material disintegrates ⟨*lead is a ~ of uranium*⟩ □ see NUCLEAR
REACTION
decay time 1 the time taken for the activity of a radioactive sub-

stance to fall to a particular value □ see HALF LIFE **2** the lifetime of a satellite which experiences a small atmospheric drag so that its ultimate fate is to burn up in the Earth's atmosphere □ see TETHERED SATELLITE

declination (symbol δ) the angular distance of a celestial body measured north or south of the celestial equator in the equatorial system of coordinates. The angle is counted positive for a celestial body north of the equator, and negative for bodies south of it ⟨ *the nearest star, Proxima Centari, has a ~ of −62°34'* ⟩ – compare RIGHT ASCENSION □ SEE CELESTIAL COORDINATES, CELESTIAL EQUATOR

decoder any device which converts information from a coded form (eg binary code) to another form (eg decimal) which is more easily understood by a person or an electronic system

decompression a reduction in air pressure in a spacecraft or other device □ see AEROEMBOLISM

decompression sickness – see AEROEMBOLISM

dedicated of or being an electronic device or a system programmed or assigned to one particular use ⟨ *a ~ flight of the Space Shuttle for experiments in the life sciences* ⟩

deep space any region of space which is not near the Earth, usu taken to mean regions more remote than the Moon □ see DEEP SPACE NETWORK

Deep Space Network (abbr **DSN**) a communications and tracking system operated by the Jet Propulsion Laboratory for all spacecraft circumnavigating the Solar System. The main 'ear' and 'mouth' of the DSN comprise a 64m diameter radio antenna, backed up by two 26m diameter antennae. With these antennae the DSN maintains contact with missions to Mars (Viking and Voyager), to Venus (Pioneer Venus), to Jupiter (Pioneers 10 and 11, Voyager, and Galileo), and to Halley's Comet. The DSN has also been used for numerous radio experiments, including radar studies of planet surfaces. □ see DATA RATE, VOYAGER, NASA, JET PROPULSION LABORATORY

degassing the deliberate process of removing gas from a material, usu by the application of heat under a vacuum

degenerate characterized by or made of atoms stripped of their electrons and packed very densely ⟨ *a neutron star is composed of ~ matter* ⟩ □ see PULSAR, NEUTRON STAR

delta (symbol δ) the fourth letter of the Greek alphabet, used to indicate (**a**) the fourth brightest star in a constellation (**b**) the declination of a celestial object in the sky. □ see DECLINATION

demodulate to recover information (eg speech) from a carrier wave (eg a radio wave) – compare MODULATE □ SEE PULSE CODE MODULATION

demultiplexing a method of recovering the individual messages from a single communications channel (eg a fibre optics cable) that carries several different messages simultaneously – compare MULTIPLEXING □ SEE BANDWIDTH

density 1 (symbol ρ) the ratio of the mass of a substance to its volume. There is a wide variation in the densities of celestial bodies ranging from about 10^{-20}kgm^{-3} for interstellar gas to more than 10^{17}kgm^{-3} for neutron stars ⟨*the ~ of water is 4200kgm^{-3}*⟩. □ see NEUTRON STAR, INTERSTELLAR MEDIUM **2** the amount of blackening of an image of a celestial object on a photographic film negative, used to estimate the relative brightness of that object □ see SCHMIDT TELESCOPE, MAGNITUDE

de-orbit burn the use of a (retrograde) rocket engine to reduce orbital velocity so that a spacecraft enters the Earth's atmosphere □ see RETROROCKET, THRUSTERS

deploy 1 to place in the appropriate position or arrange strategically ⟨*to ~ a radio telescope from the Space Shuttle*⟩ **2** to utilize or bring into effective action

descending node *also* **southbound node** the point at which a celestial body crosses from the north to the south of the celestial equator or ecliptic – compare ASCENDING NODE □ SEE CELESTIAL EQUATOR

destruct to deliberately destroy a rocket vehicle after it has been launched □ see ABORT

deuterium (symbol **D** or **H**) *also* **heavy hydrogen** a nonradioactive isotope of hydrogen having a nucleus (called deuteron) composed of one neutron and one proton. Water made from deuterium is known as heavy water and is used as a moderator in some types of nuclear reactor. Some deuterium was formed in the Big Bang which began the universe. It is also an important intermediate nucleus in the fusion reactions inside stars, where it is formed by the fusion of two protons at high temperature, followed by the conversion of one of the protons into a neutron to form the proton-neutron pair – the deuteron. A helium nucleus is then produced by a proton combining with the deuteron to produce helium-3; two of these nuclei come together to produce the stable helium-4 nucleus. Without deuterium, the chain reaction in stars involving the conversion of protons (hydrogen nuclei) into helium could not take place. And it is doubtful if stable long-lived stars could exist. The formation of deuterium in stars is slow since it is controlled by a nuclear force called the weak interaction. So stars consume their hydrogen slowly ensuring their long life and, in turn, providing stable conditions on their planets, enabling life to evolve. The fact that deuterium is produced at all appears to be a (fortuitous) accident of nature. If the strong nuclear force (the force between protons and neutrons in nuclei) were just 2% stronger, it would be possible for two protons to stick together to form a di-proton. But, fortunately, this does not occur, firstly, because two protons electrically repel each other and, secondly, because the Pauli Exclusion Principle requires that the two protons line up with their spins opposed which further reduces their attraction for each other. If di-proton nuclei were formed in stars it would mean the rapid (possibly explosive) consumption of hydrogen. Indeed, it is doubtful if any hydrogen would have survived the hot initial stage of the Big Bang. This appears to be another example of nature deliberately 'fine tuning' the laws of physics to ensure (so it would seem) that the universe could evolve with life (on Earth, at least) given priority! □ see ISOTOPE, HYDROGEN, PROTON, WEAK INTERACTIONS, BIG BANG, NUCLEOSYNTHESIS, PAULI EXCLUSION PRINCIPLE, NEUTRINO, ANTHROPIC PRINCIPLE

diamonds a pattern of shock waves, often visible in the exhaust of a rocket, which looks like a number of diamond shapes placed end to end

digital 1 of or being a circuit or system that responds to or produces electrical signals of two values: logic high (binary 1) and logic low (binary 0) ⟨*a ~ communications system*⟩ **2** of a display (*eg an instrument readout*) showing a quantity as a set of numbers – compare ANALOGUE

digital computer a device which uses digital circuits (eg gates and counters) to process data. Digital computers have largely replaced analogue computers which work on continuously variable quantities – compare ANALOGUE COMPUTER

digital display a readout of the value of something (eg the surface temperature of a spacecraft) as numbers on a digital display (eg a seven-segment display) – compare ANALOGUE DISPLAY

digitize to convert information from analogue to digital form ⟨*measurements by spacecraft are* ~ed *before transmission to Earth*⟩ □ see ANALOGUE-TO-DIGITAL CONVERTER

digitizer – see ANALOGUE-TO-DIGITAL CONVERTER

dipole aerial *also* **Yagi aerial** a radio transmitting or receiving aerial (eg a rooftop TV aerial) consisting of one or more pairs of vertical or horizontal metal rods. The lengths of the rods are shorter for higher frequency radio transmissions – compare DISH AERIAL □ SEE POWERSAT

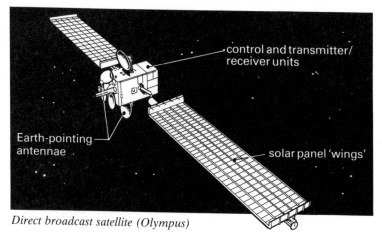

control and transmitter/ receiver units

Earth-pointing antennae

solar panel 'wings'

Direct broadcast satellite (Olympus)

direct broadcast satellite (abbr **DBS**) an Earth-orbiting artificial satellite which broadcasts TV programmes direct to homes. Direct broadcast satellites are already in use in North America and provide people in rural or thinly populated areas with a TV service. A similar service is being considered for the UK during the latter part of the 1980s, using Olympus satellites built by British Aerospace. Householders and others who want to receive programmes via a DBS would need to point a small rooftop-mounted dish-shaped aerial to a point in the sky where the satellite 'hangs' in geostationary orbit. Transmissions from direct broadcast satellites will be scrambled so that only those subscribers who have the right decoding equipment connected to their TV sets will be able to receive programmes. In America, the company RCA Astro-Electronics are designing and building two high-powered DBSs for the US Satellite Broadcasting Company (USSB) to provide the highest quality TV pictures and stereo sound over 50 states and including Puerto Rico and the Virgin Islands. Each satellite has six broadcast channels, backed up by six redundant channels, and will provide a selection of entertainment, sports, education, and public affairs programmes. The satellites will be launched by the Space Shuttle or Ariane 4. The high 240W of power on each channel means that programmes are received using only a 0.6m diameter dish-shaped rooftop antenna. □ see COMMUNICATIONS SATELLITE, TRANSPONDER, FOOTPRINT, DISH AERIAL, BRITISH AEROSPACE, BIG COMMUNICATOR, ANIK

directional suitable for transmitting or receiving signals in one direction only

directional antenna an aerial that radiates or receives radio waves in a narrow beam ⟨*a dish aerial is a* ~⟩ □ see DIPOLE AERIAL, DISH AERIAL, GAIN 2

discharge 1 a usu brief flow of electricity ⟨*a* ~ *of lightning struck the launch pad*⟩ **2** loss of power from a cell or battery with age or use

discriminator an electronic circuit that selects pulses according to their amplitude. Discriminators are the basis of some X-ray and gamma-ray sensors which determine the spread in energy of received photons □ see GAMMA-RAY OBSERVATORY

dish aerial a device with a dish-shaped collector for focussing radio signals onto an aerial in the centre of the dish. Large dish aerials are used for communicating with artificial satellites and interplanetary spacecraft since a large dish can receive a lot of weak signals and transmit a narrow beam of radio waves. Similarly, radio astronomers use large dish aerials to examine radio emissions from distant galaxies. Ground-based microwave communications links (eg between an Earth station and a centre for space research) also use dish aerials mounted on masts to send information in straight lines from one mast to another. □ see MICROWAVES, RADIO ASTRONOMY, DIRECT BROADCAST SATELLITE, APERTURE

distortion corruption of the signal travelling along a communications channel or of the message or image received, so that less information is received than is sent. Distortion can be caused by faults in the equipment used for sending and receiving the information or by outside electrical or other interference. □ see THERMAL NOISE, SIG-NAL-TO-NOISE RATIO

docking the act of joining together, in space, two or more spacecraft (space ships), or a spacecraft with a space station. The first space docking was achieved by America in December 1965, with the meeting of the two two-man spacecraft, Gemini 6 and Gemini 7, 250km above the Earth. The first (and only) Soviet-American 'space handshake' took place in 1975 when an Apollo spacecraft docked with a two-man Soyuz spacecraft. For nearly 48 hours the three US astronauts and the two Soviet cosmonauts exchanged gifts, shared meals, and spoke with the Soviet and US Presidents while the 'show' was televised for all the world to see. A similar space docking between these two countries has been mooted for the 1980s. Space docking is now routinely carried out when the Soyuz spacecraft carry a relief crew of cosmonauts to the orbiting Salyut 7 space station. □ see MAT-ING, GEMINI PROJECT, APOLLO, SOYUZ

Doppler effect a change in the apparent frequency (and hence wavelength) of sound, light, or other waves as a result of relative motion between their source and an observer. For a source (eg a star) moving away from an observer, the observed wavelength is longer than it would be if there were no relative motion. Thus light is shifted towards the red end of the visible spectrum, an effect known as red shift. If the source is moving towards the observer, the wavelengths become shorter (known as the blue shift). □ see RED SHIFT, BLUE SHIFT, HUBBLE'S LAW

drag a force that makes a body lose energy as it moves through a fluid □ see ATMOSPHERIC DRAG, DRAG PARACHUTE

drag parachute any of various types of parachute (eg a drogue parachute) deployed by an aerospace vehicle to slow it down and/or to stabilize it □ see DROGUE PARACHUTE, DROGUE RECOVERY

drift 1 the horizontal movement of an aerospace vehicle from a prescribed flight path, usu due to crosswinds **2** the slow change in the reading of an instrument, usu caused by instrument error **3** the slow change in frequency of a radio transmitter, usu due to faulty design

drogue parachute a small parachute for pulling a larger parachute from stowage or for slowing down the descent of a probe in a planet's atmosphere □ see [1]GALILEO, VENERA, DRAG PARACHUTE

drogue recovery a method of safely slowing the descent of a spacecraft through the atmosphere using parachutes. After reentry into the atmosphere, one or more small drogue parachutes are deployed to reduce the spacecraft's speed and to stabilize it. Then larger recovery parachutes are deployed to lower the spacecraft to the Earth's surface without damaging the spacecraft or its occupants. Apollo astronauts were safely returned to Earth (more accurately, the sea) in this way. And off-duty Soviet cosmonauts from the Salyut space station are brought down on Russian soil by drogue parachutes.

□ see SOLID ROCKET BOOSTER, APOLLO, SOYUZ

DSN – see DEEP SPACE NETWORK

dust – see COSMIC DUST, INTERSTELLAR MEDIUM, MICROMETEOROID

dust shield a cover or deflecting plate protecting instruments and other delicate equipment aboard spacecraft from the high-speed impact of cosmic dust and micrometeorites □ see [1]GIOTTO, [1]GALILEO, MICROMETEOROID, INTERSTELLAR MEDIUM

Eagle the lunar module which, on 20 July 1969, landed the first men (Armstrong and Aldrin) in Mare Tranquillitatis (Sea of Tranquillity) on the Moon ⟨*"The ~ has landed!"*⟩ □ see APOLLO, MOON, ARMSTRONG

Earth the third planet of the Solar System, which has a diameter of 12 756km, orbits the Sun at about 150 million kilometres, and has one satellite (the Moon). The Earth has a substantial atmosphere, mainly of nitrogen and oxygen, which supports an immense variety of life. The upper regions of the atmosphere have been studied, first by high-altitude balloons, then by sounding rockets, and more recently by artificial satellites. Two thirds of the planet is covered by water that has an average depth of 3900m; the average height of the land above the water is 860m. The Earth's interior has three main layers, the crust, mantle, and core. The crust is about 32km thick under the land and 10km under the oceans, and consists largely of sedimentary rocks, such as limestone and sandstone, having a density of about 3, and resting on a base of igneous (volcanic) rocks. The mantle extends to a depth of about 2900km where its density reaches about 5.5, and is composed largely of silicate rock. The core increases in density from about 10 where it joins the mantle, to about 13 at the centre of the Earth. The core consists largely of nickel-iron, similar in composition to the nickel-iron meteorites which contain 6% nickel. The inner core, from about 5150km to the centre, appears solid, but the outer core appears to be liquid and is probably the source of the Earth's magnetic field. The temperature inside the Earth varies from about 1200K at a depth of 100km, to about 3000K at 1000km, and 6400K at the centre. □ see EARTH'S ATMOSPHERE, GEOMAGNETISM

Earth Radiation Budget Satellite (abbr **ERBS**) a roughly triangular satellite designed to fit the payload bay of the Space Shuttle and released from *Challenger* on 5 October 1984. ERBS was released by the Remote Manipulator System into a similar orbit to *Challenger's*. ERBS then raised its orbit during several hours of low thrust engine operation into a 97-minute orbit at an altitude of about 600km, inclined at 56° to the equator. Its purpose is to study the way solar energy is received and reradiated by the Earth. □ see RIDE, REMOTE MANIPULATOR SYSTEM, SOLAR MAX, SOLAR CONSTANT

Earth Resources Technology Satellite (abbr **ERTS**) – see LANDSAT

Earth's atmosphere the envelope of gases that clings to the Earth. At sea level the atmosphere is made up of 78% of nitrogen, 21% of oxygen, 0.9% of argon and 0.04% of carbon dioxide. In addition, it contains water vapour, sulphur dioxide, carbon monoxide, dust particles, bacteria, etc. The atmosphere is divided into various layers determined by the variation of temperature with altitude. In ascending order, these layers are: troposphere, stratosphere, mesosphere, thermosphere, and exosphere. The troposphere extends to an altitude of about 8km at the poles and 18km within the tropics, and contains three-quarters of the mass of all the atmosphere. It is the region of weather systems and clouds, and is heated by infrared radiation and convection. At the top of the troposphere, the temperature falls to about −60C which marks the tropopause. Above the tropopause is the stratosphere, in which the temperature is at first steady and then increases to about 0C at an altitude of about 50km which marks the stratopause. This is a stable region in which there is no vertical air movement. The heating within the stratosphere comes from the absorption of ultraviolet radiation from the Sun by ozone molecules. Ozone is formed by the effect of ultraviolet radiation on oxygen molecules, and forms the ozonosphere within the stratosphere. Beyond the stratosphere is the mesosphere in which the temperature once again falls since ozone is less plentiful. At the top of the mesosphere, at a height of 85km, the temperature is −90C. Above the mesosphere is the thermosphere in which oxygen and nitrogen absorb solar ultraviolet radiation making the temperature rise to about 1300C at a height of 500km. However, the heating effect at this temperature is negligible since the density of the atmosphere is, on average, one million millionth of that at sea level. Shooting stars and aurorae are produced in the thermosphere. Beyond the thermosphere, the atmosphere 'leaks' away into interplanetary space. This region is known as the exosphere. It contains the Van Allen radiation belts and extends into the magnetosphere. □ see ATMOSPHERE, SHOOTING STAR, ECOSPHERE, IONOSPHERE, VAN ALLEN ZONES, MAGNETOSPHERE, AURORA

earthshine sunlight reflected from the Earth that illuminates the 'night' side of the Moon. Close to new moon this can be seen as 'the old moon in the new moon's arms'.

Earth station *also* **ground station** a base for relaying radio signals to and from artificial satellites and interplanetary spacecraft. An Earth station usu has one or more dish-shaped aerials pointing skywards. Signals sent to a satellite are said to be 'uplink' and those coming from the satellite are said to be 'downlink' ⟨*the ~ at Goonhilly Downs in Cornwall*⟩. □ see DISH AERIAL, COMMUNICATIONS SATELLITE, DIRECT BROADCAST SATELLITE, DEEP SPACE NETWORK

eccentricity (symbol **e**) a measure of how much an ellipse (eg a planet's orbit) departs from circularity. If the eccentricity of an orbit is high (as in the case of many comets), the difference between its closest approach (perihelion) and furthest approach (aphelion) to the Sun is large. The value of e is given by $c/2a$, where c is the distance between the two foci of the ellipse, and 2a is the length of its major axis. For a circular orbit $c=0$, so $e=0$. Of the planets, Pluto has the largest eccentricity (0.25) and Venus the least (0.007). □ see ELLIPSE, COMET, HALLEY'S COMET

eclipse the partial or total obscuration of a luminous body by a

Earth station

nonluminous body. It is possible to work out the diameter of an asteroid by measuring how long it takes to eclipse a star. Partial or total eclipses of the Sun occur when the Moon passes between the Earth and the Sun and casts a shadow on the Earth. Partial or total eclipses of the Moon occur when the Earth passes between the Sun and the Moon and casts a shadow on the Moon. Four eclipses are occurring in 1986, two of the Sun and two of the Moon. On 9 April, a partial eclipse of the Sun was visible in Indonesia, Australia, Papua New Guinea, south part of New Zealand, and Antarctica. On 24 April, a total eclipse of the Moon was visible in the Americas, Australasia, and Asia. On 3 October, a total eclipse of the Sun is visible in North America, north-eastern Asia, Iceland, Greenland, northern part of South America, and in Arctic regions. On 17 October, a total eclipse of the Moon is visible in Australasia, Asia, Africa, and Europe. □ see OCCULTATION

ecliptic *also* **ecliptic plane** the plane of the Earth's orbit extended to meet the celestial sphere. It is therefore taken to be the annual path of the Sun across the sky. The orbits of the Moon and planets, apart from Pluto, lie very close to the ecliptic. Spacecraft sent to explore the Solar System move close to the ecliptic plane. However, after visiting Saturn, Voyager 1 has now left the ecliptic. □ see SOLAR HELIO-CENTRIC OBSERVATORY, CELESTIAL EQUATOR

ecliptic plane – see ECLIPTIC

ecosphere a zone of space surrounding the Sun (or some other star) in which conditions are favourable for life to prosper – compare BIOSPHERE □ SEE ASTROBIOLOGY, SPACE GREENHOUSE

Ehricke, Dr Krafte (1917–84) a German engineer who emigrated to the United States with von Braun and other V2 rocket engineers to become a key figure in the American space programme. In 1954 he joined General Dynamics-Convair, San Diego, California, and worked on the development of the first multistage intercontinental ballistic missile, Atlas, which was later to become the workhorse of the space age. In the 1950s and 60s, Ehricke conceived and helped in the development of the Centaur upper stage rocket which uses liquid hydrogen and liquid oxygen propellants, and variants of this are in use in the Shuttle programme. In 1965 he joined North American Rockwell. Unlike Braun, Ehricke was a 'star man' and believed that mankind's destiny was to continue its historic expansion by basing industries in space using the resources of the Solar System. Central to this theme, which he called the 'extraterrestrial imperative', is the development of the Moon as an industrial base serving the expanding needs of the Earth, and the creation of new civilizations on the Moon. The industrialization of the Moon, the Earth's 'seventh continent', is the theme of his book *The Seventh Continent* which he was working on at the time of his death. □ see BRAUN, ATLAS, CENTAUR, MOON BASE, LUNAR SLIDE LANDER

¹Einstein – see HIGH ENERGY ASTRONOMY OBSERVATORY

²Einstein, Albert (1879–1955) a German-Swiss-American physicist whose first scientific papers (1905) plausibly explained the photoelectric effect previously observed by Planck. Einstein proposed that a particular wavelength of light was made up of packets (quanta) of fixed energy content which were absorbed by electrons in some metals, the electrons subsequently being released from the metal's surface. Light of shorter wavelength contained more energetic quanta whose absorption caused the release of electrons with a greater energy. This explanation of the photoelectric effect laid the foundations of the quantum theory. This theory and Einstein's special and general theories of relativity (1905 and 1916) have had a profound impact on our understanding of the physical world. They have helped us to understand the large scale energy changes of stars and galaxies in the universe. But they also made possible atomic bombs which produced such devastation when they were exploded above unsuspecting Japanese civilians towards the end of World War 2. After the War and to the end of his life Einstein fought stubbornly for a world agreement to end the threat of nuclear war. But his ability to revolutionize physics outweighed his ability to change power politics, and at the time of his death the peril was greater than ever before. Shortly after his death, the manmade chemical element of atomic number 99 was named einsteinium in his honour. □ see PLANCK'S CONSTANT, QUANTUM, QUANTUM THEORY, RELATIVITY, MASS-ENERGY EQUIVALENCE, NEWTON'S LAWS OF MOTION

ejection capsule 1 a compartment in some manned spacecraft which can be ejected free of an exploding launch rocket and parachuted safely to the ground **2** a boxlike module containing recording instruments (eg film from a spy satellite) that is ejected and returned to Earth by parachute

electrical of or connected with electricity ⟨an ~ discharge⟩

electric field a region in which a charged particle (eg a proton) experiences a force. Electric fields are put to good use in the cathode-ray tube and many other (electronic) instruments. □ see COSMIC RAYS, ANTIMATTER PROPULSION

electricity (the effects and use of) electrically charged particles (eg electrons and protons) at rest or in motion. An electric current is a flow of such particles. □ see ATOM, ION, PLASMA

electric propulsion the generation of thrust by a rocket engine which uses an electric field to expel high-speed ions (eg mercury or caesium atoms which are stripped of one or more of their electrons). An electric propulsion rocket requires an on-board electrical power source to generate the electric field for accelerating the positively charged ions. Solar panels, radioisotope thermoelectric generators, and fuel cells are suitable sources, but nuclear power is a proven technology that could also be employed, especially for spaceships which do not have a local star as a source of solar energy. In a nuclear electric propulsion engine, a nuclear reactor produces an ample supply of electrical power. This energy could be used to heat caesium (the propellant) so that it vapourizes and enters the rocket's combustion chamber. Here the atoms are ionized by passing them over a hot grid of platinum. The positively charged caesium ions enter the thrust chamber where they are accelerated by an electric field. The separated electrons are directed by a separate route and pass into the exhaust stream to produce an electrically neutral flow. Experimental electric propulsion rockets (though not using nuclear power) have been successfully tested in space, but their thrust is very low. They are being considered as economical alternatives to chemical rocket engines (eg liquid hydrogen/liquid oxygen engines) for long flight times. Two variants of the electric propulsion rocket are the plasma rocket engine and the antimatter rocket engine. □ see ION, PLASMA ROCKET, ANTIMATTER PROPULSION

electrode a metal conductor used to make contact with a circuit ⟨the ~ of a fuel cell⟩ □ see ANODE, CATHODE, FUEL CELL, CONTINUOUS-FLOW ELECTROPHORESIS SYSTEM

electroluminescence the conversion of electricity to light at relatively low temperatures (eg the emission of light by the phosphor of a VDU when struck by an electron beam)

electromagnetic having both electrical and magnetic properties

electromagnetic pulse (abbr **EMP**) an intense and sudden burst of electromagnetic energy produced by the explosion of nuclear weapons in the atmosphere. In addition to the appalling effects of radiation, and fire and blast damage when nuclear weapons are used, the electromagnetic pulse would cause world-wide disruption to radio communications and irreparable damage to many types of electrical equipment. The burst of radiation releases an intense shower of electrons which rains down from above and burns out all electronic equipment. Consequently military planners see fibre optics cables as one solution to the electrical problems produced by the electromagnetic pulse since laser light carrying information down a fibre optics cable is not susceptible to these effects. □ see OPTICAL COMMUNICA-

TIONS, ELECTROMAGNETIC RADIATION

electromagnetic radiation (abbr **emr**) a flow of energy which travels through a vacuum at the speed of light. Electromagnetic radiation may be considered either as wave motion (consisting of oscillating magnetic and electric fields) or as a stream of particles (quanta). Electromagnetic waves include radio, infrared, visible light, ultraviolet, X rays, and gamma rays. They are characterized by their different wavelengths (or frequencies), the means by which they are generated and detected, and the ways in which they react with matter. □ see QUANTUM, PLANCK'S CONSTANT, SPEED OF LIGHT, RED SHIFT

electromagnetic spectrum the entire range of wavelengths (or frequencies) which travel at the speed of light. The electromagnetic spectrum extends from very short wavelength gamma rays to the longest wavelength radio waves. □ see ELECTROMAGNETIC RADIATION, RED SHIFT, OPTICAL COMMUNICATIONS

electron a particle which carries a negative charge and which is one of the basic building blocks of all substances. An electron has a negative charge of 1.6×10^{-19} coulombs. An electric current through a conductor (eg copper) is the movement of a large number of electrons. □ see ELECTRONICS, ATOM, PROTON, LEPTON, IONIZATION, POSITRON

electronic of, being, or using devices (eg integrated circuits) to switch, amplify, count, and in other ways usefully control electricity

electronic engineering a branch of engineering (eg telecommunications) dealing with the applications of electronic devices and systems □ see MICROELECTRONICS

electronics the study and application of the behaviour and effects of electrons in transistors, television tubes, and other devices □ see MICROELECTRONICS

electronic system a set of electronic building blocks (eg amplifiers and switches) connected together to produce a useful function (eg a computer system) □ see SYSTEM, MICROELECTRONICS

electron volt (symbol **eV**) a unit for measuring the energy of atomic particles, esp the energy which binds together the building blocks (eg electrons and protons) of atoms. The electron volt is defined as the energy gained by an electron moving through a potential difference of 1 volt, and is equal to 1.602×10^{-19} joules. Chemical reactions, ie reactions that affect the energy distribution of electrons in orbit round the nucleus of atoms and molecules, require energy exchanges of a few electron volts. But nuclear reactions that affect the arrangement of protons and neutrons within atomic nuclei require about a million electron volts, which is why a kilogram of plutonium has roughly the explosive energy of a million kilograms of TNT. □ see PHOTON, NUCLEAR FISSION, NUCLEAR FUSION, PLANCK'S CONSTANT, STAR, SUPERNOVA, MASS-ENERGY EQUIVALENCE

element 1 any component which is part of a larger system ⟨*a transmitter is an ~ of a communications system*⟩ □ see MODULE **2** any of more than 100 natural substances (eg helium) and manmade sub-

stances (eg einsteinium) which are made up of atoms of only one type
□ see ATOM, COSMIC ABUNDANCE, NUCLEOSYNTHESIS

elementary particles *also* **fundamental particles** the simplest forms of matter and radiation. The behaviour of elementary particles is determined by a set of fundamental properties including rest mass, electrical charge, and angular momentum (or spin). Photons (quanta of electromagnetic radiation), electrons, protons, and neutrinos are the only completely stable particles. Free neutrons are unstable and disintegrate into protons and neutrinos, but are stable in the nuclei of atoms. The neutron is now regarded as the union of one 'up' quark and two 'down' quarks. When a neutron decays into a proton, one of the down quarks changes into an up quark, an electron being created to carry away one unit of electric charge. Physicists are not really sure which are the truly elementary particles, for even the very stable proton may be weakly unstable. It is possible to account for all ordinary matter with the up and down quarks, the electron, the neutrino, and the four antiparticles of these particles, together with gluons, photons, and gravitons. □ see NUCLEUS, ELECTRON, PROTON, NEUTRON, NEUTRINO, QUARK, ANTIPARTICLE, LEPTON, HADRON, BIG BANG, GENERAL UNIFIED THEORY

ellipse a smooth oval curve as followed by planet Mars round the Sun. More accurately, an ellipse is the curve generated by a point moving in such a way that the sum of its distances from two fixed points (the foci) is a constant and equal to the length of the major axis of the ellipse. Most planets in the Solar System follow near circular paths (a special case of an ellipse). □ see KEPLER'S LAWS, ECCENTRICITY

ellipsoid a surface that is obtained by rotating an ellipse about one of its axes

elliptical of or shaped like an ellipse ⟨*comets usu have highly ~ orbits*⟩

elongation the angular distance between the Sun and a planet or the Moon as seen from Earth. An elongation of 0° is called conjunction, one of 90° is called quadrature, and one of 180° is called opposition. The inferior planets, Mercury and Venus, cannot be in quadrature but reach positions of maximum angular distance (greatest elongations). Venus has a greatest elongation varying between 47° and 48°, while Mercury, which follows a more elliptical orbit, has a greatest elongation varying between 18° and 28°. □ see CONJUNCTION, OPPOSITION, ECLIPSE

emission spectrum a spectrum formed by the emission from ionized gases of electromagnetic radiation of specific wavelengths. Electrons in the atoms of a heated, or otherwise excited, gas 'jump' to higher energy levels. When these electrons fall back to lower energy levels, the atom emits radiation of a narrow wavelength and these are seen as bright emission lines in the spectrum. Since the wavelengths of these bright lines are characteristic of the heated gas, astronomers can identify the elements present in the outer atmosphere of a star or in

the cloud of dust and gas in a nebula – compare ABSORPTION
SPECTRUM □ SEE ELECTROMAGNETIC SPECTRUM, CONTINUOUS
SPECTRUM, SPECTROMETER, LASER

EMP – see ELECTROMAGNETIC PULSE

emr – see ELECTROMAGNETIC RADIATION

EMU – see EXTRAVEHICULAR MOBILITY UNIT

encounter 1 the arrival of a spacecraft at its destination ⟨*Giotto's ~
with Halley's Comet*⟩ □ see FLYBY **2** a meeting with an alien being (as
some people have claimed) ⟨*a face-to-face meeting with an alien being
is called 'an ~ of the third kind'*⟩ □ see RENDEZVOUS

engine a machine which converts heat energy into mechanical
energy ⟨*a rocket ~*⟩ □ see THERMODYNAMICS, ROCKET ENGINE

enter to go into orbit about a planet or satellite, or to pass (temporarily) into an airspace ⟨*Mariner 9 ~ed into an orbit of Mars on
November 13 1971*⟩

Environmental Science Services Administration (abbr
ESSA) – see NATIONAL OCEANIC AND ATMOSPHERIC
ADMINISTRATION

environmental testing the usu lengthy process of making sure a
spacecraft or similar vehicle can withstand the rigours of a rocket
launch, the flight through space, and the encounter with the object of
its mission. For example, the spacecraft Giotto, which is making a rendezvous with Halley's comet in 1986, was subjected to thermal tests at
the Centre Spatiale de Toulouse in July 1984. The spacecraft was spun
in a vacuum chamber for 15 days at a rate of 10 revolutions per minute
and positioned at various angles to simulate the effects of heat on the
spacecraft at launch, during the trip through space, and the encounter
with the comet. It was also subjected to vibration and acoustic tests to
assess the effects of the roar and vibration of the engines of the Ariane
rocket which launched it. Similar tests are carried out on all spacecraft
and, depending on the destination, further special tests would be carried out. The Soviet Venera spacecraft, for example, destined for the
surface of Venus, had to undergo severe tests involving temperatures
of at least 450C and a pressure of 90 atmospheres to simulate the
extreme conditions on the surface of this planet. The Galileo
spacecraft which will arrive at Jupiter in 1988 is also subjected to tests
including its resistance to the strong electrical fields produced by Jupiter. □ see [1]GIOTTO, VENERA, [1]GALILEO

equatorial 1 of, at, or in the plane of the Earth's equator □ see
CELESTIAL COORDINATES **2** *esp of a telescope* supported on two axes
so that the telescope can readily follow the path of a celestial body
□ see EQUATORIAL MOUNTING

equatorial mounting a method of mounting a telescope so that
one axis (the polar axis) is parallel to the Earth's axis, while the
second axis (the declination axis) is at right angles to it. Compared
with the altazimuth mounting, the equatorial mounting has the advantage that once the telescope is clamped in declination, it need only be
made to rotate about its polar axis once every 24 hours to follow a star

automatically. Most large telescopes (eg the Hale 5m reflector) are equatorially mounted – compare ALTAZIMUTH MOUNTING □ SEE DECLINATION, POLAR, REFLECTING TELESCOPE

equatorial orbit a satellite which continuously orbits above the Earth's, or any other planet's, equator ⟨*a geostationary orbit is an ∼*⟩ – compare POLAR ORBIT □ SEE GEOSTATIONARY

equinox either of the two times each year that occur about 21 March (the vernal equinox) and 23 September (the autumnal equinox) when the Sun crosses the celestial equator and day and night are of equal length everywhere on Earth. The seasons are reversed in the Southern hemisphere. □ see CELESTIAL EQUATOR

ERBS – see EARTH RADIATION BUDGET SATELLITE

ERTS [*E*arth *R*esources *T*echnology *S*atellite] – see LANDSAT

ES – see EXPERT SYSTEM

ESA – see EUROPEAN SPACE AGENCY

ESA Halley probe – see ¹GIOTTO

escape velocity the minimum speed an object (eg a spacecraft) must have if it is to escape from the gravitational attraction of a planet, satellite, or other body. If the object does not reach escape velocity, it will follow an elliptical orbit around the body, as do the planets, asteroids, artificial satellites, and all periodic comets. The escape velocity from the Earth's surface is 11.2kms⁻¹; from the Moon, 2.4kms⁻¹; and from the Sun, 817.7kms⁻¹. The Pioneer and Voyager spacecraft have speeds in excess of the escape velocity needed to leave the Solar System on their way to the stars. If a body is orbiting at a distance r from another body, the escape velocity, v, is given by $v = \sqrt{(2gR^2/r)}$, where g is the acceleration of gravity at the surface of the body, and R is its radius. □ see ORBITAL ELEMENTS, ORBITAL PERIOD, KEPLER'S LAWS, PIONEERS 10 AND 11, VOYAGER, EVENT HORIZON, BLACK HOLE

ESSA [*E*nvironmental *S*cience *S*ervices *A*dministration] – see NATIONAL OCEANIC AND ATMOSPHERIC ADMINISTRATION

ESTEC – see EUROPEAN SPACE AGENCY

ET 1 – see EXTERNAL TANK **2** – see ²EXTRATERRESTRIAL

Eureca [*Eu*ropean *R*etrievable *Ca*rrier] a retrievable 'space platform' which will be launched by the Space Shuttle to carry out experiments in Earth orbit. Eureca is being developed by Germany for the European Space Agency (ESA), and it will be launched for the first time in early 1988. Eureca will have its own propulsion system which will take it into an operational orbit 500km high, inclined at 18.5° to the equator. Data from the experiments will be transmitted to Earth stations. After about six months in orbit, Eureca will return to a lower orbit to meet a later flight of the Space Shuttle. The latter's Remote Manipulator System (RMS) will retrieve the platform and place it into the payload bay ready for a return to Earth. It will then be taken to Europe so that scientists can fit new experiments to the platform ready for the next mission. About 1000kg of payload space will be available on Eureca, largely for experiments in life sciences and materials processing. □ see SHUTTLE PALLET SATELLITE, GETA-WAY SPECIAL, MAUS, EUROPEAN SPACE AGENCY, COLUMBUS, LIFE SCIENCES, MATERIALS PROCESSING

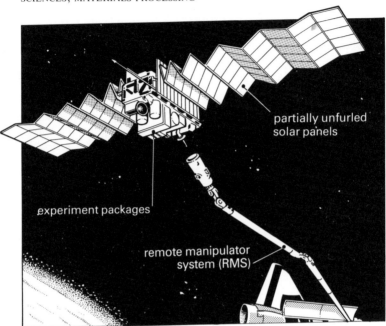

Eureca (deployed by Shuttle's Remote Manipulator System)

Europa the second in increasing distance and the smallest of Jupiter's Galilean moons. Photographs of Europa, taken by the two Voyager spacecraft which passed close by in 1979, revealed a smooth surface criss-crossed with mysterious lines thousands of kilometres long. They are believed to be refrozen cracks that open occasionally to erupt water from a liquid ocean several kilometres below the surface. Thus Europa, like its sister moon Io, may still be geologically active and continuously making a new surface, not of sulphur, as on Io, but of frozen water. This planet-sized glacier gives Europa its high albedo.
□ see JUPITER, IO, GANYMEDE, ALBEDO

European Retrievable Carrier – see EURECA

European Space Agency (abbr **ESA**) an international organization of 12 member countries (Austria, Belgium, Denmark, France, West Germany, Ireland, Italy, Netherlands, Spain, Sweden, Switzerland, and the United Kingdom) intended to give Europe an independent role in exploring space and using it for manufacturing, communications, navigation, weather forecasting, and other peaceful purposes. The European Space Agency's Research and Technology Centre (ESTEC, at Noordwijk in the Netherlands) is ESA's largest establishment. Its project teams work on the design and development of satellites. Using its own rocket, Ariane, ESA has successfully launched a number of satellites into Earth orbit for scientific purposes, such as investigating the Earth's aurorae and magnetosphere, cosmic rays from deep space, and X rays from the Sun. The main communications satellite has been Meteosat which has been Europe's contribution to the World Weather Watch Programme. Giotto, the spacecraft sent to slip through the tail of Halley's comet in 1986, was designed and built by ESA. Spacelab, a manned reusable laboratory to operate in the cargo bay of the Space Shuttle, is currently ESA's most important joint venture with NASA. And ESA is considering building Columbus which will carry experiment packages alongside the proposed US Space Station. ESA is also considering five new satellites for launch in the late 1980s and early 1990s: these include FIRST (Far Infrared and Submillimetre Space Telescope) to extend the work of the orbiting infrared satellite, IRAS; X-ray Multi-Mirror (XMM) telescope, for measuring the intensity of X rays from stars and galaxies; SOHO (Solar Heliocentric Observatory) to study the Sun's atmosphere; and AGORA (Asteroid Gravity, Optical and Radar Analysis) designed to rendezvous with and study selected asteroids in the Asteroid Belt between Mars and Jupiter. □ see ARIANE, INFRARED ASTRONOMY SATELLITE, ¹GIOTTO, SPACELAB, US SPACE STATION, HUBBLE SPACE TELESCOPE, COLUMBUS, HERMES, HOTOL, SOLAR HELIOCENTRIC OBSERVATORY, INFRARED SPACE OBSERVATORY, KEPLER 1

European Telecommunications Satellite – see EUTELSAT
European X-ray Observatory Satellite – see EXOSAT
Eutelsat [*Eu*ropean *Tel*ecommunications *Sat*ellite Organization] an organisation of European countries which commissions and manages satellites for a variety of services, including TV distribution to cable networks, telephony, data transmissions, and the Satellite Multiservice System (SMS). SMS provides a system of international links for business services (eg video-conferencing, fast facsimile, and remote printing). The European Broadcasting Union (EBU) is also using Eutelsat for the services it provides to Eurovision. Eutelsat uses two communications satellites, Eutelsat 1 and 2 launched by the European Space Agency's Ariane rocket in 1983 and 1984, respectively. These satellites are expected to provide communications services through to the early 1990s. □ see EUROPEAN SPACE AGENCY, ARIANE, COMMUNICATIONS SATELLITE

EUVE – see EXTREME ULTRAVIOLET EXPLORER
EVA – see EXTRAVEHICULAR ACTIVITY
evening star the planet Venus (or sometimes Mercury) seen shining brightly in the western sky at sunset □ see VENUS
event horizon the spherical surface surrounding a black hole, where the escape velocity equals the speed of light and hence no information can reach an external observer. The radius of the event horizon is known as the Schwarzchild radius and depends on the mass

of the black hole. Any body inside the event horizon shrinks to a singularity at the centre of the black hole. □ see BLACK HOLE, ESCAPE VELOCITY, RELATIVITY, SINGULARITY

exhaust velocity the rapid stream of gases and particles that leave the nozzle of a rocket engine. The rate of change of momentum of the exhaust stream provides the driving force for the rocket. □ see ROCKET ENGINE, LINEAR MOMENTUM, SPECIFIC IMPULSE

exobiology – see ASTROBIOLOGY

Exosat [*E*uropean *X*-ray *O*bservatory *Sat*ellite] Europe's first X-ray satellite launched in May 1983 from the Vandenberg Airbase, California, by the Delta rocket, to study X-ray emissions from celestial objects in the energy range 0.04keV to 80keV. Exosat, built by West Germany, had a mass of about 500kg and orbited between 300km and 200 000km. It could make direct observations of X-ray sources and it used lunar occultation of the X-ray sources to determine their position with great precision. □ see X-RAY ASTRONOMY, EUROPEAN SPACE AGENCY, ELECTRON VOLT

exosphere – see EARTH'S ATMOSPHERE

expanding universe the widely held view that all matter in the universe is moving further apart. This conclusion follows from the fact that light from distant galaxies shows a red shift which indicates that the galaxies are receding from us – compare STEADY STATE □ SEE RED SHIFT, BIG BANG, HUBBLE'S LAW

expedition to Mars a long-held dream, still to be realized, for a manned exploration of Mars. Though the Mariner, Viking lander, and Voyager flyby spacecraft did not find any sign of life on Mars, both the USA and the USSR would like to send astronauts and cosmonauts to Mars (and bring them back safely!) before the year 2010. Before such an exciting event can take place, the good health of the astronauts must be assured for the long journey and a suitable rocket engine must be developed. American astronauts and Soviet cosmonauts are learning how to deal with the effects on the body of lengthy periods of weightlessness. Early studies suggested that only a nuclear-powered engine would be suitable, and this type of engine has yet to be produced. However, cryogenic rocket engines fuelled by liquid hydrogen and oxygen are well developed and could be used. Dr Bob Parkinson (Study manager for Space Station Studies at British Aerospace) has proposed the following plan for a five-person expedition using cryogenic rocket engines. Three spaceships are locked together for the journey to Mars. Two of the spaceships are identical to provide a backup should there be a serious failure to one of them: these are the Mars orbiters which return to Earth. The third vehicle is the Mars lander and does not return. Each orbiter has three rocket stages: the first stage sends the ships on a trajectory to Mars; the second stage is used for manoeuvres at Mars; and the third stage is used for the return flight. The lander has the first and second stages only, and these are based on the orbital transfer vehicle being developed in support of the US Space Station. The ships are fuelled in

Earth-orbit before the major part of their journey. The ships go into orbit round Mars on 10 June 1995 and spend 45 days surveying the surface for a suitable landing site. During this phase there is the possibility of a side trip to the moon Phobos. Having decided on a site, the Mars surface party lands in the lander, explores, collects samples, and finally returns to the orbiters using only the centre part of the lander. The spaceships, again locked together, leave Mars on 25 July 1995 for the journey back to Earth. Venus is used for a gravity-assist on the return leg to reduce travel time. The ships reach Earth-orbit on 16 May 1996 for recovery by the Space Shuttle. □ see MARS, SPACE MEDICINE, ELECTRIC PROPULSION, ORBITAL TRANSFER VEHI-CLE, KEPLER 1, PHOBOS AND DEIMOS

expendable not intended to be reused ⟨*Saturn 5 was an ~ launch vehicle*⟩ – compare REUSABLE

expert system (abbr **ES**) computer software which holds information on some area of human expertise. An expert system can help a doctor to diagnose an illness, a chemist to understand chemical reactions, or a geologist to search for oil. An expert system does not just act as a database to provide answers to questions entered at the keyboard. It is designed to follow a line of enquiry and therefore provide as much useful information as possible in the shortest time. An expert system could help engineers troubleshoot problems that sometimes arise when loading liquid hydrogen into the external tanks of the Space Shuttle. To avoid costly launch delays, the expert system would capture the expertise of the launch team and make rapid decisions if an emergency arose. □ see SUPERCOMPUTER, ARTIFICIAL INTELLIGENCE, SPACE SHUTTLE

Explorer 1 the USA's first successful Earth-orbiting satellite launched on 31 January 1958. Electronic instrumentation aboard Explorer measured the intensity of cosmic rays and of the Earth's radiation belts, the temperature of the upper atmosphere, and the frequency of collisions with micrometeorites. Like Sputnik, launched a year earlier, Explorer had no means of storing data and therefore transmitted information to Earth continuously. □ see SPUTNIK 1, TELEMETRY, ARTIFICIAL SATELLITE

explosive bolt a metal bolt incorporating an explosive which, when detonated, destroys the bolt. Explosive bolts are used to separate a satellite from a launch vehicle. □ see SOLID ROCKET BOOSTER

exponent a number or symbol written at the upper right of another number to show that it is to be multiplied by itself a certain number of times. Thus, 10^6 light-years means 1 million light-years where 6 is the exponent of 10.

external tank (abbr **ET**) the large fuel tank containing liquid hydrogen and oxygen for the three main engines of the Space Shuttle attached to it. It contains 720 tonnes of liquid propellant which are used up in the first 8.5 minutes of the flight into space. At the end of this time, the Space Shuttle is at a height of 113km, at which point the empty tank separates and burns up in the atmosphere over the Indian

extraterrestrial

Ocean. Each tank is made largely of aluminium alloys and is 47m long and 8.5m in diameter. It is designed to absorb the enormous thrust (28.6 million newtons) generated by the three main engines and the two solid rocket boosters. The external tank is more than just a fuel tank: it contains a fuel feed system to deliver fuel to the Shuttle's engines; a pressurization and vent system to regulate the pressure in the tank; an environmental control system to regulate the tank's temperature and make its atmosphere inert; and an electrical system to deliver power and instrumentation signals. The skin of the external tank is sprayed with a 2.5mm thick layer of polyisocyanurate which protects it from the effects of excessive aerodynamic heating, and minimizes ice formation as it soars up through the atmosphere riding 'piggy back' on the Space Shuttle. □ see SOLID ROCKET BOOSTER, SPACE SHUTTLE, THERMAL CONTROL SYSTEM

¹extraterrestrial originating, existing, or occurring beyond the Earth ⟨~ *lifeforms*⟩ □ see ASTROBIOLOGY

²extraterrestrial (abbr **ET**) an (intelligent) lifeform from beyond the Earth □ see SEARCH FOR EXTRATERRESTRIAL INTELLIGENCE, ASTROBIOLOGY

extravehicular activity (abbr **EVA**) any activity carried out by astronauts and cosmonauts outside a space station. For such activities, astronauts wear an EVA spacesuit. □ see SPACEWALK, SPACESUIT, EXTRAVEHICULAR MOBILITY UNIT

extravehicular mobility unit (abbr **EMU**) an elaborate spacesuit worn by astronauts who work outside the Space Shuttle. The EMU comprises a spacesuit, a portable life-support subsystem (PLSS), displays and controls module (DCM), a manned manoeuvering unit (MMU), emergency life support and rescue equipment, and several other items for the comfort of the wearer. The EMU enables an astronaut to leave the spacecraft to operate payload systems, to install, remove, and transfer film cassettes on payload experiments, and to repair and replace modular equipment within the payload area. The EMU would also be used to transfer astronauts between a disabled Space Shuttle and a rescue craft. Unlike the spacesuits worn by Apollo astronauts, the EMU is not customized for the wearer: instead it is made up of selected differently-sized component parts. Basic to the spacesuit is a liquid cooling and ventilation garment (LCVG) which is zippered for front entry and which passes excess body heat to the PLSS. Underneath the LCVG is a urine collection device which receives and stores up to 950 millilitres of urine for transfer to the Shuttle's waste management system. An in-suit drink bag provides the astronaut with up to 0.6 litres of drinking water. The spacesuit has bearings in the shoulder and arm joints so that the astronaut can bend, lean, and twist with ease. Fitted to the helmet is a visor which gives protection against micrometeoroids, and ultraviolet and infrared radiation from the Sun. An electrical harness routes biomedical instrumentation signals (respiration and heartbeat rates), two-way communications, and caution-and-warning signals to the

extravehicular communicator. The latter is carried on the upper part
of the back with the DCM mounted on the chest. The DCM contains
a set of electrical and mechanical controls, a microprocessor, and
LED display, and provides the astronaut with a way of controlling the
PLSS and reading the status of its various functions. The PLSS is carried on the back and provides a constantly refreshed atmosphere for
the astronaut to breathe. It includes oxygen bottles, a fan/separator/
pump assembly, valves, and sensors. The MMU snaps onto the back
of the PLSS and allows an astronaut to carry out inspections, and service, adjust, and repair in orbit. It can also be used as a portable space
workstation for it has attach points and power outlets for lights,
cameras, and power tools. Indeed, the MMU enables astronauts to
rescue satellites which have malfunctioned, as happened during the
Discovery flight in November 1984 when astronauts, Allen and
Gardner, recovered the failed Palapa and Westar satellites which had
been launched the previous February. In order to help astronauts service satellites, the various functions of the MMU might be controlled
by spoken commands using computerized voice recognition. And
instead of reading typed instructions on the sleeve of the spacesuit,
instructions could be held in a computer and transmitted to a 25mm
screen in the suit's helmet. Another possibility is to equip the helmet
with a see-through 'head-up' display to project instructions on the
inside of the visor. □ see EXTRAVEHICULAR ACTIVITY, SOLAR MAX,
SPACE INSURANCE, REMOTE MANIPULATOR SYSTEM, SPACE SHUTTLE,
MICROMETEROID, RESCUE BALL

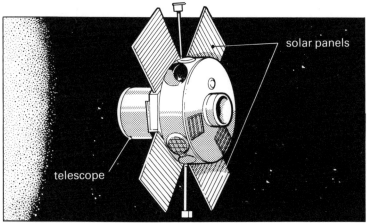

Extreme Ultraviolet Explorer

extreme ultraviolet astronomy the study of celestial objects
which emit radiation between the X-ray and ultraviolet regions of the
electromagnetic spectrum, ie between about 10nm and 100nm (100

and 1000 angstroms) □ see ULTRAVIOLET, ULTRAVIOLET ASTRONOMY, NANOMETRE

Extreme Ultraviolet Explorer (abbr **EUVE**) a US satellite which will be launched by the Space Shuttle in 1988 to make the first all-sky map in the extreme ultraviolet (EUV) which lies between the ultraviolet and X-ray regions of the electromagnetic spectrum. The EUV is the last 'unexplored' region of the electromagnetic spectrum and this study will be the first in-depth investigation. EUVE will enable astrophysicists to study several aspects of the nature and evolution of stars, especially the physical processes taking place in the hot gaseous corona surrounding them. It will gather data from hot white dwarf stars which have exhausted their nuclear fuel and emit EUV as they cool. Also it will make a study of the clouds of interstellar gas within a few hundred light-years of the Sun. Since these clouds absorb EUV, it will be possible to map their presence by observing the absorption of EUV from objects behind them. EUVE will orbit at 550km and will use four 400mm diameter telescopes to make the all-sky survey, and a spectrometer for detailed observations of sources. □ see EXTREME ULTRAVIOLET ASTRONOMY, ULTRAVIOLET ASTRONOMY, ROSAT, ASTROPHYSICS, SOLAR MAX, INTERNATIONAL ULTRAVIOLET EXPLORER

eyeballs in, eyeballs out a term used by pilots to describe the forces to which they are subjected during acceleration and deceleration. Forces due to the launch of a rocket cause 'eyeballs in'; and forces due to the firing of retrorockets cause 'eyeballs out'. □ see CENTRIFUGE

eyelens – see EYEPIECE

eyepiece *also* **eyelens** that part of a telescope or binoculars that is nearest to the eye. The eyepiece produces the magnified image that the eye looks at. □ see OBJECTIVE, TELESCOPE

F

failsafe ensuring no loss of function or danger when failure occurs ⟨*an electrical fuse is a ~ device*⟩ □ see MALFUNCTION, BUG, TROUBLESHOOTING

Faint Object Camera – see HUBBLE SPACE TELESCOPE

fairing any smooth part of an aerospace vehicle designed to reduce air resistance by smoothing out the flow of air across it □ see NOSE CONE, STREAMLINE FLOW

Far Infrared and Submillimetre Space Telescope (abbr FIRST) – see EUROPEAN SPACE AGENCY

fatigue the tendency of a material to weaken or deteriorate under repeated or continuous stress ⟨*metal ~ in aerospace components*⟩

FDS – see FLIGHT DATA SUBSYSTEM

feed to put information into a device ⟨*to ~ data into a computer*⟩

feeder an electrical conductor that sends signals from a radio transmitter to a radio antenna □ see COAXIAL CABLE, DISH AERIAL, RADIO ASTRONOMY, WAVEGUIDE

ferry to take goods to and from an orbiting spacestation (eg by Space Shuttle or Soyuz) □ see PROGRESS, ORBITAL TRANSFER VEHICLE

FGS – see FINE GUIDANCE SENSORS

fibre optics the use of hair-thin transparent glass or plastics fibres to transmit light by repeated internal reflections. Future spacestations may well utilize fibre optics to collect and distribute solar radiation throughout the spacestation for illumination, recreation, and health. □ see OPTICAL FIBRE, OPTICAL COMMUNICATIONS

fibre optics communications – see OPTICAL COMMUNICATIONS

field 1 a group of computer instructions dedicated to a particular purpose (eg reading data) **2** the property of the space surrounding a body which causes one body to act on another ⟨*gravitational ~*⟩ **3** a set of scanning lines which builds up a complete picture on a VDU line-by-line

fifth generation computer a supercomputer which can acquire knowledge and use it intelligently. The fifth generation computer was first proposed as a 10-year project by Japan in 1981 and subsequently this challenge has been taken up by other countries. Its success, however, depends on finding solutions to five problems: first, how to build a computer on a chip which is considerably more complex then today's very large scale integration devices; second, how to devise a language able to handle over 100 million instructions per second so that it can make intelligent use of information fed to it; third, how to provide the computer with an enormous and complex knowledge base which it can draw upon for its decision-making tasks; fourth, how to design a computer architecture which performs many tasks at once so that it acts in a way more closely resembling the human brain in making inferences from the information it receives; fifth, how to improve existing software, such as the language Prolog, so that the computer can recognize human speech. Designers of fifth generation computers want people to be able to converse with the computer so that they are able to consult it as an expert in any chosen field of knowledge. It will make possible intelligent robots which can see and hear spoken commands, and which can be involved in decision-making in government, industry, and the home. The fifth generation computer will bring closer science fiction's image of a talking computer (eg HAL in the film *2001: A Space Odyssey*) which takes over the 'housekeeping' duties of astronauts and cosmonauts on long trips through space. The fifth generation computer will enable an unmanned robot to make intelligent decisions when exploring the surface of a distant planet (eg whether to explore an interesting canyon on Mars if it then loses contact with its parent spacecraft in orbit round Mars). □ see SUPERCOMPUTER, EXPERT SYSTEM, ARTIFICIAL INTELLIGENCE

figuring the process of grinding and polishing the surface of a lens or mirror to give it the shape (eg a paraboloid mirror) required in a telescope or other optical instrument

filter 1 a device for controlling the range of frequencies which passes through an electronic circuit ⟨*a band-pass* ~⟩ □ see FREQUENCY DIVISION MULTIPLEXING **2** a thin film of material which allows light of one colour (or a narrow band of wavelengths) to pass through it ⟨*an infrared* ~⟩ □ see PHOTOMETER, BANDWIDTH

fin 1 a fixed or adjustable aerofoil projecting from an aerospace vehicle to provide in-flight control ⟨*a stabilizing* ~⟩ **2** a flat or shaped (eg corrugated) surface that helps to dissipate heat ⟨*a cooling* ~⟩

fine guidance sensors (abbr **FGS**) guidance sensors to enable the Hubble Space Telescope (HST) to point accurately at stars. A star or object to be studied by the HST must be adjacent to two 'guide stars'. These guide stars must reach stringent criteria such as brightness, colour, position, and lack of stellar companions. Using the guide stars, the FGS will enable the HST to lock onto faint stars for up to 24 hours with a movement of only a few milliseconds of arc, which is equivalent to focussing on a 15mm circle at a distance of 400km. □ see HUBBLE SPACE TELESCOPE

finite 1 having definable limits ⟨*light has a* ~ *range of frequencies*⟩ **2** completely determined by counting or measurement ⟨*the* ~ *speed of light*⟩ – compare INFINITE

fire 1 to ignite a rocket engine ⟨*the test rocket was* ~d⟩ **2** to launch a rocket vehicle ⟨*to* ~ *a ballistic missile*⟩ □ see THRUSTERS, ROCKET ENGINE

FIRST [*F*ar *I*nfrared and *S*ubmillimetre Space *T*elescope] – see EUROPEAN SPACE AGENCY

fission – see NUCLEAR FISSION

flare – see SOLAR FLARE

flight control room a place from which the countdown, launch, and subsequent operations of an aerospace vehicle are monitored and controlled through telemetry and voice communications □ see CON-

SOLE, BLOCKHOUSE, JOHNSON SPACE CENTRE

flight crew astronauts or cosmonauts who command, pilot, navig-
ate, or in some other way operate a manned spacecraft □ see
COMMANDER, PILOT, MISSION SPECIALIST, PAYLOAD SPECIALIST,
SPACE SHUTTLE, FLIGHT DECK, MID DECK

Flight Data Subsystem (abbr **FDS**) a computer which carries out
two basic functions aboard the Voyager spacecraft: it commands the
11 instruments on the spacecraft in their task of recording scientific
data; and it processes this data in a form suitable for transmission back
to Earth. At the Jet Propulsion Laboratory (JPL), two 'navigators',
Dick Rice and Ed Blizzard, reprogram the FDS continuously from
Earth so that Voyager is able to adapt to the changing environment as
it approaches the outer regions of the Solar System. From Jupiter, the
FDS was able to send data back to Earth at a rate of 115.2 kilobits per
second. When Voyager 2 reaches Uranus in 1986 it will be able to
return data at 30 kilobits per second. Each TV picture taken by the
Voyager cameras is made up of 800 lines with 800 picture elements in
each line. The brightness of each pixel is described by 8 bits.
Therefore each TV picture consists of $800 \times 800 \times 8 = 5$ million bits.
Hence it would take about 170 seconds to send back a single TV picture of Uranus. However, it is possible for FDS to send back the data
at twice this rate by measuring only the brightness difference between
one pixel and its neighbour. This technique is called data compression.
Voyager 2 has two nearly identical FDS computers, the second being a
spare in case of component failures (not a bad idea since the memory
of one FDS computer failed on Voyager 1 in 1981). A Dual Processor
Program (DPP) has been developed to handle both the FDS computers on Voyager 2: it instructs one FDS to handle compression of
TV image data, while the other controls and reads the instruments.
This real-time programming of the Voyager computers illustrates the
level of sophistication of these remarkable spacecraft which act as an
extension of our senses deep in interplanetary space. The Voyager
spacecraft will become the most remote scientific instruments to be
operated as they leave the Solar System in the 1990s. □ see VOYAGER,
SUPERCOMPUTER

flight deck the portion of the Space Shuttle above the mid deck
where the flight crew operate control and navigation equipment. The
commander and pilot occupy standard left and right positions respectively, on the flight deck. There are two other seats on the flight deck:
one is occupied by the 'engineer' who is usually Mission Specialist 2,
and he/she is an integral part of the flight crew for ascent/descent
operations; the other seat is usually occupied by Mission Specialist 1,
who is responsible for on-orbit operations. □ see FLIGHT CREW, MID
DECK, SPACE SHUTTLE, MISSION SPECIALIST

flight path the track made or to be followed by an aerospace vehicle through the air or space ⟨*the Shuttle* ∼⟩ □ see TRAJECTORY, ORBIT

flight simulation – see SIMULATION

flight test a test of the performance of an aerospace vehicle, or of a

flow

component carried by it in flight □ see HOT TEST

flow a stream of air or liquid (or of a plasma) ⟨*a ~ of propellant*⟩ – compare FLUX □ see STREAMLINE FLOW

fluid a substance which does not have a shape of its own and adapts to the shape of its container. Water (a liquid) and air (a gas) are fluids. In zero gravity (eg when the Space Shuttle is in orbit) liquids tend to gather in spherical globules owing to the small attractive forces between the molecules of the liquid. □ see PLASMA

flux 1 the rate at which electromagnetic energy flows ⟨*the ~ of solar radiation*⟩ **2** the intensity of nuclear radiation measured by the number of particles passing per second through a given area ⟨*neutron ~*⟩ – compare FLOW

fly 1 to operate or travel in an aircraft or spacecraft ⟨*an astronaut flies the Space Shuttle*⟩ **2** to cause to fly ⟨*ESA flies the Spacelab aboard the Space Shuttle*⟩

flyby a path which takes a spacecraft close to a planet or satellite without landing on it ⟨*the Voyager ~ of Neptune*⟩ – compare LANDER □ see GRAVITY ASSIST, RENDEZVOUS

footprint 1 an area on the Earth's surface covered by the transmissions of a communications satellite **2** an area within which a spacecraft is expected to land **3** the space taken up by a desk-top computer, typewriter, or other (electronic) equipment

force any physical property (eg gravity) that changes the speed of a body, or distorts the body if it is at rest. A force of 1 newton (N) is defined as the force required to give a mass of 1kilogram an acceleration of 1 metre per second. At launch, the Space Shuttle is accelerated into space by a force of 28.5 million newtons generated by its rocket engines. □ see GRAVITY

four-dimensional requiring four coordinates to describe the position of an object in spacetime – compare THREE-DIMENSIONAL □ see SPACETIME

Fraunhofer lines the most prominent absorption lines in the spectrum of starlight or sunlight. These lines were first studied by Joseph von Fraunhofer in 1814 in the spectrum of the Sun's photosphere. Over 25 000 of these lines have been identified in the Sun's spectrum. The lines are due to the presence of calcium, hydrogen, sodium, magnesium, and iron atoms in the Sun's corona which selectively absorbs light originating lower down in the photosphere. □ see ABSORPTION SPECTRUM, SPECTROMETER

free fall the motion of a freely falling body acted on by gravity. During free fall, a person feels weightless though gravity is present – it's just that he/she is not being restrained against the pull of gravity. An astronaut in an orbiting satellite is falling towards the Earth, spacecraft and all, and therefore does not feel any weight. □ see ZERO-G, MICROGRAVITY, SATELLITE

free flight flight through the air without engine assistance ⟨*~ tests of the Space Shuttle*⟩ □ see BALLISTIC MISSILE

free-flying not attached to another spacecraft for electrical power or other services ⟨*Eureca is designed for ~ operation*⟩

frequency the number of times per second that a periodic event (eg the reversal of the mains alternating current) repeats itself. Frequency is measured in hertz ⟨*the ~ of the pulsar in the Crab nebula is 30Hz*⟩. □ see PERIOD, ELECTROMAGNETIC WAVE

fuel 1 a substance used to provide heat and/or power by burning in oxygen ⟨*liquid hydrogen is the ~ in a cryogenic engine*⟩ □ see OXIDIZER **2** a substance that releases energy when its atoms are split ⟨*uranium is a ~ in a nuclear reactor*⟩ □ see NUCLEAR FISSION

fuel cell a device that changes chemical energy into electrical energy by the reaction between two chemicals (eg hydrogen and oxygen) continuously supplied to it. A fuel cell is actually made up of several individual cells stacked together, the number of cells determining the voltage produced by the fuel cell. Fuel cells were used in the Apollo flights to the Moon, and are the source of power on the Space Shuttle. A by-product of the reaction between oxygen and hydrogen is drinkable water which is used by the astronauts. □ see SPACE SHUTTLE

fundamental particles – see ELEMENTARY PARTICLES

fuselage the central part of an aerospace vehicle designed to accommodate passengers and cargo. The forward fuselage of the Space Shuttle contains the cockpit, living quarters, and experiment control station.

fusion – see NUCLEAR FUSION

g 1 the symbol for the prefix giga meaning one thousand million
□ see GIGAHERTZ **2** the symbol for the acceleration due to gravity,
equal to about 9.8ms^{-2} at sea level □ see GRAVITY
G the symbol for the gravitational constant □ see GRAVITATIONAL
CONSTANT

Gagarin, Yuri Alekseyevich (1934–68) Soviet pilot who became the
world's first cosmonaut/astronaut in 1961 when he was sent into orbit
for 90 minutes. Gagarin did not live to see Armstrong land on the
Moon in 1969 for he died, tragically, in a plane crash in 1968. □ see
ARMSTRONG, TERESHKOVA

gain 1 the increase in power, voltage, or current of a signal as it
passes through an amplifier **2** a measure of how well a radio aerial
receives or transmits signals in one direction ⟨*the spacecraft Voyager
uses a high-*gain *antenna for communications with Earth*⟩ □ see DISH
AERIAL, APERTURE

galactic 1 of a galaxy, esp the Milky Way galaxy **2** pertaining to the
galactic system of coordinates ⟨*~ longitude*⟩ □ see GALAXY,
INTERGALACTIC

galaxy a conspicuous usu disc-shaped structure made up of a slowly
rotating assembly of stars, gas, dust, and radiation. Our own Milky
Way galaxy is a typical galaxy, about 10^5 light-years across and containing about a hundred billion (10^{11}) stars. Galaxies are distributed
more or less evenly throughout space, though they tend to cluster
together in groups ranging from a few dozen to many thousands. The
centrifugal effect of the rotation of each galaxy counterbalances its
tendency to collapse due to its own gravity. At the moment individual
galaxies and clusters of galaxies are moving away from each other,
which saves the universe from collapsing due to its own gravity. Most
of the observed galaxies fall broadly into three categories according to
their general shapes: elliptical, spiral, and irregular. The Milky Way
galaxy is an example of a spiral galaxy. An analysis of 1000 of the
brightest galaxies shows that about 75% are spirals, 20% are elliptical,
and 5% are irregular. Since they all appear to have an age of about 12
billion (12 × 10^9) years, it is thought that these shapes indicate different species of galaxies rather than one species in different stages of
evolution. Calculations suggest that these various shapes arose from
the clumping together of matter in the first millionth of a second after
the universe came into being. □ see HUBBLE'S LAW, SPIRAL GALAXY,
SEYFERT GALAXY, CLUSTER 2, MILKY WAY, NEUTRINO, SUN, STAR,
POPULATIONS OF STARS, BIG BANG, QUASAR

Galaxy – see MILKY WAY 1

Galilean of, discovered, or developed by Galileo Galilei, the
founder of experimental physics and astronomy □ see GALILEAN
SATELLITES, GALILEAN TELESCOPE, JUPITER, ^{2}GALILEO

Galileo

Galilean satellites the four larger moons (Io, Europa, Ganymede, and Callisto) of Jupiter discovered by Galileo in 1610. They can be seen with binoculars from Earth on a clear night, and the Voyager spacecraft sent back fascinating close-up pictures of these four moons during the flyby of Jupiter in 1979. □ see JUPITER, IO, EUROPA, GANYMEDE, CALLISTO, VOYAGER, ²GALILEO

Galilean telescope the first astronomical telescope developed by Galileo in 1609. It is made up of a single long-focus converging object lens and a shorter focal length diverging eyepiece. This design produces an upright image and is used in the modern opera glass. Though Galileo's telescope had a magnification of only about 30, and produced a poor image, it opened up a completely new way of thinking about the universe. Galileo made these notes on 7 January 1610: "... when I was viewing the heavenly bodies with a spyglass, Jupiter presented itself to me; and because I had prepared a very excellent instrument for myself, I perceived ... that beside the planet there were three little stars, small indeed but very bright". These were three of the four inner (Galilean) satellites of Jupiter. □ see ²GALILEO, TELESCOPE, JUPITER

¹Galileo a 2550kg spacecraft due to be carried into Earth orbit by Mission 61G of the Space Shuttle in May 1986 and then launched on a trajectory to the planet Jupiter where it is due to arrive in late 1988 for a 20-month study of this planet and its moons. Galileo's launch craft for its 800 million kilometre journey is the first of the 'wide body' Centaur upper-stage rockets which carry 50% more fuel than previous Centaurs. There are two main components to the Galileo spacecraft: an orbiter which will go into orbit round Jupiter; and a probe which is designed to enter Jupiter's atmosphere. The 1.5m wide and 0.9m long probe will separate from the orbiter 5 months before its entry into Jupiter's atmosphere. The probe will be stabilized by spinning it before release from the orbiter section. The probe will enter Jupiter's atmosphere at a speed of 185 000kmh – fast enough to cross the Atlantic in 90 seconds! On entering Jupiter's atmosphere, it will be slowed down by atmospheric drag so rapidly that it will experience the enormous deceleration of 250g. The sensitive instruments inside the probe will be protected from the intense heat generated on the surface of the probe by a massive carbon phenolic heat shield comprising the Deceleration Module. Two minutes after entry, drogue parachutes will open, pulling away the Deceleration Module and allowing the instrument section to sink slowly towards the brilliant cloud tops by the main parachute. Measurements will be radioed to the orbiter flying on a parallel course 190 000km above, for relay to Earth. These instruments (which amount to only 28kg of the probe's total mass of 331kg) will relay data about the chemical composition of the atmosphere, including minor constituents and the relative abundance of hydrogen and helium; the way the density and temperature of atmosphere varies with altitude; and, in particular, the origin and nature of Jupiter's intense electrical

Galileo

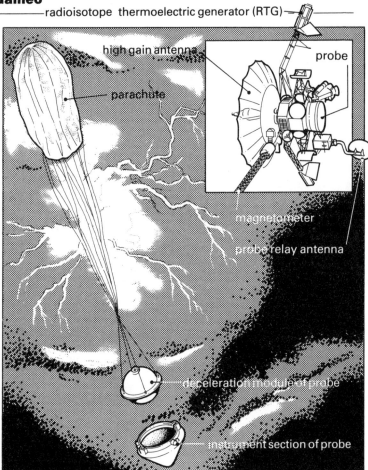

high gain antenna

probe

parachute

magnetometer

probe relay antenna

deceleration module of probe

instrument section of probe

Galileo

activity. The probe's tasks will be completed in 60 to 75 minutes, and the orbiter will reposition itself for a close flyby of Jupiter's satellite Io, before insertion into orbit about Jupiter. During the minute or so measurements are being taken, electrical power will come from a lithium battery which will have been dormant for more than two years during the trip to Jupiter. The probe will be destroyed once the atmospheric pressure has reached about 25 bar and the temperature has risen to about 450K. During the 20-month mission, the orbiter will make 11 targeted close encounters of the Galilean satellites – Io, Europa, Ganymede, and Callisto. These approaches (at times as close as 200km from a satellite's surface) will require a series of complex manoeuvres involving gravity-assists from individual satellites and the firing of the orbiter's engine. High resolution pictures of Jupiter and its satellites are expected. En route to Jupiter, the Galileo spacecraft will be taking a close look at the asteroid Amphitrite which is in a near-circular orbit at a distance of 2.5AU from the Sun. Amphitrite is about 200km in diameter and is one of the larger minor planets of the Solar System. Observations will reveal its rotation rate, shape, mass, density, and mineral composition to help scientists decide whether Amphitrite was made at the time the Solar System was formed, or whether it is a fragment of a broken-up planet. This asteroid, the 29th to be discovered, was detected by Albert Marth in London on 1 March 1854 and named after one of the wives of the mythical god Neptune. □ see JUPITER, IO, SOLAR SYSTEM, ASTEROID, CENTAUR, HEAT SHIELD, SPACE SHUTTLE

²Galileo [Galileo Galilei] (1564–1642) Italian astronomer and physicist who is always known by his first name, perhaps because of the Tuscan habit of using a variation of the last name as the first name of the oldest son. Galileo was born 3 days before Michelangelo died; and, incidentally, he died in the same year Newton was born. A good experimenter and theoretician, Galileo made original discoveries about the pendulum (by observing a swinging chandelier) when he was a student of medicine at Pisa. But his real contribution to the nature of gravity began when he observed the way different objects fell (some of these experiments took place from the top of the leaning tower of Pisa). He concluded that, in the absence of air resistance, all bodies reach the ground at the same time. At the end of the 16th century, Galileo had moved to Padua and was lecturing (brilliantly it seems) to students about a 'new' physics. His views about the structure of the universe were similar to those of Copernicus, the central idea of which is that Earth revolves around the Sun, not the Sun round the Earth. In 1609, using lenses manufactured in Holland, Galileo devised a telescope which had a magnifying power of about 30 – and thus began the age of telescopic astronomy. With the telescope, he proved that the Sun had spots and Jupiter had moons. The markings on the Sun were at odds with accepted dogma that the Sun, and all other celestial bodies, were perfect – sunspots were not allowed! And the rotation of Jupiter's moons showed that not all bodies rotated round the Earth. His views were generally greeted with delight, but not everybody was pleased with the idea of a Sun-centred universe for it demoted the Earth from its privileged position. To make matters worse, in 1632 Galileo announced his 'Copernican' views in a book *Dialogue on the Two World Systems*. In 1633 Pope Pius V declared Galileo's beliefs a heresy. Soon after, at nearly 70 years of age, Galileo, was brought before the Inquisition and forced (under threat of torture) to renounce his views – though legend has it that, having completed his renunciation, he muttered "Eppur si muove!" ("And yet it moves!"), referring to the Earth. But Galileo was silenced until his death, and his book *Dialogue* was not removed from the Catholic list of prohibited books until 1835. The spacecraft Galileo, due for a close-up study of Jupiter

gallium arsenide

in 1988, has been named in Galileo's honour. □ see ²KEPLER, NEWTON, GALILEAN TELESCOPE

gallium arsenide (chemical formula **GaAs**) a crystalline material which, like silicon, is used to make diodes, transistors, and integrated circuits. Its main claim to fame is that semiconductor devices made from it are able to conduct electricity five to ten times faster than those made from silicon. This property makes GaAs of interest to manufacturers of computer memory devices since there is an ever-increasing need for computers to process information faster. Data held in a computer memory device made from GaAs can be stored and retrieved more rapidly than with any other type of memory, so this material is in great demand for military, and subsequently civilian computers. However, there are a few drawbacks to the use of GaAs, one of which is that gallium and arsenic, the two elements from which it is made, are in short supply, mainly being found as impurities in aluminium and copper ores respectively. On the other hand, silicon is plentiful being found in silicates such as sand. The present cost of GaAs is about thirty times that of silicon, and the cost is increased because about 90% of GaAs chips are rejected after production. Furthermore, it is not so easy to manufacture integrated circuits from GaAs as it is from silicon, since it does not form a protective layer of oxide to resist the diffusion of a dopant during the process of photolithography. Thus it is unlikely that there will be a rapid rise in the use of GaAs-based semiconductors in the foreseeable future except for specialist applications where cost is not a major consideration. Manufacturers of GaAs have turned to space for help. NASA has signed a joint agreement with Microgravity Research Associates of Miami to develop a furnace for processing GaAs in the hard vacuum and zero gravity of space. The prototype furnaces are being flown aboard Shuttle flights and the Miami-based company is expected to sell GaAs material at $1M a kilogram, presumably to weapons manufacturers. □ see MATERIALS PROCESSING

gamma (symbol γ) the third letter of the Greek alphabet, used to identify the third brightest star in a constellation (eg Orion)

gamma-ray astronomy the study of celestial objects from their gamma-ray emissions at wavelengths 1000 times shorter than those of visible light. This shorter wavelength makes gamma rays the most energetic of all electromagnetic radiation. Like X-ray astronomy, which detects radiation 100 times less energetic than gamma rays, gamma-ray astronomy was born when it became possible to send instruments into space aboard satellites and spacecraft. The first satellite to detect gamma rays from deep space, ie from beyond the Solar System, was the Orbiting Solar Observatory-3 (OSO-3) launched in 1968. OSO-3 used an instrument called a scintillation counter which found that the gamma rays were particularly strong within about 30° of the centre of our galaxy. Indeed it is only because gamma rays are so penetrating that we can 'see' what is going on in the centre of our galaxy through all the gas and dust in the way. Later satellites, such as SAS-2 (1972), used instruments called spark chambers to detect gamma rays with enormous energies in the band 20 to 100 million electron volts (20 to 100MeV). And, most significantly, some of these gamma rays came from the Vela and Crab pulsars. The European Space Agency's COS-B satellite (1975) used a large spark chamber of improved resolving power to confirm that these pulsars and about twenty other sources, including the quasar 3C 273, produced gamma rays. One interesting discovery has been that there is a diffuse background of gamma rays; it has no specific direction, and could originate from the interaction of cosmic radiation with the interstellar medium. The origin of gamma rays is particularly interesting to astrophysicists, for should any radiation be found with energies in the range 0.5 to 3MeV it would suggest that heavy elements are made in supernovae explosions. □ see GAMMA RAYS, SUPERNOVA, SCINTILLATION COUNTER, ELECTRON VOLT, HIGH ENERGY ASTRONOMICAL OBSERVATORY, GAMMA-RAY OBSERVATORY, NUCLEOSYNTHESIS, PULSAR

Gamma-ray Observatory (abbr **GRO**) a 15 tonne Earth-orbiting observatory designed to investigate gamma rays from stars and galaxies and due to be launched by the Space Shuttle in 1988. The Shuttle will also retrieve and service the GRO from time to time. The GRO will contain four different instruments to pinpoint and measure the intensity and wavelength of gamma rays from celestial sources such as pulsars and quasars which are believed to be associated with black holes. One way of identifying a black hole is to look for gamma rays of particular energies which are known to be produced in its neighbourhood. One of the instruments aboard the GRO will look for the source of puzzling gamma ray 'bursters'. Something is occasionally emitting bursts of gamma-ray energy, but so far no source has been identified with any other known objects. One of these 'bursters' was detected in 1979 and showed a series of bursts of gamma rays at 8-second intervals: the bursts came from the direction of the Large Magellenic Cloud, and if the calculations are correct, the source emitted more energy in one tenth of a second than the Sun does in ten thousand years. □ see GAMMA RAYS, CLUSTER 3

gamma rays the most energetic electromagnetic radiation in the universe, having wavelengths less than about 10^{-12}m – shorter even than X rays. Because of their extremely short wavelength, gamma rays are spoken of in terms of their energy, measured in electron volts (eV). Their energies are higher than a tenth of a million electron volts and they have been observed with energies of hundreds of millions of electron volts. Gamma rays are the by-products of some types of nuclear reactions (eg those taking place inside of a nuclear reactor). They are also produced by the extremely violent reactions taking place at the centre of our Galaxy, and in black holes, pulsars, and quasars. Gamma rays from space are being studied by instruments aboard artificial satellites. □ see GAMMA-RAY ASTRONOMY, GAMMA-RAY OBSERVATORY, ANNIHILATION, COSMIC RAYS, X RAYS, ELECTRON VOLT

gantry a (massive) frame comprising elevated platforms for servicing a rocket as it rests on its launch pad □ see MOBILE LAUNCH PLATFORM

Ganymede the third in increasing distance and the brightest of Jupiter's Galilean moons, slightly larger than the planet Mercury. Photographs taken by the Voyager spacecraft showed Ganymede's surface to be torn and twisted, parts looking like ploughed fields. But these features are tens of kilometres wide and hundreds of kilometres long and are probably caused by deformations in Ganymede's icy surface. The surface appears to be a mixture of dirty ice and rocks which gives this moon a low albedo. Bright patches on the surface could be underlying 'clean' ice which has been thrown up by meteor impacts. Few large craters and huge fault lines are seen on the surface of this remarkable and unique satellite. □ see JUPITER, CALLISTO, IO, EUROPA, VOYAGER, ALBEDO

gas matter which is able to fill any volume it is put into since its molecules move independently of one another □ see FLUID, PLASMA

GAS – see GETAWAY SPECIAL

gas laser a device that excites a gas to produce an intense and narrow beam of radiation continuously. There is considerable interest in the military applications of gas lasers by both the USA and USSR, President Reagan having committed the USA to developing laser-beam weapons in the interests of defence against enemy missiles – known as the Star Wars programme. Many gases (eg carbon dioxide, argon, and krypton) can be made to lase. In the gas-dynamic laser, the lasing action takes place in a chamber called the optical cavity through which a gas (eg carbon dioxide) is pumped at supersonic speed and at a pressure below atmospheric pressure. Electrodes feed electrical power into the gas to stimulate the lasing action. Nevertheless most of the electrical power is wasted as heat which is carried off by the gas. A further development of the gas-dynamic laser is the chemical laser which has the advantage that it requires no external source of power. The energy to pump the lasing gas is provided by a chemical reaction between carbon monoxide and nitrous oxide which produces carbon dioxide with its electrons already in an excited state and able to produce photons of light. Mirrors at each end of the optical cavity reflect the emitted photons to enhance the lasing action. The hot gas is exhausted from the laser at supersonic speed and in this respect resembles a jet engine. Gas lasers are not so much designed as discovered: the physics is not completely understood and a new gas combination or a new mirror design may produce a radical change in the laser's performance. □ see LASER, STAR WARS

geiger counter an instrument for detecting and measuring the intensity of radiation (eg gamma rays and alpha particles). Its sensor consists of a cylindrical geiger tube which contains two electrodes, an anode and a cathode, between which is maintained a direct current voltage of a few hundred volts. The tube contains a small quantity of gas at low pressure. Any radiation entering the tube causes the gas to ionize and brief small pulses of current are produced which are counted by electronic circuits to give a reading of the strength of the radiation (eg of cosmic rays near a spacecraft). □ see GAMMA RAYS, SCINTILLATION COUNTER

Gemini project a series of two-man spacecraft that followed the one-man Mercury project in the 1960s, and preceded the Apollo missions to the Moon. Gemini 3 was the first manned flight launched in 1963 (in which John Young was reprimanded for taking along a corned beef sandwich without permission!). Astronaut Ed White made America's first spacewalk aboard Gemini 4 in June 1965. Geminis 6 and 7 rendezvoused in December 1965, and Gemini 8 successfully docked with an unmanned spacecraft in March 1966. Gemini 12 completed the series in November 1966. The Gemini project showed that two men could function efficiently for long periods in weightless conditions, work effectively outside the spacecraft, and control the docking of two spacecraft – all important exercises for the Moon landings accomplished later by the Apollo project. □ see MERCURY PROJECT, WHITE, APOLLO, YOUNG

general relativity (abbr **GR**) a theory proposed by Albert Einstein in 1916 showing that a gravitational field causes space to be curved and time to run slow. The direct consequence of general relativity (GR) is that a strong gravitational field (such as that surrounding black holes and neutron stars) upsets the normal laws of physics. For example, light should bend when it passes close to a massive body. And this has been verified by measuring the delay in the signals grazing the Sun from Viking and other spacecraft on the opposite side of the Sun: the measurement agrees with the prediction of Einstein's GR to within 3%. Strong gravitational fields should make atoms oscillate more slowly so that atomic clocks run slower. Astrophysicists are looking to space for further verification of the predictions of GR: the properties of the black hole itself; the marked curvature of space-time near to massive black holes; the possible existence of gravitational waves; and the value of the gravitational constant, are all under scrutiny by instruments launched into space aboard spacecraft and satellites. □ see EVENT HORIZON, BLACK HOLE, EINSTEIN, SPECIAL RELATIVITY, SPACETIME

General Unified Theory (abbr **GUT**) a single mathematical theory capable of explaining the four fundamental forces of nature: gravity, electromagnetism, and the weak and strong nuclear forces. No successful GUT has yet been able to incorporate all four forces. However, it has been shown that the electromagnetic and weak nuclear forces, though very different in their operation, are two aspects of a single electroweak force, and recent advances suggest that the strong nuclear force can be brought into the scheme. Gravity alone remains to be incorporated in the theory. □ see MAGIC NUMBER, EINSTEIN, GRAVITY

geo – see GEOSTATIONARY ORBIT

geocentric having Earth as a centre 〈*the Moon is in ~ orbit about* *the Earth* ⟩ – compare AEROCENTRIC

GEODSS – see GROUND-BASED ELECTRO-OPTICAL DEEP SPACE SURVEILLANCE

geomagnetism the Earth's magnetic field. Its origin is believed to involve currents flowing in the rotating liquid outer core of the Earth which is composed of molten nickel-iron. The magnetic field above the Earth is much like that produced by a giant bar magnet at the centre of the Earth and inclined at about 14 degrees to the axis of rotation. Fossil evidence indicates that the Earth's magnetic field completely reverses (magnetic north becomes magnetic south) every few hundred thousand years, but the cause is unknown. However, the Earth's magnetic field interacts with the solar wind to produce the magnetosphere, and magnetic storms (violent fluctuations in the strength of the field) are linked with solar flares on the Sun. □ see MAGNETOSPHERE, EXPLORER 1, AURORA, ACTIVE MAGNETOSPHERIC PARTICLE EXPLORER

geophysics the physics of the Earth and the air and space surrounding it. Though originally including the study of the history, movement, and composition of planet Earth (ie disciplines of geology, oceanography, seismology, hydrology, etc), geophysics now extends to meteorology, geomagnetism, and astrophysics. This extension of the meaning of geophysics comes from our greater understanding of the origin and composition of the Solar System afforded by satellites and spacecraft. □ see MAGNETOSPHERE, SATELLITE LASER-RANGING

GEOS – see GEOSYNCHRONOUS ORBIT SCIENTIFIC SATELLITE

geosphere the solid and liquid parts of the Earth above which is the atmosphere □ see EARTH'S ATMOSPHERE, BIOSPHERE

geostationary *also* **geosynchronous** of or being an Earth-orbiting artificial satellite remaining in one part of the sky as seen from the Earth's surface □ see GEOSTATIONARY ORBIT, CLARKE

Geostationary Operational Environmental Satellite (abbr **GOES**) a series of US weather satellites designed to monitor weather conditions over the United States and the Pacific Ocean from Alaska to New Zealand. GOES-6 launched in April 1983 was the third in the series and forms part of the World Weather Watch project. At the heart of a GOES is an instrument called a radiometer which senses the infrared radiation produced by the Earth and clouds. The satellite spins at 100 revolutions per minute to provide gyroscopic stability, while its antenna remains fixed and pointing towards the Earth. The radiometer is housed in the spinning portion and builds up an image of the Earth line by line, rather like the way a TV picture is built up. On each scan, a scan mirror steps 1/100th of a degree and after 18 minutes and 1800 scans a complete picture is ready for transmission to an Earth station. The image gives information about temperature and moisture distribution from the cloud tops to the ground. □ see NATIONAL OCEANIC AND ATMOSPHERIC ADMINISTRATION, RADIOMETER, SPIN STABILIZED

geostationary orbit (abbr **geo**) *also* **geosynchronous orbit, Clarke**

orbit an orbit of an artificial satellite moving west to east above the equator that has a period of 23 hours, 56 minutes, and 4.1 seconds and therefore stays in the same place above the Earth. Geostationary orbits are preferred for communication and navigation satellites since they appear to 'hang' in the sky at one point and are therefore always accessible from an Earth station. Geostationary orbits are at an altitude of 36 900km above the equator, and are difficult to achieve since they require a high orbital speed. Three communications satellites in geostationary orbit provide global coverage. □ see ORBIT, GEOSYNCHRONOUS ORBIT SCIENTIFIC SATELLITE, COMMUNICATIONS SATELLITE

geosynchronous – see GEOSTATIONARY

Geosynchronous Orbit Scientific Satellite (abbr **GEOS**) the first all-scientific satellite built for the European Space Agency by British Aerospace and launched into geostationary orbit to measure electric and magnetic fields, as well as particle streams, in the Earth's magnetosphere. GEOS-1 was launched in 1977 but failed to go into geostationary orbit and was eventually placed in an elliptical orbit. GEOS-2 was successfully launched in 1978 and its seven instruments proved to be so useful that its mission was extended until 1983. With the growing number of satellites in geostationary orbit, there is an increasing risk of collisions between working and deactivated satellites. Thus, in January 1984, the orbit of GEOS-2 was raised by 270km where the satellite now drifts at a rate of about 3.5° longitude per day. GEOS-2 becomes visible to the Earth station at Michelstadt in West Germany for 4 weeks every 3.5 months. From its new vantage point, GEOS-2 was able to provide useful data in support of the three satellites in the Active Magnetospheric Particle Explorer (AMPTE) project which was patrolling the fringes of the magnetosphere in 1985. □ see GEOSTATIONARY ORBIT, MAGNETOSPHERE, BRITISH AEROSPACE, GEOPHYSICS, ACTIVE MAGNETOSPHERIC PARTICLE EXPLORER

Getaway Special (abbr **GAS**) any of several small self-contained experiment packages which are accommodated in the payload bay of the Space Shuttle as volume and weight become available. These packages are generally flown on a first-come first-served basis, and are intended to stimulate and encourage the use of space by researchers, educational institutions, and private individuals. For example, on mission STS-7 launched on 18 June 1983, *Challenger's* payload bay contained seven GAS payloads of which one, GAS-G012 contained a colony of ants in a small canister. This experiment was developed by the New Jersey High Schools of Camden and Wilson and was aimed at examining the behaviour of ants in zero gravity. Students at the Ashford High School for Girls, England, have won a competition with their proposal to investigate the growth of a chemical garden in weightlessness. It should be launched by the Space Shuttle in 1986. The GAS payloads have also been used to deploy two small satellites. The Global Low-Orbiting Message Relay Satellite (GLOMR) and the

Northern Utah Satellite (NUSAT) were launched from mission 51B in May 1985. ☐ see PAYLOAD SPECIALIST, SHUTTLE PALLET SATELLITE

giant planets the four outer planets of the Solar System: Jupiter, Saturn, Uranus, and Neptune: their diameters are between 3.8 and 11.2 times the diameter of the Earth. ☐ see JUPITER, SATURN, URANUS, NEPTUNE

giant star a large, ageing, and very conspicuous star which has exhausted most of its supply of hydrogen to produce a dense core of helium. Hydrogen continues to burn in a vast and very tenuous expanding shell. The helium core continues to contract and heats up. If the helium begins to burn, a shell of burning helium spreads outwards leaving behind a carbon-oxygen core. Further nucleosynthesis may take place involving heavier elements depending on the mass of the star. Capella is a bright yellow giant and is the brightest star in the constellation Auriga; its spectrum shows its atmosphere has a lot of lithium in it. Arcturus is a red giant and the brightest star in the constellation Bootes; its diameter is about 30 times larger than the Sun's. Betelgeuse is the second brightest star in the constellation Orionis; it has a diameter about 800 times larger than the Sun's. Our Sun will eventually become a similar distended red giant completely incinerating all the planets from Mercury to Pluto. ☐ see STAR, SUPERNOVA, HERTZSPRUNG-RUSSELL DIAGRAM, NUCLEOSYNTHESIS

gigabit a quantity of binary data equal to 1 thousand million (10^9) bits. ☐ see MEMORY, SUPERCOMPUTER

gigabyte (symbol **GB**) a quantity of computer data equal to 1 thousand million (10^9) bytes ☐ see FIFTH GENERATION COMPUTER

gigahertz (symbol **GHz**) a frequency equal to 1 thousand million (10^9) hertz which for radio waves is known as ultra-high frequency ⟨*radar waves are in the ~ range*⟩ ☐ see UHF, X-BAND

gimbal a mechanical device on which a rocket engine, nozzle, or other device is mounted, that has two mutually perpendicular and intersecting axes of rotation to allow the device to move freely in any direction ☐ see GYROSCOPE, NOZZLE

[1]Giotto *also* **ESA Halley Probe** a spacecraft launched by ESA's rocket Ariane 1 from the Kourou launch site in French Guiana, South America, on 2 July 1985, and destined for a rendezvous with Halley's comet in March 1986. Following launch, Giotto was parked in geostationary orbit. After a few revolutions in this orbit, and when the spacecraft was near its perigee position, its solid propellant motors were fired to put it into a heliocentric trajectory on target for the meeting with Halley's comet one month after its closest approach to the Sun. Giotto passed within 500km of the comet's nucleus on the sunward side travelling at 68kms⁻¹. At this speed, cometary dust struck Giotto with explosive force so it was protected by (bumper) dust shields. Since the exact position of the nucleus was not known, data from the earlier Soviet Vega probes were used to make fine adjustments to Giotto's path using its thrusters. Giotto carried twelve instruments, one of which – an Ion Mass Spectrometer (IMS), which was looked after by a team from NASA's Jet Propulsion Laboratory, enabled the scientists to study the types of ions created by the solar wind on the material from which the nucleus is made. For this purpose, the IMS was split into two instruments, HIS and HERS. HIS (High Resolution Spectrometer) looked at what was happening within the inner portion of the coma. HERS (High Energy Range Spectrometer) measured the trajectory and speed of the electrically

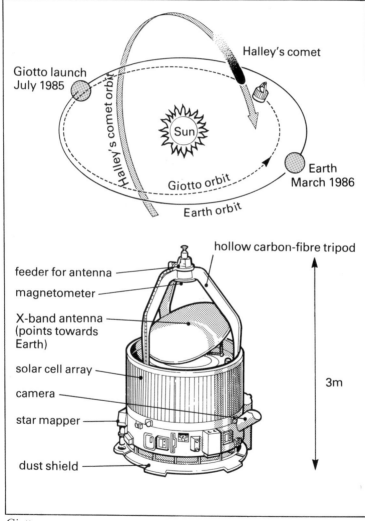

Giotto

charged ions along with their mass-to-charge ratio. Giotto was also equipped with a camera to send back colour pictures of the nucleus.
□ see HALLEY'S COMET, ARIANE, EUROPEAN SPACE AGENCY, NASA, DEEP SPACE NETWORK, JET PROPULSION LABORATORY, PARKING ORBIT, PERIGEE, PLANET-A, SPECTROMETER, AUSTRALIAN NATIONAL RADIO ASTRONOMY OBSERVATORY

²Giotto, [Giotto di Bondone] (1266–1337) Florentine artist and one of the founders of modern realistic painting who (probably) witnessed the apparition of (Halley's) comet of 1301, and included it in his fresco 'Adoration of the Magi' which is in the Scrovegni Chapel in Padua. In this painting, the comet takes the place of the (presumably) non-miraculous star of Bethlehem. In the seventeenth century, Edmund Halley realized that the comet of 1301 was one of the many appearances of a comet which later became known as Halley's comet. The European Space Agency's spacecraft, Giotto, is named in honour of the Italian painter. □ see ¹GIOTTO

glide to descend without the use of engines

glide path the path of descent of an aerospace vehicle before landing ⟨*the Space Shuttle's* ∼⟩

glitch 1 a sudden small increase in the speed of rotation of a pulsar, thought to be due to a fracture in the crust of the pulsar's spinning neutron star, that makes it shrink slightly and speed up □ see PULSAR, ANGULAR MOMENTUM **2** a brief but often recurring distortion to the shape of an electrical signal as seen on a cathode-ray tube or other recording device □ see BLIP

Global Navigation Satellites System (abbr GLONASS) – see GLOBAL POSITIONING SYSTEM

Global Positioning System (abbr **GPS**) *also* **Navstar** a military satellite system operated by the US Air Force that enables someone to find their position on the Earth's surface to within a few metres. By the end of the 1980s, GPS will have 18 satellites operating in 12-hour, 20 000km high orbits. Each satellite continually transmits its position and the precise time of transmission. A passive receiving set on the ground, in an aircraft, on a ship, or carried by a soldier, receives signals from four satellites. This data is processed by the receiver. The receiver displays the user's position to a few metres, speed to an accuracy of a few kilometres per hour, and time to within a few millionths of a second. Though principally intended for military use (eg the precise delivery of weapons on a target), air traffic control, Earth-mapping surveys, search-and-rescue operations, and transatlantic navigation are some other uses for GPS. The rubidium atomic clocks aboard the GPS satellites are so accurate that the effect of time dilation (predicted by the special theory of relativity) has to be taken into account when receiving signals from the satellites. In October 1982, the Soviet Union launched a similar system of navigation satellites called GLONASS (Global Navigation Satellites System). These satellites share much the same orbit as the GPS satellites and were part of the Cosmos series designated Cosmos 1413, 1414, and

1415. □ see INMARSAT, ATOMIC TIME, SPECIAL RELATIVITY, TIME DILATION

globular cluster – see CLUSTER 2

GLONASS [*Glo*bal *Na*vigation *S*atellites *S*ystem] – see GLOBAL POSITIONING SYSTEM

gluon – see QUARK

GMT – see GREENWICH MEAN TIME

Goddard, Robert Hutchings (1882–1945) American physicist who became interested in rocketry as a teenager, and launched a 1.2m long liquid-fuelled (gasoline and liquid oxygen) rocket in 1926. In Massachusetts he was banned from experimenting with rockets, but the philanthropist Daniel Guggenheim was persuaded to award Goddard $50 000 to continue his experiments in New Mexico. By 1935 he was designing and launching rockets that travelled at 800kmh⁻¹ to an altitude of several kilometres. Some of these were multistage rockets and had gyroscopic stabilization. But the US Government was not interested in these developments, at least until World War 2 when it became known that Germany was developing rockets as powerful weapons. Goddard, neglected in his lifetime, died before America entered the space age, but the Goddard Spaceflight Centre in Maryland is named in his honour. □ see TSIOLKOVSKY, V-2, BRAUN

GOES – see GEOSTATIONARY OPERATIONAL ENVIRONMENTAL SATELLITE

Gorizont a series of Soviet communications satellites providing round-the-clock telephone, telegraph, and TV links through the 'Orbita' receiving stations both within the USSR and abroad. These 2000kg cylindrical satellites have a pair of solar panels and a dish aerial at one end. Unlike the Molniya communications satellites, the Gorizont satellites operate in geostationary orbits of between about 140 and 180°E longitude to be accessible to ground stations in the Soviet Union. □ see MOLNIYA

GPS – see GLOBAL POSITIONING SYSTEM

Grand Tour a NASA plan to send spacecraft on a speedy journey to the giant planets of the Solar System between 1977 and 1982. The tour would have taken advantage of a rare alignment of Jupiter, Saturn, Neptune, and Pluto on the same side of the Sun, but the plan was cancelled because it was too costly. The Grand Tour would have used gravity assist to give each spacecraft an energy boost as it swept by a planet to speed them on their way to the next planet. The Voyager 2 spacecraft is now doing a similar tour of the outer planets but the journey is taking much longer since it cannot take advantage of the special alignment of the planets. Perhaps by the next time a similar alignment takes place, astronauts will have the stamina to take the route Voyager 2 is now taking. □ see GRAVITY ASSIST, VOYAGER

gravitation – see GRAVITY

gravitational collapse the sudden contraction of a star which has exhausted its nuclear fuel. The contraction occurs when internal gas pressure is insufficient to support the star's weight. There are three likely end-points of gravitational collapse: the star becomes a black hole, a neutron star, or a white dwarf. □ see SUPERNOVA, BLACK HOLE, NEUTRON STAR, WHITE DWARF

gravitational constant (symbol **G**) a universal constant (ie assumed to be the same everywhere in the universe) which is defined as the force of attraction between two bodies of mass 1kg separated by by a distance of 1m. G is equal to 6.672×10^{-11} Nm^2kg^{-2}. Theories that the universe is expanding suggest that the value of the gravitational constant should be decreasing very slightly with time, but astronomical observations have so far not confirmed this prediction. The very small value of G has had important consequences for the evolution of the universe. For example, if G were 10% larger, stars would consume their hydrogen faster, and our Sun would have long ago become a red giant and vaporized the Earth. Indeed, the universe might long ago have ceased to expand and would by now have collapsed in on itself. □ see GRAVITY, NEWTON'S LAW OF GRAVITATION, EXPANDING UNIVERSE, BIG BANG, MAGIC NUMBER, WEAK INTERACTION, ANTHROPIC PRINCIPLE

gravitational force – see GRAVITY

gravitational waves waves in the gravitational field which are analogous to light waves in the electromagnetic field, and which travel at the same speed as light waves. There is no direct experimental evidence for gravitational waves, but their existance is predicted by Einstein's general theory of relativity. Gravitational waves would be produced by the collapse of massive stars into black holes (eg supernovae explosions). The waves would carry energy away from the object in packets (quanta) called gravitons. Since gravitational waves should carry away energy when two massive stars are orbiting close to each other, astronomers are searching for binary pulsars. Observations of the only known binary pulsar indicate that its orbital period is slowing down by about 10^{-4} seconds per year. This agrees very well with the prediction of relativity theory. □ see GRAVITON, RELATIVITY, PULSAR, ELECTROMAGNETIC WAVE

graviton a hypothetical quantum which is involved in exchanges in a gravitational field, just as the photon is the quantum of the electromagnetic field □ see PHOTON, QUANTUM THEORY

gravity *also* **gravitation, gravitational force** the familiar force of attraction which acts universally between two bodies (eg between us and the Earth). Astronauts who visited the Moon, were the first human beings to be subjected to the gravity of another celestial body – it felt the same only weaker! Gravity is one of the four fundamental forces of nature, the other three being the electromagnetic force, and two nuclear forces, the strong and the weak. Gravity is the weakest of these natural forces but it acts over vast distances. It shapes the orbital motions of planets in the Solar System, and it largely determines the structure and development of the universe. In the 17th century, Sir Isaac Newton studied gravity and produced the famous Law of Gravitation which showed that the more massive a body the more strongly it pulls another body towards it. The strength of this pull falls off as the square of the distance from the centre of the body. This (universal) law appears to hold throughout the universe, but Einstein suggested that objects vast distances apart might experience a gravitational repulsive force, and this has implications for the long-term development of the universe. It is not possible to control the force of gravity (except by moving away from the body producing it), or to neutralize it – at the moment an antigravity engine does not exist! However, on long trips through space, gravity can be produced artificially by subjecting astronauts to a centrifugal force in a rotating chamber. In this way astronauts can remain fit and strong ready for their return to Earth or for their exploration of another planet. □ see NEWTON, NEWTON'S LAW OF GRAVITATION, GRAVITATIONAL CONSTANT, ZERO-G, ARTIFICIAL GRAVITY, SPECIAL RELATIVITY, SPACE MEDICINE

gravity assist a technique for giving a spacecraft an energy boost by planning a trajectory that passes close to a planet or satellite. Gravity assist is an example of how the theory of astrodynamics is used to enable spacecraft to operate more effectively as remote gatherers of information. The spacecraft gains energy at the expense of the gravitational and orbital energy of the planet which therefore slows up imperceptibly. This 'gravitational sling shot' technique was first used on the Mariner 10 mission to Venus and Mercury in 1974, and by Pioneers 10 and 11, and Voyager spacecraft to the outer planets. Gravity assist is sometimes called planet hopping. □ see GRAND TOUR, VOYAGER, ULYSSES, MARINER

grazing incidence X-ray telescope a reflecting telescope in which X-rays are brought to a focus after 'grazing' a reflecting surface. It is because the energy of X-ray photons is so high compared with light photons that they are reflected at such a small angle. This angle is about 1° for 0.3nm X rays reflected from a gold surface. Though costly, X-ray mirrors are used in X-ray telescopes orbiting the Earth. □ see X RAYS, X-RAY ASTRONOMY, HIGH ENERGY ASTRONOMICAL OBSERVATORY, NANOMETRE

Great Red Spot a long-lived large mobile oval feature in Jupiter's atmosphere, close to the equator, that has been studied for more than 100 years. Its colour ranges between pale pink and brick red and may come from the condensation of phosphorus at the cloud tops, or from contamination by organic molecules such as nitriles produced by thunderstorms in Jupiter's atmosphere. Recently, the Great Red Spot has been examined at close range by the Pioneer and Voyager spacecraft, and more information is expected from the probe to be released into Jupiter's atmosphere by the Galileo spacecraft. □ see JUPITER, PIONEERS 10 AND 11, VOYAGER, ¹GALILEO

greenhouse effect the warming of the lower atmosphere of a planet (or the air in a greenhouse) by the trapping of infrared radiation. The greenhouse effect is only effective if the atmosphere contains carbon dioxide which is opaque to long wavelength infrared radiation. Sunlight that reaches the surface of a planet is absorbed and reradiated at (longer) infrared wavelengths. The carbon dioxide in the atmosphere absorbs and re-emits the infrared. A balance between the influx of solar radiation and the escape of infrared radiation only happens when the surface and atmosphere have both reached a higher temperature than would have been reached in the absence of the carbon dioxide. The very high temperature (about 450C) on Venus is largely due to a marked greenhouse effect caused by an atmosphere containing over 98% carbon dioxide. The greenhouse effect also operates on Earth. □ see VENUS, SPACE GREENHOUSE, SPACE SETTLEMENT, EARTH RADIATION BUDGET SATELLITE, SOLAR CONSTANT

Greenwich Mean Time (abbr **GMT**) *also* **universal time** the basis for measuring time throughout the world. It is measured with reference to the Greenwich meridian, ie the meridian of zero longitude which passes through the Borough of Greenwich in London. GMT is determined from the apparent motion of the Sun and is equal to the mean solar time. It is counted in hours from midnight. □ see MEAN SOLAR TIME

GRO – see GAMMA-RAY OBSERVATORY

ground-based of or being on the Earth's surface ⟨~ *astronomy*⟩ – compare SPACE-BASED

Ground-based Electro-optical Deep Space Surveillance (abbr **GEODSS**) a network of optical telescopes operated by the US Air Force to track spacecraft orbiting at altitudes of between 800 and 36 000km. By January 1987, the network will have five observatories around the world: White sands, New Mexico; Mount Haleakala, Hawaii; Taegu, South Korea; Diego Garcia island, Indian Ocean; and, provisionally, Portugal. The system sweeps the skies watching for hostile satellites and monitoring orbital spacings to prevent collisions. The system also helps to forecast where decaying satellites will fall. Each installation has two primary 1m telescopes, and a separate 0.4m auxiliary telescope to track fast-moving satellites in low-earth orbit. The images produced by these telescopes fall onto high-sensitivity electronic sensors (eg charge-coupled devices), which produce images displayed on a TV screen. The telescopes track the background stars so that the satellites show up as streaks on the screen. A computer works out the coordinates of a streak, and calculates its position and velocity. The object can then be identified using an electronic-zoom feature which can distinguish between steady or tumbling satellites, spent-rockets, or debris. GEODSS was used to search for the disabled Palapa-B2 and Westar-6 communications satellites which failed to reach orbit after launch from the Space Shuttle in February 1984. The Soviet Union operates a similar satellite surveillance system. □ see CHARGE-COUPLED DEVICE, SPACE INSURANCE

ground station – see EARTH STATION

g-suit a suit that presses on the abdomen and legs and prevents blackouts by retarding the collection of blood below the chest. An astronaut would wear a g-suit if high accelerations are expected. □ see BLACKOUT 2, SPACE MEDICINE

guidance the process of keeping an aerospace vehicle on course. Guidance employs one or more of the following techniques: in preset guidance, the vehicle (eg a cruise missile) follows a path set and controlled by on-board systems; in command guidance, the vehicle follows a path controlled by external (radio) signals; in beam-rider guidance, the vehicle follows a beam of radiation (eg infrared or radar); in terrestrial-reference guidance, the vehicle's path is determined by some manmade or natural property of the Earth; celestial guidance uses the Sun or stars as reference points; homing guidance responds to information from the destination; in active homing guidance, the homing signal is modified by information sent from the vehicle; corrections to a course made after launch is called mid-course guidance; and terminal guidance takes place in the later stages of a journey. □ see FINE GUIDANCE SENSORS, TELEMETRY, GYROSCOPE

guided missile any missile carrying a warhead which is capable of being guided, or which guides itself, towards a target after launch ⟨*a cruise missile is a ~*⟩ □ see GUIDANCE, GYROSCOPE

GUT – see GENERAL UNIFIED THEORY

gyro – see GYROSCOPE

gyroscope *also* **gyro** a device, based on a rapidly spinning rotor, to provide a constant reference direction for the purposes of navigation and control. Since a gyroscope resists any force tending to change its axis of rotation, it is used as the basis of a gyrocompass in which the horizontal axis of the gyroscope points to true north. If the gyrocompass is carried by a moving vehicle, the forces developed when the vehicle tries to change direction can be used to keep the vehicle on course. □ see ANGULAR MOMENTUM, SPIN STABILIZED

hadron any elementary particle that can participate in strong interactions (eg forces between particles in the nuclei of atoms), in weak interactions, and, if charged, in electromagnetic interactions. There are two classes of hadrons: baryons (neutrons and protons) which obey the Pauli Exclusion Principle, and mesons which do not. The two baryons and the two much lighter leptons, the electron and the electron-neutrino, are sufficient to build a universe closely resembling the one that exists. □ see ELEMENTARY PARTICLES, NEUTRON, PROTON, MESON, LEPTON, PAULI EXCLUSION PRINCIPLE

Hale Observatories a group of astronomical observatories comprising the Mount Wilson Observatory, the Big Bear Solar Observatory, and the Palomar Observatory, all of which are in California, and the Las Campana Observatory in Chile. The group is named after George Ellery Hale who helped to set up the Mount Wilson and Palomar Observatories and, in particular, the 5m (200in) Hale telescope on Mount Palomar which took the first photographs in 1948. This reflecting telescope was the world's largest until the Soviet Academy of Sciences built a 6m instrument. The observatories have a number of smaller instruments, including the 1.2m Schmidt telescope with which high-resolution photographs of the sky have been obtained. These Earth-based telescopes will continue to play an important part in astronomical research, even though the orbiting Hubble Space Telescope enables astronomers to see deeper into space than ever before. □ see BETA PICTORIS, HUBBLE, SCHMIDT TELESCOPE, REFLECTING TELESCOPE

half-life the time taken for half of the atoms of a radioactive substance to disintegrate

Halley, Edmund (1656–1742) English astronomer, a well-to-do and staunch friend of Newton, interested in astronomy from his school days. He founded the first observatory in the Southern hemisphere on the island of St Helena in the South Atlantic. While professor of geometry at Oxford, Halley (with Newton's help) compiled records of numerous comets and computed their paths across the sky. One of the comets he recorded was the bright comet of 1682, and he was struck by the similarity of this comet with those that appeared in 1456, 1531, and 1607, for these four had come at intervals of between 75 and 76 years. Halley thought these comets were one and the same, and predicted the next appearance in 1758. He did not live long enough to witness its return, but it has become known as Halley's comet ever since. □ see HALLEY'S COMET, COMET, NEWTON

Halley's comet a comet which has a uniquely extensive history of appearances, some of which have coincided with notable historical events thus causing it to have been regarded as portentous (it appears on the Bayeux Tapestry which depicts the Battle of Hastings of 1066).

It was named after the astronomer Edmund Halley who used Newton's Law of Gravitation to predict its return in 1758 after noticing that the orbits of the bright comets in 1531, 1607, and 1682 were very similar. Halley's comet has a period of 76 years, its orbit is retrograde, ie it moves round the Sun in the opposite direction to that of the planets, and its most recent appearance in the Spring of 1986 was the subject of an intensive programme of Earth-based and space-based investigation. However, Halley's comet was very unfavourably placed in relation to the Earth when it was at perihelion (closest distance to the Sun) in early February 1986. It is at this time that comets produce their longest tails. Unfortunately, Halley's comet was on the far side of the Sun and its tail pointed away from the Earth. Its closest pre-perihelion distance to the Earth was 73×10^6km on the 27 November 1985, and observers with binoculars were able to see it throughout November just below the star cluster Perseus, in the constellation Taurus. From mid-March to mid-April 1986, Halley's comet was visible in the early morning sky in Britain as it passed on its way out of the Solar System. The further south the observer, the higher in the sky the comet and the easier it was to see. At a latitude of 40°S, Halley's comet was 70° above the horizon and showed off a tail 10° long in a dark sky. Anyone journeying to the south of the equator was rewarded with a fine sight. □ see COMET, [1]GIOTTO, PLANET-A, VEGA, OORT CLOUD

halo the approximately spherical region of space, above and below the plan of a galaxy, which is largely free of dust and populated by older stars and globular clusters □ see GALAXY, OORT CLOUD, HELIOPAUSE

hard of or being radiation of relatively high penetrating power ⟨~ X rays⟩

hard landing the (destructive) impact of a spacecraft on the surface of a planet or one of its moons, when not softened by parachutes or descent rockets. An instrument package on a spacecraft might be designed to withstand a hard landing – compare SOFT LANDING

hard vacuum the vacuum of space (which can be reproduced in the laboratory) in which the pressure exerted by any gases is less than 10^{-10}Nm^{-2}. The atmosphere at sea level produces a pressure of 10^5Nm^{-2}. □ see NEWTON

harvest moon the full moon nearest the time of the September equinox. At this time the Moon appears to rise at about the same time each evening because it is low down on the horizon. □ see EQUINOX

hatch a small door or opening in the hull of a spacecraft, that is usually tightly sealed when closed to prevent the escape of air to the outside vacuum □ see AIRLOCK

head the coma and the nucleus of a comet from which extends a long tenuous tail when the comet approaches perihelion □ see COMET

HEAO – see HIGH ENERGY ASTRONOMICAL OBSERVATORY

heat the form of energy associated with random movement of atoms and molecules, and transferred by conduction, convection, or radiation. Heat is measured in joules. – compare TEMPERATURE

heat death – see SECOND LAW OF THERMODYNAMICS

heat exchanger a device (eg in an air conditioning system) that transfers heat from one liquid or gas to another without their mixing □ see SPACE RADIATOR

heat pipe a device for efficiently transferring heat from one place to another using the evaporation and condensation of a liquid in a sealed tube. The liquid evaporates and draws heat from the source in contact with it. The vapour moves rapidly to the other end of the tube where it condenses and gives up the heat to a radiator and is lost. The condensed liquid is drawn along a wick and returns to the source of heat. The wick can be made of felt or foam, or can consist of grooves machined into the walls of the tube. The latter design is used in spacecraft where gravity cannot be relied upon to return the liquid to the source of heat. Heat pipes can conduct heat many hundreds of times better than copper. Different liquids, including water, are used depending on the application. The biggest application of heat pipes is for cooling electronic components (eg transistors), and they are being developed for handling waste heat in space stations and spacecraft. □ see SPACE RADIATOR

heat shield a device which protects something from heat. A heat shield is necessary to protect the Space Shuttle, or another spacecraft, from the effects of heat generated by passing through the Earth's atmosphere at high speed, or on missions near the Sun. □ see THERMAL CONTROL SYSTEM, ABLATING MATERIAL, ATMOSPHERIC DRAG, HEAT PIPE

heavy hydrogen – see DEUTERIUM

heliocentric having the Sun as centre ⟨*Halley's comet is in ~ orbit*⟩ – compare GEOCENTRIC

heliopause the predicted boundary region beyond Pluto's orbit where the Sun's solar wind and magnetic field gives way to the dust, particles, and magnetic fields between the stars. The heliopause might lie between 50 and 100AU from the Sun (1AU = 150 × 10^6km). Voyager 1 is now about 25AU from the Sun (above and between the orbits of Uranus and Neptune). This spacecraft has for some time been detecting radio waves at a frequency of about 3kHz which could represent emission from the heliopause. Some scientists have speculated that Voyager 1 could cross the heliopause in the early 1990s – compare MAGNETOPAUSE □ see INTERSTELLAR MEDIUM, PIONEERS 10 AND 11, VOYAGER

helium (chemical symbol **He**) the second lightest and second most abundant chemical element in the universe after hydrogen. A neutral atom of helium, in its most abundant form, has a nucleus of two protons and two neutrons around which orbit two electrons. The nucleus, called an alpha particle, is exceptionally stable and is emitted by some radioactive substances. All but about 1% of the estimated 25% of helium in the universe is believed to have been synthesized in the first few minutes after the Big Bang. A small proportion is synthesized from hydrogen by nuclear fusion processes in the centres of some stars. However, helium is not easy to detect spectroscopically, not even in our Sun. One of the UK experiments aboard Spacelab 2 in July 1985 was CHASE (Coronal Helium Abundance Spacelab Experiment) which used a new technique to measure the amount of helium in the corona of the Sun. □ see SPACELAB, BIG BANG, HYDROGEN, DEUTERIUM, MISSION 51F

helium burning a process that takes place in the centre of a star which has used up all its hydrogen and involves the fusion of three helium nuclei to form a carbon nucleus. This reaction occurs late in the life of a star and produces only about 10% of the energy as hydrogen burning. □ see HYDROGEN BURNING, NUCLEOSYNTHESIS, DEUTERIUM, GIANT STAR

Hermes a small space shuttle which has been designed but not built by France as a member of the European Space Agency. This four-man vehicle would be launched by Ariane 5 and would be capable of landing at the Kourou launch site in Guiana or at other European sites. It could be used in association with the US Space Station in the 1990s, and its main purpose would be to carry materials back and forth to a space platform used for making new products. □ see ARIANE, EUROPEAN SPACE AGENCY, US SPACE STATION

hertz (symbol **Hz**) The SI unit of frequency equal to 1/second (s^{-1}). The frequency of the mains alternating current is 50Hz. □ see FREQUENCY, WAVELENGTH

Hertzsprung-Russell diagram (abbr **H-R diagram**) a graph that shows the relationship between the temperature of stars (T) (horizontal axis), and their luminosity (L) (vertical axis). About 90% of all stars lie on this T-L graph which comprises a diagonal band with the hottest brightest stars at the upper left and the coolest faintest stars at the lower right. This grading, in descending order of temperature, is known as the main sequence and from it the evolution of stars can be derived. Stars which are just beginning to form from clouds of gas and dust, and which are too cool for nuclear reactions to take place in them, lie far to the right of the main sequence. But the more massive the star, the more quickly it contracts under gravity and the sooner it moves on to the main sequence. Also, the more massive the star, the faster it squanders its energy reserves and the shorter the time it spends on the main sequence. The H-R diagram was developed independently by the Danish astronomer E Hertzsprung in 1911, and by the American astronomer H N Russell in 1913. In about 4000 million years time the Sun will move off the main sequence to commence the giant phase of its existence. By then it may be possible to replace the Earth's ageing Sun by a younger star! □ see STAR, SUPERNOVA, WHITE DWARF, GIANT STAR, PROTOSTAR, NUCLEOSYNTHESIS, HYDROGEN BURNING, HELIUM BURNING, CARBON, UNIVERSE, ANTHROPIC PRINCIPLE

high 1 *of technology* up to date and advanced in development and concept ⟨*the Voyager spacecraft is ~ technology*⟩ **2** of or being the larger of two voltage levels in a digital logic circuit ⟨high-*level logic is given the binary number 1*⟩ – compare LOW

High Energy Astronomical Observatory (abbr **HEAO**) a series of three large (about 3000kg) Earth satellites which measured the strength of X-ray sources and cosmic rays in space. HEAO-1 was launched in 1977 and from a 400km high orbit carried out a survey of celestial sources of X rays using detectors more sensitive than those carried by two previous satellites, Uhuru and Ariel 5. The second, HEAO-2 (now renamed Einstein), was launched in November 1978 into a 540km high orbit inclined at 23.5° to the equator. The task of this 3175kg satellite was to pinpoint X-ray sources with a resolution comparable with that obtained with optical telescopes (ie 1 to 2 arc-seconds) , and to measure the distribution of energies in the X rays they emitted. The heart of the observatory was a grazing-incidence X-ray telescope which focussed X rays onto a solid-state spectrometer. This device used silicon-germanium crystals which, when struck by X-ray photons, produced electrical pulses enabling the energy and wavelength of the X rays to be found. Einstein was stabilized in orbit by gyroscopes and reaction jets, and data was transmitted at a rate of 128 kilobits per second to the Goddard Space Flight Centre via the Deep Space Network. HEAO-3 was launched on 30 September 1979. It carried two cosmic ray telescopes, one to study heavy nuclei and the other to identify their isotopes. ☐ see X-RAY ASTRONOMY, SPECTROMETER, GRAZING INCIDENCE X-RAY TELESCOPE

Hipparcos a 500kg satellite being developed by Matra of France for launch into geostationary orbit in 1987 from where it will produce the first accurate map of the positions of 100 000 selected stars. Hipparcos will be the first satellite to make astrometric measurements from space. It is named after the famous Greek astronomer (c190–120BC) who measured the lunar parallax and hence the distance from the Earth to the Moon. ☐ see ASTROMETRY, PARALLAX

Hohmann transfer an orbit requiring the minimum energy to transfer a spacecraft from one celestial body to another ☐ see TRANS-FER ORBIT

hold a temporary stoppage of a countdown (eg of a launch) ⟨"T minus fifty and ~ing"⟩ ☐ see COUNTDOWN

homing guidance – see GUIDANCE

homogeneity the generally assumed property of the universe that at a given time the universe appears the same to all typical observers wherever they may be ☐ see COSMOLOGICAL MODEL, BIG BANG

horizon 1 *in visual astronomy* the great circle where an observer's horizontal plane meets the celestial sphere ☐ see CELESTIAL SPHERE, CELESTIAL COORDINATES, EVENT HORIZON **2** *in cosmology* the distance from which no light would have yet had time to reach us. For a universe of finite age, the horizon is of the order of the age of the universe times the speed of light. ☐ see COSMOLOGY, BIG BANG, EVENT HORIZON, HUBBLE'S LAW

HOTOL [*H*orizontal *T*ake-*o*ff and *L*anding] a pilotless and reusable aircraft-type vehicle which could take off from and land on runways similar in length to those used by Concorde, and be capable of putting payloads up to 7000kg into low-Earth orbit. The HOTOL concept is being studied by British Aerospace and Rolls-Royce. Unlike the Space Shuttle, which has to carry the oxygen necessary to burn its hydrogen fuel, the HOTOL engines would 'breathe' oxygen from the atmosphere the vehicle passes through. By not carrying liquid oxygen,

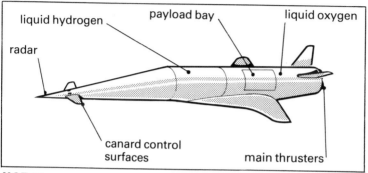

HOTOL

HOTOL would reach low-Earth orbit without having to jettison booster rockets on the way up. HOTOL would take off from its trolley undercarriage after a 2300m run, and climb to an altitude of 26km in 9 minutes when its speed would be Mach 5. From its operational orbit at an altitude of about 300km, HOTOL would launch a 7 tonne satellite payload into geostationary orbit using an upper stage boost motor in its payload bay. If HOTOL were converted to carry passengers in airliner-style comfort it could journey between London and Sidney in 67 minutes! □ see BRITISH AEROSPACE, SPACE SHUTTLE

hot test *also* **captive firing, captive test** test firing of a rocket engine while holding it down in a test stand. Instruments attached to the engine provide data to evaluate performance prior to possible modifications and eventual flight test. □ see FLIGHT TEST

housekeeping the routine tasks that have to be done in order for something to work properly. It is usual for a computer to monitor and control the environment aboard a spacecraft or space station so that equipment and astronauts can do their jobs efficiently.

Hoyle, Fred (*b*1915) English astronomer who became professor of astronomy at the University of Cambridge in 1945 after working on radar developments in World War 2. In the late 1940s Hoyle, together with Hermann Bondi and Thomas Gold, put forward the view that the universe has always been just about the same as it is now. This is the Steady State model of the universe, and Hoyle has explained it in books for the layman. Indeed, Hoyle has written a number of science fiction books under his own name. The Steady State model is less popular than the Big Bang model which proposes that the universe had a definite beginning, or a number of beginnings. □ see STEADY STATE, PANSPERMIA

H-R diagram – see HERTZSPRUNG-RUSSELL DIAGRAM
HST – see HUBBLE SPACE TELESCOPE
Hubble, Edwin Powell (1889–1953) American astronomer who, after service in World War 1, worked at Mount Wilson Observatory studying the patches of luminous 'fog' or nebulae in the night sky. Using the largest telescope of its day, the 2.5m reflector, he was able to make out stars in the Andromeda galaxy. Some of the stars were Cepheid variables, and, using the period-luminosity law of Shapley and Leavitt developed for stars in the Milky Way, he was able to conclude that Andromeda is about 800 thousand light years away (later shown to be an underestimate). Hubble's observations of Andromeda, and of a number of extragalactic nebulae, proved that there were other star systems (galaxies) besides our own Milky Way galaxy. Thousands of millions of such systems are now known to exist. In 1929 he analysed the speeds of recession of a number of galaxies and showed that the speed at which a galaxy speeds away from us is proportional to its distance (Hubble's Law). The simplest explanation of this recession is that the universe began with a Big Bang. □ see CEPHEID VARIABLE, GALAXY, HUBBLE'S LAW, RED SHIFT, BIG BANG
Hubble's constant – see HUBBLE'S LAW
Hubble's Law a law stating that the speed at which two typical galaxies are moving apart is proportional to the distance between them. This law can be written speed = H × distance where H is known as Hubble's constant. Its value is generally quoted as 50 kilometres per second per magaparsec. This means that two galaxies,

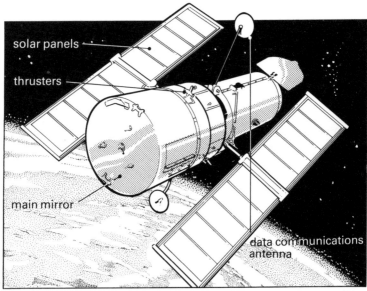

Hubble Space Telescope

say 10 megaparsecs (about 30 million light-years) apart, are moving apart from each other at about 500kms⁻¹. This law was first proposed by the American astronomer Edwin Hubble in 1929 who observed that galaxies appeared to be rushing away from each other. Hubble used the law to measure the distance of galaxies from their speeds of recession obtained by the red-shift in their spectra. Hubble's law and the later detection of the microwave background radiation, has given rise to the notion that the universe has evolved from a single cataclysmic Big Bang. Hubble's Law enables us to calculate when this explosion occurred: since speed = distance/time, the reciprocal of Hubble's constant tells us when the expansion began – about 10 thousand million years ago. However, the speed at which the galaxies move apart must be slowing down as the intergalactic gravitational forces try to restrain the expansion (thus H is not actually a constant as its name suggests!). If we take into account the idea of a decelerating expansion, then the creation of the universe works out at about 18 thousand million years ago when it was infinitely compressed and expanding infinitely rapidly. □ see HUBBLE, RED SHIFT, BIG BANG, GALAXY

Hubble Space Telescope (abbr **HST**) a multipurpose optical telescope scheduled to be launched by the Space Shuttle *Atlantis*. The HST has a Cassegrain system with a 2.5m diameter primary mirror and a 0.3m secondary mirror. Though the primary mirror is smaller than that of many Earth-based telescopes, it will be able to see seven times deeper into space – perhaps even to the edges of the visible universe. From this distance light takes 14 billion years to reach us so the pictures will reveal some of the early history of the universe. Because the HST will not be affected by poor seeing caused by the Earth's atmosphere and its structure not subject to bending due to gravity, it will be able to look at quasars and galaxies, and search for evidence of planets round distant stars which are 50 times fainter than those visible from the Earth's surface. The main 2.5m mirror is ground to such a precise shape that it will operate at its theoretical maximum resolving power; its surface is so smooth that if it were scaled up to the size of America the bumps on it would be no larger than an upturned saucer! The HST will be pointed extremely accurately at these celestial objects using an attitude control system capable of keeping the telescope on target to within 0.007 arc seconds; this will enable the telescope to identify a dime in Los Angeles from San Fransisco, 440 miles away. Power for the instruments and control systems will come from solar panels which will supply a total of 4.4kW at 34V. Though most of the HST was built by American companies, the two solar panels and one of the instruments (the Faint Object Camera) have been made by the European Space Agency. In fact, the two solar panels were made by British Aerospace, and contain a total of 48 760 solar cells (each 20 × 40mm) spread over two 3 × 12.5m 'wings'. The Faint Object Camera uses an image intensifier based on four charge-coupled devices each of 800 × 800 pixels and cooled by heat pipes to enable objects down to magnitude 28 to be detected. The other four instruments are: the Faint Object Spectrograph to study the light from quasars and comets; a high resolution spectrograph for exploring the physical characteristics of interstellar gas clouds and exploding galaxies; a photometer for measuring the light intensity and time variations in the light received from these distant objects; and a wide-field planetary camera to provide high-resolution photographs of planets, and sweeping views of galaxies and star fields. Data from these instruments will be transmitted to Earth stations via the orbiting Tracking Data and Relay Satellites (TDRS). Astronauts from the Space Shuttle will service the telescope from time-to-time, and it may be necessary to bring it back to Earth every five years to give it a complete overhaul. The HST has a mass of 11 tonnes, a length of 13.1m, and diameter of 4.3m so that it can fit in the Space Shuttle's cargo bay. □ see CASSEGRAIN TELESCOPE, REFLECTING TELESCOPE, BRITISH AEROSPACE, EUROPEAN SPACE AGENCY, RESOLVING POWER, SEEING, ATTITUDE CONTROL, BETA PICTORIS, TRACKING AND DATA RELAY SATELLITE, CHARGE-COUPLED DEVICE, HEAT PIPE

hydrogen (chemical symbol **H**) the lightest and the most abundant chemical element in the universe. The nucleus of ordinary hydrogen consists of a single proton, but there are two much less common heavier forms (isotopes): deuterium, which has a nucleus consisting of one proton and one neutron; and tritium, one proton and two neutrons. Tritium is a radioactive isotope of hydrogen. Neutral atoms of any sort of hydrogen consist of a hydrogen nucleus and one electron in orbit about it. It is generally agreed that all the hydrogen in space was created when the universe began (eg in the Big Bang). Deuterium, too, was formed then, but most of it would have been converted to helium in the first few minutes. Clouds of neutral hydrogen occur throughout interstellar and intergalactic space. Astronomers can detect it as it emits radio waves with a 210mm wavelength. The large quantity of hydrogen in the Sun and most other stars is being converted into helium by fusion reactions. Molecular hydrogen is also present in large quantities in the atmospheres of the outer planets Jupiter, Saturn, Uranus, and Neptune. One important use of hydrogen is as a rocket fuel: liquid hydrogen burns easily in (liquid) oxygen. Perhaps future spacecraft will be able to replenish their fuel supplies by dipping into the hydrogen-rich atmospheres of the outer planets! □ see BIG BANG, SUN, STAR, NUCLEOSYNTHESIS, LIQUID HYDROGEN, HYDROGEN BURNING, HELIUM, DEUTERIUM

hydrogen bomb – see THERMONUCLEAR

hydrogen burning the process by which energy is produced by all main sequence stars, such as the Sun, for the greater part of their lives. Hydrogen burning takes place at high temperatures and pressures when four protons (the nuclei of hydrogen) fuse to produce nuclei of helium. □ see HELIUM BURNING, NUCLEOSYNTHESIS, HERTZSPRUNG-RUSSELL DIAGRAM

hydrolaboratory a building which houses a large tank of water so

that astronauts can practise extravehicular activities in simulated conditions of weightlessness underwater. The astronauts (and
cosmonauts) are suitably weighted so that they neither rise nor fall in
the water and they carry out repair and maintenance tasks similar to
the ones they are to carry out in Earth orbit. □ see EXTRAVEHICULAR
ACTIVITY, JOHNSON SPACE CENTRE

hyperbola a smooth open curve with two branches, so shaped that
the difference between the distances from any point on the curve and
two fixed points (foci) is constant. If the path of a body (eg some comets) round the Sun is a hyperbola, that body will not return to the
Solar System. □ see CONIC SECTION, ELLIPSE, ORBIT

hyperboloid a surface that is obtained by rotating a hyperbola
about its axis □ see HYPERBOLA, CASSEGRAIN TELESCOPE

hypergol – see HYPERGOLIC FUEL

hypergolic self-igniting ⟨*a* ~ *fuel*⟩

hypergolic fuel *also* **hypergol** a rocket fuel that ignites on contact
with an oxidizer □ see OXIDIZER, STORABLE PROPELLANT

hypersonic of or being a speed more than five times the speed of
sound ⟨*the Space Shuttle reenters the Earth's atmosphere at* ~ *speed*⟩
– compare SUPERSONIC □ see MACH NUMBER

hypersonics the branch of aerodynamics which deals with the
effects of air flowing over surfaces at speeds much greater than the
speed of sound

hyperspace a realm of space that extends above the space and time
of our physical world and which might be exploited by commanders of
future starships making rapid long distance trips to stars and galaxies.
Hyperspace adds at least one more dimension to the familiar length,
breadth, and depth of three-dimensional space which, together with
one-dimensional time, form the concept of four-dimensional
spacetime embodied in Einstein's general theory of relativity. The
exploitation of hyperspace for space travel supposes that it is possible
for a spaceship to travel faster than light, an impossible objective if we
are to believe one of the conclusions of Einstein's theory. But in the
same way that today's aircraft (eg Concorde) can travel at faster-than-
sound (supersonic) speeds, tomorrow's starships might be able to
travel at faster-than-light (superluminary) speeds. Thus a future
interstellar starship would first accelerate to the speed of light at which
point it would disappear from sight as it entered the realm of
hyperspace: it would have 'jumped' into hyperspace and exist 'above'
the four-dimensional realm of spacetime. During the starship's jump
into hyperspace, the passage of time would be negligible compared
with the much longer passage of time during its travel in four-dimensional space. In order to ensure accurate navigation to a distant star,
the starship's commander would need precise information from
on-board instruments about his/her position and that of the target star
in space-time if he/she is to execute a proper hyperspace jump and not
end up at the wrong star – or in another universe! □ see SPACETIME,
STARSHIP, ELECTRIC PROPULSION, ANTIMATTER PROPULSION

IC – see INTEGRATED CIRCUIT

ICBM – see INTERCONTINENTAL BALLISTIC MISSILE

ICE – see INTERNATIONAL COMETRY EXPLORER

ignition the start of self-sustaining combustion of propellants in a rocket motor □ see HYPERGOLIC FUEL, CRYOGENICS

IGY – see INTERNATIONAL GEOPHYSICAL YEAR

image a likeness to an object created optically or electronically ⟨*the Voyager spacecraft transmitted ~s of Saturn's rings*⟩ □ see TELESCOPE

image converter a device for changing an invisible image (eg infrared light from stars) into one which is visible □ see CHARGE-COUPLED DEVICE, IMAGE INTENSIFIER, LANDSAT

image intensifier *also* **image tube** an electronic device for making a faint optical image brighter. One type of image intensifier is an evacuated device in which the weak radiation falls on a light-sensitive cathode. Electrons are emitted from this photocathode and are accelerated by an electric field to fall on a positively charged phosphor screen and become visible. Some image intensifiers use a number of photocathodes one after the other (a multistage tube) to intensify the image further. Image intensifiers are used at the focus of optical telescopes to make faint images visible. □ see CHARGE-COUPLED DEVICE

image recognition the identification of an object by using computers linked to TV cameras and/or other light sensors. Image recognition is being developed so that robots aboard spacecraft can identify and control the assembly of structures and control of instruments in space. □ see ROBOT, CHARGE-COUPLED DEVICE, TELEPRESENCE

image sensor any device for detecting an image and converting it into a pattern of electrical signals □ see CHARGE-COUPLED DEVICE, VIDEO CAMERA

image tube – see IMAGE INTENSIFIER

Improved Stratospheric and Mesospheric Sounder (abbr **ISAMS**) – see UPPER ATMOSPHERE RESEARCH SATELLITE

inclination (symbol **i**) the angle between the plane of the orbit of a planet, comet, or satellite and the equatorial plane ⟨*standard Space Shuttle missions have orbital ~s of 28.5° and 57°*⟩ □ see ORBIT

Indian Remote Sensing satellite – see REMOTE SENSING

Indian satellite (abbr **INSAT**) – see INDIAN SPACE RESEARCH ORGANIZATION

Indian Space Research Organization (abbr **ISRO**) an Indian group which builds, tests, and launches Indian satellites, mainly from the island of Shriharikota (100km north of Madras) which is connected to the Indian mainland by a 6km-long causeway. The Rohini series of satellites, built and launched in India, includes Rohini-D2 launched on 17 April 1983 to send back photographs and other information about the Earth's surface. India's main launch vehicle is the 23m-long 17 tonne SLV-3 rocket which is capable of putting a 41kg satellite into near-Earth orbit. But India plans to build rockets rivalling Europe's Ariane launcher and able to launch 1000kg satellites to investigate India's agriculture, forestry, meterology, hydrology, and fish reserves. The economic benefits of their findings are expected to pay for the cost of launching the satellites within two to three years. The Indian space programme involves collaboration with both Russia and America. On 3 April 1983, a Soviet rocket carried India's first cosmonaut, Rakesh Sharma, and two Russian cosmonauts up to the orbiting space station Salyut-7 to join the resident crew of three. Sharma carried out a number of experiments including one that produced a material called metallic glass that is highly resistant to radiation, high temperature, and chemical attack. Indian materials scientists hope to use the microgravity environment of space to manufacture this valuable material. Sharma's most publicized experiment involved yoga exercises. Since yoga enables an astronaut to control respiration and blood circulation during long periods of weightlessness, it could be extremely useful for the crew of a permanently manned space station. India's American connection was the release of the weather and communications satellite, INSAT-1B, from the Space Shuttle *Challenger*, on 30 August 1983. INSAT-1C was launched in June 1986 from another Space Shuttle, together with an Indian payload specialist. India thus became the second nation, following France, to fly astronauts under both the US and Soviet space programmes. □ see SOYUZ, REMOTE SENSING, MATERIALS PROCESSING

inertia the property of a body which makes it resist a change to its velocity. The inertia of a body is directly related to its mass. □ see NEWTON'S LAWS OF MOTION, MACH'S PRINCIPLE

inertial navigation a method of navigation of an aircraft or spacecraft by using a preprogrammed on-board computer to adjust acceleration and give distance travelled and current position □ see GYROSCOPE

inertial upper stage (abbr **IUS**) a two-stage solid-fuel rocket engine used to boost satellites into higher Earth-orbits or into interplanetary trajectories, from the payload bay of an orbiting Space Shuttle. On the STS-6 flight in April 1984, the second stage failed to complete the burn-time needed to put a Tracking and Data Relay Satellite (TDRS-A) into geostationary orbit. The fault developed when a seal collapsed because of overheating caused by hot gases leaking through the nozzle thermal protection system. Subsequent IUS rockets have been redesigned to prevent this happening. □ see TRACKING AND DATA RELAY SATELLITE, PAYLOAD ASSIST MODULE, CENTAUR

inferior planets the planets Venus and Mercury which are closer to the Sun than the Earth – compare SUPERIOR PLANETS

infinite 1 extending indefinitely ⟨an ~ universe⟩ **2** greater than the largest number chosen ⟨an ~ number of stars⟩ – compare FINITE

◻ see INFINITY, COSMOLOGY, UNIVERSE

infinity (symbol ∞) **1** any time, space, or quantity which is immeasurably large **2** a distance so great that the rays of light from a point source (eg a star) at that distance may be considered parallel
◻ see TELESCOPE, BIG BANG

information processing – see DATA PROCESSING

information technology (abbr **IT**) the gathering, processing, and circulation of information by combining the data-processing power of the computer with the message-sending capability of communications. Nowadays, thousands of ordinary people are able to access and manipulate information from their offices, schools, and homes in a way previously only available to a few specialists in commerce, industry, and research institutions. Microelectronics is providing the major stimulus to the development and use of information technology. People worldwide can gain access to this information through the increasing use of communications satellites. ◻ see MICROELECTRONICS, COMMUNICATIONS SATELLITE, DIRECT BROADCAST SATELLITE

infrared of or being electromagnetic radiation having wavelengths between the red end of the visible spectrum and the microwave region, ie wavelengths between about 700 nanometres and 1 millimetre, respectively. Infrared rays are invisible to the eye but they are being increasingly used in electronic systems (eg TV and video remote control systems), and tiny infrared-emitting lasers are used in the developing areas of optical communications, the compact disk, and the video disk. Infrared-sensitive telescopes aboard artificial satellites are helping astronomers to understand the birth of stars from clouds of dust and gas in the Milky Way galaxy. ◻ see ELECTROMAGNETIC RADIATION, INFRARED ASTRONOMY SATELLITE, HUBBLE SPACE TELESCOPE, OPTICAL COMMUNICATIONS, LASER

infrared astronomy the study of celestial objects from their infrared emissions in the wavelength band 0.8μm (800nm) to 1mm. Though ground-based observations can be made through several atmospheric windows up to about 17μm, measurements at longer wavelengths must be made from balloons, high-altitude rockets, or satellites. Recently, the Infrared Astronomy Satellite (IRAS) has made some very exciting discoveries from Earth orbit, the best vantage point of all for infrared telescopes. ◻ see INFRARED ASTRONOMY SATELLITE, EUROPEAN SPACE AGENCY, INFRARED SPACE OBSERVATORY, UK INFRARED TELESCOPE, HUBBLE SPACE TELESCOPE

Infrared Astronomy Satellite (abbr **IRAS**) an Earth-orbiting satellite which used a 600mm diameter telescope and a sensor cryogenically cooled with liquid helium to about −257C to detect the faint infrared radiation from clouds of dust in interstellar space. IRAS exhausted its helium coolant in late 1983, but by that time it had gathered, over a period of 300 days, sufficient data to enable a comprehensive map of the universe at infrared wavelengths to be obtained. A catalogue of about a quarter of a million infrared sources has been published. Since infrared radiation passes more easily than visible light through clouds of dust, IRAS has provided a new view of the structure of the centre of our galaxy, the Milky Way. It has also discovered a ring of 'dust' particles surrounding the first magnitude star Vega. IRAS did not have the resolving power to determine whether these particles are of planetary size, which is a pity since a closer study of the distant Vega system may give us a clue to our own origins. □ see INFRARED, CRYOGENICS, BETA PICTORIS, INFRARED SPACE OBSERVATORY

Infrared Space Observatory (abbr **ISO**) an Earth-orbiting telescope which the European Space Agency (ESA) is planning to build and launch in the early 1990s. The telescope would have a 0.6m-diameter infrared telescope and its cryogenically cooled detector would be used to study infrared emissions from galactic objects (eg new-born stars) in the 5 to 100 micron (μm) wavelength band. □ see INFRARED ASTRONOMY, INFRARED ASTRONOMY SATELLITE, EUROPEAN SPACE AGENCY, MICRON

inject 1 to boost a spacecraft into a particular orbit or trajectory ⟨*to ~ the spacecraft Giotto into a heliocentric orbit*⟩ **2** to introduce fuel, a fuel/air mixture, oxidizer or another substance into the combustion engine of a rocket or other engine

Inmarsat [*In*ternational *Ma*ritime *Sat*ellite Organization] an organization which commissions and administers telephone, telex, and facsimile services via satellite communications to world-wide shipping and offshore industries. The system can also be used for transmitting data on ship-engine performance, fuel consumption, position, weather conditions, etc. About 2500 ships are now fitted with a special antenna which always points at a satellite however much the ship heaves and rolls. Inmarsat has 44 member countries and uses US and European satellites in geostationary orbits. For example, in 1984 the European Space Agency leased their communications satellite Marecs B2 to Inmarsat: this satellite was launched by Ariane 3 on 9 November 1983 from Kourou, French Guiana and placed in geostationary orbit at 177.5° E. Marecs is close to Marisat, positioned at 176.5° E, which it has replaced. It is able to handle 50 telephone calls simultaneously compared with Marisat's 10. □ see GLOBAL POSITIONING SYSTEM, INTELSAT, LANDSAT, EUROPEAN SPACE AGENCY, ARIANE

inner planets *also* **terrestrial planets** the planets Mercury, Venus, Earth, and Mars which lie within the asteroid belt – compare OUTER PLANETS □ see ASTEROID

¹**input 1** the act or process of putting information into a computer or other data processing machine ⟨*a keyboard is an ~ device*⟩ **2** the point (eg a socket) at which information is fed into a machine **3** information fed into a data processing machine – compare ¹OUTPUT

²**input** to put information into a device ⟨*to ~ a computer program*⟩ – compare ²OUTPUT

INSAT [*In*dian *Sat*ellite] – see INDIAN SPACE RESEARCH ORGANIZATION

insolation

insolation *also* **solar flux** the amount of solar radiation falling on a unit area per unit time (ie Wm⁻²). Outside the Earth's atmosphere, insolation is equal to the solar constant. Its value is used by designers of the solar cells that provide the electrical energy for spacecraft.
□ see SOLAR CONSTANT, EARTH RADIATION BUDGET SATELLITE

Institute of Space and Astronautical Science (abbr **ISAS**) a Japanese space research group that designs, develops, tests, and flies its own launchers and scientific satellites using its Kagoshima Space Centre. Prior to 1981, it successfully launched 12 satellites using the Lamda or Mu solid propellant rockets. Five of the satellites had technological payloads to test the performance of the various development stages of the Mu vehicles and seven were scientific satellites to study the ionosphere, magnetosphere, X-ray bursts, and solar flares. Tenma (which means Pegasus) was a spin-stabilized, 218kg X-ray satellite launched on 20 February 1983 to study X-ray emissions from neutron stars and other celestial objects. ISAS's current launcher is the 61 tonne MU-3SII which was used in 1985 to launch an engineering test spacecraft, Tansei-5, into solar orbit. This spacecraft is a replica of the Planet-A spacecraft destined for a rendezvous with Halley's comet in 1986. □ see PLANET-A, X-RAY ASTRONOMY, NATIONAL SPACE DEVELOPMENT AGENCY

instrumentation the design, construction, and use of (electronic) devices for making measurements ⟨~ *on a weather satellite*⟩ □ see SENSOR, SPECTROMETER, RADIOMETER, THERMOMETER, GEIGER COUNTER, ALTIMETER, PHOTOMETER, CHARGE-COUPLED DEVICE, ANALOGUE-TO-DIGITAL CONVERTER

insulator a material which does not allow electricity, heat, or sound to pass through it ⟨*PVC tape is an electrical* ~⟩ □ see THERMAL TILES, THERMAL CONTROL SYSTEM

integrated circuit (abbr **IC**) an electrical circuit comprising transistors, resistors, diodes, and sometimes capacitors formed and connected together on a single small chip of silicon □ see SILICON CHIP, MICROELECTRONICS, RADIATION HARDENING

Intelsat [*In*ternational *Tele*communications *Sat*ellite] an organization which masterminds international communications by satellite. At the moment, Intelsat has over 20 operational satellites parked in geostationary orbit. The most recent of these, the Intelsat-5 series, have a mass of about 1100kg and are the largest, most complex, and most powerful communications satellites yet flown in space. They have a box-shaped body and two solar panels with a total span of about 16m, and are launched by ESA's Ariane and the US Atlas Centaur launchers. Each Intelsat-5 can handle the equivalent of 12 000 telephone calls and 2 colour television channels. Fifteen satellites are planned for the present series which after 1986 will be followed by the drum-shaped spin-stabilized Intelsat-6 series. □ see COMMUNICATIONS SATELLITE, MOLNIYA, DIRECT BROADCAST SATELLITE, ARIANE, ATLAS, C-BAND, K-BAND

intensity 1 the strength of a signal ⟨*the* ~ *of a laser beam*⟩ □ see

STAR WARS **2** the brightness of a source of radiation ⟨*the ~ of a quasar*⟩ □ see JANSKY, INSOLATION, MAGNITUDE

intercontinental ballistic missile (abbr ICBM) a ballistic missile with a range of a few thousand miles. The US Minuteman and the Soviet SS-16 are examples of solid-fuelled ICBMs. □ see BALLISTIC MISSILE, LASER, STAR WARS

interference anything (eg unusual atmospheric conditions) which interferes with the transmission of signals □ see STATIC 1, SEEING

interferometer an instrument for measuring the diameters of sources of radiation (eg giant stars or radio galaxies). It works by receiving radiation from separated antennae (as in radio astronomy), or by splitting starlight into two or more parts (as in a stellar interferometer), and bringing them together to form an interference pattern. The pattern is analysed to provide the angular diameter of the object. If its distance is known, its physical size can be measured. In this way, it is possible to say that the red giant star, Betelgeuse, has a diameter of 800 million miles, but the technique can only be used on relatively close and large stars. □ see APERTURE SYNTHESIS, VERY LARGE ARRAY, QUASAT

intergalactic located or taking place between the galaxies ⟨*~ space*⟩

intergalactic medium the matter between the galaxies. Little is known about the intergalactic medium. In clusters of galaxies, the intergalactic medium is found to be completely ionized and probably amounts to a significant proportion of the total mass of the cluster. □ see STAR, HALO, GALAXY

International Cometry Explorer (abbr ICE) a spacecraft launched in 1978 as ISEE-3 (International Sun-Earth Explorer 3) mainly to measure the properties of the solar wind and solar flares. In 1982, a complex series of manoeuvres was started by controllers at the Goddard Space Flight Centre using on-board propellant and gravity assists from the Earth and the Moon to move (the now renamed) ICE from its solar monitoring post to the region through which the comet Giacoboni-Zinner passed on its orbit round the Sun. ICE plunged through the tail of this comet on 11 September 1985 to make measurements of the interaction of the solar wind with the comet's tail, and also the distribution of dust within its tail. No pictures were returned to Earth by ICE as there are no cameras on board. Comet Giacoboni-Zinner is followed in its orbit by a large amount of debris and the passage of the Earth through its tail on 8 October 1985 produced a magnificent display of shooting stars in the night sky over Japan. The manoeuvring of ICE into a new orbit to intercept this comet has been a triumph of space technology and celestial mechanics, and ICE was the first spacecraft ever to meet a comet. Since ICE is working far beyond its designed distance from Earth, the large 64m diameter antenna at the Jet Propulsion Laboratory's (JPL's) Deep Space Network is being used to track it. After the rendezvous with comet Giacobini-Zinner, ICE took measurements of the solar wind upstream of Halley's comet in March 1986. □ see GRAVITY ASSIST, COMET, SOLAR WIND, PIONEER VENUS, DEEP SPACE NETWORK, HALLEY'S COMET

International Geophysical Year (abbr **IGY**) a period from July 1957 to December 1958 when many countries cooperated in a study of the effects of solar activity on the Earth. IGY was timed to coincide with a peak of solar activity. □ see SPUTNIK 1, EXPLORER 1, VANGUARD

International Maritime Satellite Organization – see INMARSAT

International Solar Polar Mission (abbr **ISPM**) – see ULYSSES

International Sun-Earth Explorer (abbr **ISEE**) – see INTERNATIONAL COMETRY EXPLORER

International Telecommunications Satellite – see INTELSAT

International Ultraviolet Explorer (abbr **IUE**) a 671kg satellite launched into geostationary orbit on 26 January 1978 to investigate ultraviolet radiation from stars, quasars, and galaxies. IUE was equipped with a pair of ultraviolet spectrometers at the focus of a 0.45m aperture Cassegrain telescope to detect radiation in the 115 nanometre (nm) – 320nm wavelength band. The telescope was a joint project between NASA, ESA, and the UK's Science Research Council. Two Earth stations at Maryland and Madrid can point IUE at any object using onboard radio-controlled hydrazine thrusters. As well as providing ultraviolet images of many cosmic objects, including gas streams around binary star systems and the effects of interstellar gas and dust on starlight, IUE discovered that the planet Uranus shone brightly in ultraviolet. At first these emissions were thought to be caused by hydrogen trapped in the planet's magnetosphere. But preliminary results from the Voyager 2 spacecraft which passed the planet in January 1986 cast doubt on this idea. Instead it seems that the ultraviolet glow from the planet is caused by solar particles smashing into molecules in its upper atmosphere. The structure of this unique planet will be made clearer when all of Voyager 2's data from its flyby of Uranus is looked at more closely. □ see ULTRAVIOLET ASTRONOMY, URANUS, EXTREME ULTRAVIOLET EXPLORER, VOYAGER

interplanetary operating between the planets ⟨~ *spacecraft*⟩

interplanetary probe – see SPACECRAFT

interstellar located or taking place between the stars ⟨~ *spaceflight*⟩ □ see STARSHIP, COSMIC DUST, WORLD SHIP

interstellar dust – see INTERSTELLAR MEDIUM

interstellar medium the gas and dust which exists between the stars in the Galaxy. This is largely confined to the spiral arms and contributes about 10% to the mass of the Galaxy. Clouds of several different gases, including hydrogen, have been detected by radio astronomers, in cool regions of space. Close to stars, these gases and others are heated to high temperatures. Small particles of dust composed of carbon and silicates, sometimes with an icy covering, are also found. This dust causes dimming and reddening of starlight. □ see

GALAXY, COSMIC DUST, COSMIC RAYS, SPIRAL ARMS, INFRARED
ASTRONOMY SATELLITE, HYDROGEN

Io the innermost of the four large (Galilean) moons of Jupiter. Its
rotation is synchronous and one face is kept permanently towards
Jupiter (as our Moon keeps one face towards the Earth). The Voyager
spacecraft found Io to be the most volcanic body in the Solar System,
and, unlike the Moon, it is certainly not geologically dead: the
cameras revealed a yellow-brown surface mottled with hot-spots which
spew umbrella-shaped plumes of molten sulphur 100km high above
Io's surface. These emissions of sulphur are more characteristic of
geysers than volcanoes, rather like those of the much smaller geysers
in the Yellowstone National Park. One of Io's volcanoes, called Pele,
was erupting when Voyager 1 passed by in 1979. It sent plumes of
sulphur across 1400km of Io's surface. Io's mottled appearance is
caused by the different temperatures to which the sulphur is exposed
as it lies on the surface. Io's surface is constantly being renewed by
these volcanoes so it does not have the ancient cratered appearance
that most moons have. It is suggested that Io is so hot inside because it
is caught in a constant gravitational tug-of-war between Jupiter and
Io's sister moon, Europa, 250 000km further out. The constant tidal
tug produces internal heating on a massive scale. A lot of sulphur,
potassium, and sodium ions form an ionosphere round Io which
interacts with Jupiter's magnetosphere producing violent electrical
behaviour which Voyager detected and which is also detectable from
Earth. □ see VOYAGER, JUPITER, ¹GALILEO, ION, IONOSPHERE, MAG-
NETOSPHERE, SYNCHRONOUS ROTATION

ion an atom, or group of atoms, which has lost or gained one or
more electrons and which therefore carries a positive or negative
electrical charge □ see IONIZATION

ion engine – see ELECTRIC PROPULSION

ionization the separation of one or more electrons from their
parent atoms to create free electrons and positively charged atoms.
When these ions recombine, electromagnetic radiation (eg light and
ultraviolet) is emitted. Thus the Earth's aurora is caused by recom-
bination of the ions after ionization by the solar wind. A study of the
wavelengths of the light from ionized gases in the solar system and the
depths of space provides information about the types of gases present
there. □ see SOLAR WIND, IONOSPHERE, MAGNETOSPHERE, PLASMA,
ELECTRIC PROPULSION, ANTIMATTER PROPULSION

ionosphere the portion of the Earth's atmosphere which extends
from about 50km above the Earth's surface. The ionosphere contains
electrons and ions created by the impact of the Sun's ultraviolet radiation and the solar wind on air molecules in the upper atmosphere. The
ionosphere has several distinct layers and their degree of ionization
varies with the time of day, season, latitude, and the state of the solar
activity. The ionosphere reflects radio waves of frequencies less than
about 30MHz allowing long-distance radio communication by successive reflections between it and the ground. However, the ionosphere

Ion Release Module

makes radio astronomy impossible for frequencies less than about
10MHz. The planets Jupiter, Mars, and Venus, and Jupiter's moon Io,
also have an ionosphere. □ see IONIZATION, RADIO ASTRONOMY,
MAGNETOSPHERE, SOLAR WIND, ULTRAVIOLET, JUPITER, IO, PLASMA,
ACTIVE MAGNETOSPHERIC PARTICLE EXPLORER

Ion Release Module (abbr **IRM**) – see ACTIVE MAGNETOSPHERIC
PARTICLE EXPLORER

IRAS – see INFRARED ASTRONOMY SATELLITE

IRS [*I*ndian *R*emote *S*ensing (satellite)] – see REMOTE SENSING

ISA – see ISRAEL SPACE AGENCY

ISAMS [*I*mproved *S*tratospheric *a*nd *M*esospheric *S*ounder] – see
UPPER ATMOSPHERE RESEARCH SATELLITE

ISAS – see INSTITUTE OF SPACE AND ASTRONAUTICAL SCIENCE

ISEE [*I*nternational *S*un-*E*arth *E*xplorer] – see INTERNATIONAL
COMETRY EXPLORER

ISO – see INFRARED SPACE OBSERVATORY

isotope any of two or more species of atoms of a chemical element
(eg hydrogen) in which the nuclei contain different numbers of
neutrons but the same numbers of protons. Thus isotopes have the
same atomic number but different mass numbers. □ see ATOM,
HYDROGEN

ISPM [*I*nternational *S*olar *P*olar *M*ission] – see ULYSSES

Israel Space Agency (abbr **ISA**) an organization formed in 1984
to plan and implement a number of space projects. ISA plans to fly a
life sciences experiment (involving a hornet's nest) aboard the Space
Shuttle. An Israeli-made astrophysical X-ray sensor will also be carried by the Shuttle. In the long term, ISA plans to operate its own
Earth resources satellite (1988), and a weather satellite and communications satellite (1993), the latter perhaps launched by ISA's own
rocket which may be built by then.

ISRO – see INDIAN SPACE RESEARCH ORGANIZATION

IT – see INFORMATION TECHNOLOGY

IUE – see INTERNATIONAL ULTRAVIOLET EXPLORER

IUS – see INERTIAL UPPER STAGE

J

jacket a protective or (thermally) insulating cover ⟨*a ~ for preventing micrometeroid penetration of a spacecraft*⟩ □ see DUST SHIELD

jansky (symbol **Jy**) the unit used by astronomers for measuring the radio flux density from radio sources, equal to 10^{-26} watts per square metre per hertz □ see RADIO ASTRONOMY

jet 1 a long thin stream of material emitted from a compact object such as a radio galaxy □ see VIRGO A, RADIO ASTRONOMY **2** a forceful stream of fluid discharged from a narrow opening or a nozzle, moving in a narrow duct ⟨*an air ~*⟩ **3** a nozzle or other narrow opening for emitting a jet of fluid ⟨*gas ~s for attitude control*⟩ **4** an aircraft powered by a jet engine

jet propulsion propulsion of an object in one direction by forcing a fluid in the opposite direction to provide a reaction force □ see FLUID, ROCKET ENGINE, SPECIFIC IMPULSE

Jet Propulsion Laboratory (abbr **JPL**) the centre of NASA's Deep Space Network which maintains communication with interplanetary spacecraft. It is based at Goldstone in California and its main aerial is a 64m-diameter radio dish. □ see DEEP SPACE NETWORK, MISSION CONTROL, EARTH STATION

jet stream strong winds in the upper atmosphere (the troposphere or stratosphere) of the Earth blowing from west to east and often reaching speeds of 400kmh⁻¹ □ see EARTH'S ATMOSPHERE

jettison to discard unwanted or encumbering material ⟨*during ascent the Space Shuttle ~s its solid rocket booster*⟩ □ see SOLID ROCKET BOOSTER

jitter unwanted and irregular back-and-forth movement of an image displayed on a VDU or cathode-ray oscilloscope – compare JUDDER

Jodrell Bank the site in Cheshire, England of the Nuffield Radio Astronomy Laboratory. Its main antenna is a 75m fully steerable paraboloidal dish which was completed in 1957. A smaller 38m elliptical dish was completed in 1964. These antennae are part of the Multi Telescope Radio Linked Interferometer (MTRLI) which provides high-resolution maps of cosmic radio sources. The other dishes making up the interferometer are, two 25m dishes within a few kilometres of Jodrell Bank, a 24m dish at Wardle near Nantwich, the 25m dish at the Royal Radar Establishment, Malvern, and a 25m dish at Knockin, Oswestry. □ see MULLARD RADIO ASTRONOMY OBSERVATORY, VERY LARGE ARRAY, APERTURE SYNTHESIS

Johnson Space Centre (abbr **JSC**) a complex of facilities at Houston in Texas, USA, for the development, management and control of US space flights. Construction of JSC began in April 1962 and personnel began moving into it during March 1964. For the first decade of operations, it was known as the Manned Spacecraft Centre but on 17 February 1973 it was renamed the (Lyndon B) Johnson

Space Centre after the President who recommended a manned lunar landing in response to the Soviet space successes. Its most famous building is Building 30, better known as 'Mission Control', which takes over communications with spacecraft once they have lifted off from Cape Canaveral. Near to this room full of consoles are the support rooms where experiments and payloads can be monitored and specialists can work out solutions to problems that may arise during a flight. Mock-ups of spacecraft are housed in other buildings – for example Building 9A contains several Space Shuttle simulators including the Remote Manipulator System and the forward fuselage of the Space Shuttle with an attached payload bay. These are used for astronaut training, and for real-time evaluation of problems arising during a flight (eg repair of Solar Max during Mission 41C). Other buildings house a water tank for extravehicular activity training, and vacuum chambers to simulate the heat of the Sun and the chill of space. Part of JSC serves as a museum with exhibits such as the lunar module, the Apollo 17 command module, Mercury 9, Gemini 5, Saturn 5, and Skylab. JSC is also involved with the Tracking Data and Relay Satellites (TDRS), and with the nearby Ellington Air Force Base which flies the KC-135 aircraft used for zero-g training. It will be involved with the US Space Station due to begin operations in 1992 in time for the 500th anniversary of Columbus's arrival in the New World.

joule (symbol **J**) the SI unit of energy equal to 1 newton metre (Nm). The heat energy required to raise the temperature of 1kg of water by 1K (1C) is equal to about 4200J. □ see SI UNITS, POWER, WATT, KELVIN

Jovian of, resembling, or characteristic of the planet Jupiter ⟨*Saturn is a ~ planet*⟩ □ see SOLAR SYSTEM, JUPITER

JPL – see JET PROPULSION LABORATORY

JSC – see JOHNSON SPACE CENTRE

judder unwanted and jerky vibrations of an image displayed on a VDU or cathode-ray oscilloscope, in which parts of the image temporarily overlap – compare JITTER

Jupiter the fifth and largest planet of the Solar System, which has a diameter of about 142 000 kilometres, orbits the Sun at about 778 million kilometres, and has at least 15 moons. Though Jupiter has a volume about 1000 times that of the Earth, its density is less than a quarter of the Earth's because it is made up largely of hydrogen and helium. Close-up observations by Pioneers 10 and 11 spacecraft in the mid 1970s showed more clearly than observations from Earth that the atmosphere has light and dark bands called zones and belts formed by Jupiter's high-speed rotation. For a planet, Jupiter is radiating a lot of energy, some of which comes from violent electrical storms in its atmosphere. The famous Red Spot in Jupiter's atmosphere seems to be related to this activity and its dynamics have been monitored by Voyagers 1 and 2 which completed successful flybys of Jupiter in 1979. The next close-up observation of Jupiter will take place when the spacecraft Galileo reaches the planet in 1988. □ see PIONEERS 10 AND 11, VOYAGER, ¹GALILEO, GREAT RED SPOT

k the symbol for the prefix kilo meaning one thousand □ see
KILOMETRE, KILOGRAM, SI UNITS

K the symbol used for the degree kelvin □ see ABSOLUTE TEMPERA-
TURE SCALE ·

Kagoshima Space Centre – see INSTITUTE OF SPACE AND
ASTRONAUTICAL SCIENCE

K-band a range of radio waves having frequencies in the range about
11 to 15GHz. The K-band is one of the bands used by the Intelsat-5A
communications satellites operating at 14GHz uplink and 11GHz
downlink. □ see BAND, X-BAND, C-BAND, INTELSAT

kelvin (symbol **K**) a unit of temperature on the absolute temperature scale ⟨*the temperature of liquid hydrogen is 20K*⟩ □ see
ABSOLUTE TEMPERATURE SCALE, CRYOGENICS

Kennedy Space Centre – see CAPE CANAVERAL

¹Kepler a spacecraft under consideration by the European Space
Agency (ESA) for a close-up study of Mars in the 1990s. Kepler
would be built by European firms, launched by Ariane, and tracked
by European Earth stations. Kepler's study of Mars would be a
natural follow-on to the Viking lander and Voyager fly-by missions,
and add considerably to our knowledge of the atmosphere and surface
conditions of Mars. These observations would also pave the way for
robots to explore Mars and return soil samples to Earth. It is possible
that Kepler would be part of a dual mission to Mars: NASA is also
considering sending the spacecraft Aeronomy Orbiter to Mars in the
1990s. Two spacecraft orbiting Mars simultaneously could take stereo
pictures of the surface of Mars, and it would be possible to determine
the strengths of the magnetic field produced from within Mars and that
produced by its magnetosphere. And when one spacecraft keeps track of
the other, it is possible to map Mars's gravitational field. □ see MARS,
VIKING, VOYAGER, EXPEDITION TO MARS, PHOBOS AND DEIMOS

²Kepler, Johann (1571–1630) German astronomer who used the data
acquired by Tycho Brahe to conclude that Mars's orbit was shaped
like an ellipse. The first two of Kepler's Laws (of planetary motion)
were published in his book *Astronomia Nova* in 1609; the third law
appears in a second book published in 1619. Kepler was in touch by
letter with Galileo until 1610, and then the relationship seems to have
ended, possibly because Galileo could not come to terms with
Kepler's strong interest in astrology and his casting of horoscopes. (In
1620 Kepler's mother was arrested as a witch; though not tortured,
she did not long survive her release for which Kepler fought so hard.)
However, Kepler did receive one of Galileo's telescopes with which
he, too, observed Jupiter's moons (and, incidentally gave them the
name 'satellites' from a Latin term for the hangers-on of a powerful
man). His interest in lenses led him to design a telescope which used

two convex lenses, rather than one convex and one concave as in Galileo's design. He computed tables of planetary motions using the newly invented logarithms of Napier and published these in 1627. Kepler also wrote a story *Somnium* about a man who travelled to the Moon in a dream – perhaps the first authentic science fiction story to be published. He appears to have regarded space travel as a real possibility for in a letter to Galileo in April 1610, he wrote: "let us create vesels and sails adjusted to the heavenly ether, and there will be plenty of people unafraid of the empty wastes". □ see KEPLER'S LAWS, [2]GALILEO, NEWTON'S LAW OF GRAVITATION, NEWTON

Kepler's Laws three fundamental laws which describe the motion of the planets about the Sun, moons about a planet, and artificial satellites about the Earth. The three laws (originally formulated for the planets) are: 1 the orbit of each planet is an ellipse with the Sun at one focus of the ellipse; 2 as each planet revolves around the Sun the (imaginary) line connecting the planet to the Sun sweeps out equal areas in equal times – thus a planet's speed decreases as it moves further from the Sun; 3 the squares of the sidereal periods of revolution of any two planets are proportional to the cubes of their mean distance from the Sun. These laws were published in 1609 and 1619 and were based on detailed observations of the planets made by Tycho Brahe with whom Kepler worked. The correct basis for Kepler's Laws was not understood until Newton put forward his Law of Gravitation. □ see NEWTON'S LAW OF GRAVITATION, [2]COPERNICUS

kerosene *also* **paraffin** a flammable rocket fuel which will burn in liquid oxygen, as in the first-stage engines of Saturn 5 □ see SATURN 5, APOLLO, GODDARD

keyboard a group of keys arranged in a certain pattern, by which information is entered into an electronic system

keypad a small keyboard with a restricted range of keys

kick to sharply increase the speed of a spacecraft or satellite ⟨*the rocket engine ~ed the satellite into geostationary orbit*⟩ □ see THRUSTERS, APOGEE MOTOR

kilobit a unit for measuring binary data equal to either 1000 or 1024 bits □ see DATA RATE

kilogram *also* **kilogramme** (symbol **kg**) the basic unit of mass in the SI system. It is the mass of a platinum-iridium cylinder kept at the International Bureau of Weights and Measures at Sevres, Paris. Structures in the universe range from the very small (eg nuclear particles which have a mass of about 10^{-27}kg) to galaxies which have a mass of about 10^{43}kg. □ see SI UNITS, METRE, SECOND

kilohertz (symbol **kHz**) a frequency equal to 1000 (10^3) hertz □ see FREQUENCY

kilometre (symbol **km**) a distance equal to 1000 (10^3) metres. The kilometre, rather than the basic unit the metre, is generally used to express large distances: the distance from the Earth to the Sun is about 150 million kilometres (1.5×10^8km). □ see METRE, ASTRONOMICAL UNIT

kinetic energy the energy a body has as a result of its motion. An artificial satellite gains kinetic energy (and potential energy) from the chemical energy of the rocket fuel in the launching rocket. If a mass m (kg) moves with a speed v (ms^{-1}), its kinetic energy is equal to $\frac{1}{2}mv^2$ joules – compare POTENTIAL ENERGY □ see ESCAPE VELOCITY

Kitt Peak National Observatory (abbr **KPNO**) an observatory near Tucson, Arizona, at an altitude of 2060m that has a number of optical and radio telescopes. The main optical telescopes are a 4m reflecting telescope (the Mayall reflector), a 2m reflecting telescope, and a large solar telescope (the McMath solar telescope). A 36m radio telescope is also on the site.

Komarov, Vladimir (1927–67) the first Russian cosmonaut to die when his Soyuz 1 spacecraft crashed. The Russian space programme suffered a second setback when the three-man Soyuz 11 spacecraft crashed in 1971. □ see WHITE, SOYUZ

KPNO – see KITT PEAK NATIONAL OBSERVATORY

Kuiper Airborne Observatory – see URANUS

L

LAGEOS [*La*ser *Geo*dynamics *S*atellite] – see SATELLITE LASER-RANGING

Lagrangian points points in space where a body of small mass, under the gravitational pull of two much more massive bodies, remains motionless relative to them. In theory there are five such points but only two are stable. Each of these stable Lagrangian points is found at one corner of an equilateral triangle formed by the three bodies. For example, the massive planet Jupiter and the Sun occupy two corners of an equilateral triangle and the third – the Lagrangian point in Jupiter's orbital path – is occupied by the Trojan asteroids. Lagrangian points also occur 60° ahead and 60° behind the Moon in its orbit about the Earth. These points might be where clouds of dust and gas have concentrated, but they have been proposed as sites for future space settlements and solar power stations. The Lagrangian points are named after the French mathematician Joseph Louis Comte de Lagrange who first predicted these points in 1772. □ see ASTEROID, SPACE SETTLEMENT, TROJAN GROUP

Laika a 5kg Samoyed husky bitch who became the first living creature to orbit the Earth on 3 November 1957 in Sputnik-2, a month after Sputnik-1. Accustomed to confinement, the little dog was strapped into an air-conditioned chamber with a food supply suspended in gelatin. Sensors wired to her body and connected to radio transmitters provided information about her pulse and respiration rates, and her movements in weightless conditions to which, it is said, she adapted well. When the oxygen supply gave out 10 days later, Laika died. □ see ABLE AND BAKER, ABREK AND BION

lander a spacecraft, or part of a spacecraft, which makes a controlled descent to the surface of a planet or satellite ⟨*the Viking* ∼⟩ – compare ORBITER □ see VIKING, VENERA

Landsat [*Land Sat*ellite] any of a series of satellites for the remote sensing of Earth resources. Landsat 1 (formerly ERTS-1) was launched into a 918km polar orbit on 23 July 1972, its main purposes being to look for new mineral resources and diseases in crops, and to keep a check on oil pollution (eg oil spillage at sea). Its main instrument was a radiometer called the multispectral scanner (MSS) whose measurements, after computer processing, produced detailed colour images of the Earth's surface. The images did not give true colours but showed heathland in red-brown, forests in scarlet, towns in dark blue, clear water in blue, polluted water in light blue, and so on. Landsat 1 circled the Earth 14 times a day and returned to its initial orbital path every 18 days. It could therefore show changes in the surface conditions (eg the spread of crop diseases or oil spillages at sea). Landsats 2 and 3 were launched in 1975 and 1978, respectively. On 9 July 1981, a new and more powerful Earth resources satellite, Landsat 4, was launched into a 705km orbit. In addition to an improved MSS, Landsat 4 carried a thematic mapper (TM) which gave a resolution of 30m. The TM was able to photograph to a considerable depth in clear water, and, using seven wavelength bands, produced images of the Earth in more natural colours. However, the X-band transmissions from the TM failed, though it was possible to receive data on the Ku-band (15.35GH$_z$ to 17.25GH$_z$) via a satellite relay. In August 1983, a fault in its solar panels made the satellite useless. NASA plans to use one of the flights of the Space Shuttle *Discovery* in 1986 to rescue Landsat 4 and perhaps return it to Earth for repair. On 1 March 1984, the 1950kg back-up satellite Landsat 5 was launched from the Vandenberg Air Force Base, California into a 705km near-polar orbit. □ see SULLIVAN, RADIOMETER, REMOTE SENSING, SPACE SHUTTLE, SPOT

lase to function as a laser by emitting coherent light □ see LASER

laser [*l*ight *a*mplification by *s*timulated *e*mission of *r*adiation] an electro-optical device which produces an intense narrow beam of light or infrared radiation over a narrow frequency band. A material (eg a ruby crystal) lases when some of its electrons are raised to a higher energy level by 'pumping' the material with a strong light source or a strong electrical current. This higher energy level is an excited state for the electrons, and when they fall back to their normal state, they emit light photons. At one end of the material is a mirror so that photons are reflected back into the lasing material where they stimulate the emission of more photons from the excited electrons; at the other end is a partially reflective mirror so that most of the beam which is generated passes through to the outside world. The material may be a solid (eg a ruby crystal) a gas (eg carbon dioxide), and the beam can be pulsed or continuous. An intense and narrow beam of coherent light leaves the laser – coherent means all the light waves are in step (ie in phase) with each other. This coherent light will travel a long way without spreading much, so that it can be concentrated on a small target area. Thus lasers are used for surgery (eg 'spot welding' a detached retina) and for reading information on video and compact disks. The single wavelength and high frequency of laser light means that a (gallium arsenide) laser can carry a considerable amount of information along optical fibres in optical communications systems. For space communications, lasers offer much higher data rates than radio since the beam of energy carrying the data can be tightly focussed. For example, while transmitting data from Saturn, Voyager's high-gain antenna spread the signal over a million times the Earth's area: a laser system of much smaller aperture would spread over little more than one Earth area. Since lasers focus their energy in a tight beam, they are being developed for destroying hostile missiles and satellites – the 'death ray' of science fiction come true. □ see OPTICAL COMMUNICATIONS, BANDWIDTH, SATELLITE LASER-RANGING, TEAL RUBY, STAR WARS, GAS LASER, WELLS, MASER, GALLIUM ARSENIDE

laser diode a compact solid-state device (as opposed to a gas-filled laser) which produces a narrow beam of light all of one wavelength (or frequency). The light from a laser diode is used in optical communications systems as the carrier wave for messages. □ see OPTICAL COMMUNICATIONS, COHERENT LIGHT

Laser Geodynamics Satellite (abbr **LAGEOS**) – see SATELLITE LASER-RANGING

launch 1 to send off a rocket from the Earth's surface, or from the surface of another planet or moon, under its own power ⟨*to ~ a Space Shuttle*⟩ □ see LIFT-OFF, LAUNCH PAD **2** to boost a space probe from its launch rocket into Earth orbit or into interplanetary flight □ see TRANSFER ORBIT

launch pad *also* **launch site** the base, platform, or area from which an aerospace vehicle or rocket is launched ⟨*the ~ at Cape Canaveral*⟩ □ see COSMODROME, CAPE CANAVERAL

launch vehicle *also* **rocket** a system for giving a spacecraft the mechanical energy needed to put it into orbit or to send it on an interplanetary mission. The energy is obtained by burning a chemical fuel (eg liquid oxygen and hydrogen) inside the launch vehicle's combustion chamber. The hot combustion gases are then ejected at great speed through a nozzle to produce the thrust needed to drive the launch vehicle and its payload forward. □ see ROCKET ENGINE, SPECIFIC IMPULSE, MULTISTAGE ROCKET, SATURN 5, ARIANE, SPACE SHUTTLE, CENTAUR

launch window – see WINDOW 2

LCD – see LIQUID CRYSTAL DISPLAY

LDEF – see LONG DURATION EXPOSURE FACILITY

Leasat any of a series of communications satellites built by Hughes Aircraft Company, leased to the US Navy and deployed from the Space Shuttle. Leasats 1, 2, and 4 (also called Syncom IV-1, etc) were deployed successfully into low-Earth orbit from Missions 41D, 51A, and 51I, respectively. However, Leasat 4, had to be abandoned after it had reached geostationary orbit due to the failure of a transmissions cable between the UHF multiplexer and transmitter on board the satellite. There was trouble, too, as Leasat 3 was rolled out of the Shuttle's payload bay during Mission 51D; an arming lever failed to activate Leasat's electrical systems. It should have spun-up to about 33rpm, and 45 minutes after leaving the Space Shuttle, its engines would have fired to send the satellite into geostationary orbit. Leasat 3 was left in orbit 'electrically dead', but was later repaired on Mission 51I by astronauts van Hoften and Fisher working outside the Space Shuttle. 'Ox' van Hoften, 1.93m tall and 89kg mass then spun the satellite manually to stabilize it. □ see SPACE SHUTTLE, SPACE INSURANCE

LED – see LIGHT-EMITTING DIODE

lens 1 a curved piece of glass for focussing visible light or infrared ⟨*a telescope ~*⟩ □ see REFRACTING TELESCOPE **2** a set of coils and/or electrodes for focussing a beam of electrons ⟨*a magnetic ~*⟩

lepton any elementary particle that can participate in weak interactions and/or electromagnetic interactions but not in strong interactions. Leptons include the electron, mu meson (muon), and neutrino. Two leptons, the electron and the electron-neutrino, are sufficient to build a universe closely resembling the one that exists. □ see ELEMENTARY PARTICLES, ELECTRON, MESON, NEUTRINO, HADRON, QUARK

life sciences the branches of science (eg biology or medicine) that deal with living organisms and life processes □ see SPACELAB, SPACE SICKNESS, SPACE MEDICINE

lift-off *also* **blast-off** the vertical ascent of a rocket, missile, or aircraft from a launch pad ⟨*people went to see the ~ to Mars*⟩ – compare TAKE-OFF □ see COUNTDOWN

light electromagnetic radiation to which the eye is sensitive. The eye and optical telescopes respond to light which lies between the infrared (about 750nm) and ultraviolet (about 360nm) regions of the electromagnetic spectrum. In this range, the eye is most sensitive to greenish-yellow light of wavelength about 550nm. It is no accident that the Earth's atmosphere is transparent to wavelengths in this range, and optical astronomy makes use of this 'window'. But the development of instruments which detect radiation outside this range has revolutionized our knowledge about the universe. □ see ELECTROMAGNETIC RADIATION, SPECTRUM, NANOMETRE, OPTICAL ASTRONOMY

light-emitting diode (abbr **LED**) a small semiconductor diode which glows when current passes through it from its anode to its cathode terminal. LEDs are used to display information in all types of electronic equipment. They take a smaller current than a filament lamp, so are used in all types of portable battery-operated equipment. Infrared LEDs are used in optical communications systems – compare LIQUID CRYSTAL DISPLAY

light flash a brief flash of light which an astronaut sometimes 'sees' when a cosmic ray or other ionizing particle passes through his/her closed eyes □ see COSMIC RAYS, IONIZATION

light pollution manmade light (especially that from street lights) that is scattered in the upper atmosphere and limits the study of stars using optical telescopes. The problem is being solved by putting telescopes in orbit round the Earth. □ see HUBBLE SPACE TELESCOPE, SEEING, SHUTTLE GLOW

light-year the distance light travels in one year, used as a convenient unit for measuring the vast distances between cosmic objects in the universe. One light year is equal to approximately 9.5 million million kilometres (9.5×10^{12}km). The nearest star, Proxima Centauri, is about 4 light-years from Earth. □ see PARSEC

linear momentum (symbol **p**) the property of a body moving in a straight line. It is defined as the product of the mass, m, and velocity, v, of a body, ie p = mv – compare ANGULAR MOMENTUM □ see LAUNCH VEHICLE, SPECIFIC IMPULSE

liquid crystal display (abbr **LCD**) a display used in a wide range

of digital devices (eg portable computers and watches) which operates
by reflected light rather than by producing light as in the light-emitting
diode (LED). LCDs use less current than LEDs so they are used in all
types of portable battery-operated equipment which have to operate
for long periods without attention – compare LIGHT-EMITTING DIODE

liquid hydrogen (abbr LH$_2$) the fuel used in cryogenic rocket
engines which burns in liquid oxygen (the oxidizer) to produce water
vapour. On the Space Shuttle, liquid hydrogen is kept at a temperature of 20K (−253C). During launch, liquid hydrogen is fed to the
main engines at the rate of 12 084 kilograms per minute where it
mixes and burns in liquid oxygen. □ see CRYOGENICS, LIQUID
OXYGEN, EXTERNAL TANK, ROCKET ENGINE

liquid oxygen *also* **lox** (abbr LO$_2$) oxygen in liquid form used to
oxidize liquid hydrogen in a cryogenic rocket engine, ie the hydrogen
burns in the oxygen to produce heat and water vapour. On the Space
Shuttle, liquid oxygen is kept in a tank at a temperature of 90K
(−183C). During launch, liquid oxygen is fed to the main engines at
the rate of 72 340 kilograms per minute where it mixes and burns the
liquid hydrogen. □ see CRYOGENICS, LIQUID HYDROGEN, HYDROGEN,
EXTERNAL TANK, ROCKET ENGINE

little green men (intelligent) alien visitors conceived by science
fiction writers and believed to inhabit outer space □ see SEARCH FOR
EXTRATERRESTRIAL INTELLIGENCE, ENCOUNTER 2, UFO

Local Group – see CLUSTER 3

Long Duration Exposure Facility (abbr **LDEF**) a large satellite
that was first released into Earth orbit by Mission 41C of the Space
Shuttle and which acts as a mounting platform for experiments requiring a long-term exposure to space □ see SPACE SHUTTLE, MISSION 41C

Long March 2 and 3 rockets used by China for launching remote-
sensing and communications satellites. Long March 2 is a two-stage
31.65m-high liquid-fuelled (but not cryogenic) rocket which is capable
of putting a 2.2 tonne satellite into a 400km-high orbit. The 43.25m-
high Long March 3 rocket comprises a Long March 2 rocket with a
cryogenic (ie liquid hydrogen and oxygen) third stage, and this can put
a 1.4 tonne satellite into geostationary orbit. The first flight of Long
March 2 was in 1974 and on 21 October 1985 it was used to launch
China 17, a recoverable remote-sensing satellite, from a launch site in
the Gobi desert. Only two Long March 3 launches have taken place.
The first flew in 1984 with partial success, and the second, on 8 April
1984, resulted in the placing of China's test communications satellite,
STW-1, into geostationary orbit. China is now offering to launch
rockets for other nations on a commercial basis at prices competitive
with Ariane and Shuttle launches. □ see GEOSTATIONARY ORBIT,
ARIANE, REMOTE SENSING

long waves radio waves having wavelengths between 1000 and
10 000 metres, ie frequencies between 300kHz and 30kHz □ see
MEDIUM WAVES, SHORT WAVES, MICROWAVES, X-BAND, FREQUENCY,
ELECTROMAGNETIC RADIATION

low 1 *of technology* not very advanced in development and concept ⟨*a paper aeroplane is ~ technology*⟩ **2** of or being the smaller of two voltage levels in a digital logic circuit ⟨low-*level logic is given the binary number 0*⟩ – compare HIGH

lox – see LIQUID OXYGEN

LRV [*Lunar Roving Vehicle*] – see LUNAR ROVER

LSL – see LUNAR SLIDE LANDER

lunar of the Moon ⟨~ *dust*⟩

lunar base – see MOON BASE

lunar module – see APOLLO

Lunar Orbiter probes a series of five spacecraft placed in orbit round the Moon between August 1966 and August 1967 to take and transmit both wide angle and close up pictures of the Moon in preparation for the Apollo Moon landings. The first three Lunar Orbiters were placed in equatorial orbits to search for suitable landing sites, and the last two in polar orbits to provide global pictures. Once their mission was complete, they were crashed into the Moon's surface (to avoid a possible collision with the command and service module on subsequent Apollo missions?). □ see LUNIK PROBES, RANGER, SURVEYOR, ZOND

Lunar Rover *also* **Lunar Roving Vehicle** (abbr **LRV**) the battery-driven, four-wheeled vehicle used by the astronauts on Apollo missions 15, 16, and 17 (July 1971, April and December 1972) to explore a larger area of the Moon's surface than was possible on foot. This vehicle accompanied the astronauts to the surface folded up in the descent stage of the lunar module. The 213kg Lunar Rover had a remote-controlled television camera mounted on the front, and pictures, together with the astronaut's commentary, could be beamed direct to Earth using the on-board transmitter and high-gain dish aerial. The Lunar Rovers were left behind after use. □ see LUNOKHOD, APOLLO

Lunar Roving Vehicle – see LUNAR ROVER

Lunar Slide Lander (abbr **LSL**) a lunar landing vehicle (eg a ferry between an Earth-orbiting space station and a Moon base) which would skid to rest on flat stretches of dust in the 'seas' (mares) on the Moon's surface. The LSL would have a specially designed braking system which would tolerate wear and the high temperatures generated in bringing the LSL to rest on prepared stretches of dust. The linings for the braking surfaces could be manufactured from materials found on the Moon (eg aluminium oxide doped with zirconium particles). The LSL, which was proposed by the late Krafte Ehricke, would provide an economical method of landing astronauts and supplies on the Moon without polluting its residual atmosphere with gases from rocket exhausts. □ see EHRICKE, MOON BASE

Lunik probes a series of Soviet spacecraft which reached the Moon between January 1959 and August 1976. Lunik 1 (January 1959) was the first spacecraft to reach the vicinity of the Moon but missed it by 6000km; Lunik 2 (September 1959) was the first manmade object to

crash-land on the Moon's surface; Lunik 3 (October 1959) took the first pictures of the far side of the Moon; Lunik 9 (January 1966) made the first soft landing on the Moon; Lunik 10 made the first scientific experiments from lunar orbit; Lunik 16 (September 1970) made the first sample-return mission; and Lunik 17 (November 1970) placed the first unmanned roving vehicle, Lunokhod, on the Moon's surface.
□ see LUNOKHOD, ZOND, LUNAR ORBITER PROBES, LUNAR ROVER, SURVEYOR, RANGER

Lunokhod a Soviet unmanned eight-wheeled roving vehicle soft-landed on the Moon by Lunik probes 17 and 21 in 1970 and 1973, respectively. Lunokhod 1 carried a laser reflector so that its position could be determined exactly from Earth, and for 10 months it travelled 10km across Mare Imbrium taking pictures, and making measurements of the magnetic field, cosmic-ray intensity, and chemical composition of the soil. Lunokhod 2 travelled 37km in 4 months on the borders of Mare Serenitatis. □ see LUNAR ROVER, LUNIK PROBES, SURVEYOR

M

M 1 the symbol for the prefix mega meaning one million □ see MEGABIT, MEGAPARSEC **2** the symbol for Mach number □ see MACH NUMBER **3** the symbol for absolute magnitude □ see MAGNITUDE

m 1 the symbol for the prefix milli meaning one thousandth of (10^{-3}) □ see MILLIMETRE **2** the symbol for metre, the SI unit of distance □ see METRE, SI UNITS **3** the symbol for mass □ see MASS **4** the symbol for apparent magnitude □ see MAGNITUDE

machine a device for performing some useful and usu complex service ⟨*spacecraft are* ~s⟩ □ see ROCKET ENGINE, SUPERCOMPUTER, SPACECRAFT

machine vision any system comprising sensors (eg TV cameras) and programmable devices which enable a machine, esp a robot, to 'see' and act 'intelligently' □ see ROBOTICS, IMAGE RECOGNITION, TELEPRESENCE

Mach number (symbol **M**) a number representing the ratio of the speed of a body (eg an aerospace vehicle) to the speed of sound in the surrounding air. It is named after an Austrian scientist, Ernst Mach (1838–1915). A Mach number of 1.00 means the body is moving at the speed of sound through the surrounding air. Subsonic speeds have Mach numbers less than 1; supersonic speeds have a Mach number greater than 1. □ see SHOCK WAVE

Mach's principle the idea proposed by Ernst Mach in the 1870s that the origin of inertia (the reluctance of a body to change its velocity) is somehow attributable to the influence of distant galaxies. Astrophysicists do not have any idea what this influence might be, but if such a link exists it would mean that a totally isolated body would not have any inertia. □ see COSMOLOGY, UNIVERSE, ANTHROPIC PRINCIPLE, INERTIA

MAD – see MUTUAL ASSURED DESTRUCTION

magic number a large number, esp the enormous number 10^{40}, which arises in several different and unrelated contexts when examining the space-time structure of the universe. For example, the ratio of the age of the universe (as measured from the current rate of expansion, and called the 'Hubble time') to the time taken for light to travel across the proton (called the 'nuclear time') is about 10^{10} years/10^{-24} seconds, or about 10^{40} when both are measured in seconds. This large number alone is not significant but it also crops up in the following contexts: the strength of the electromagnetic force between protons is much stronger than the gravitational force between them, and the ratio is about 10^{40}; the number of charged particles (protons and electrons) in the universe is about 10^{80} (ie the square of 10^{40}); and further mathematical relationships occur which involve the 'magic' number 10^{40}. There appears to be no reason why the age of the universe should be related numerically to the number of particles in it, and some physicists have attached deep physical significance to it, as well as to the other relationships where this number occurs. It may be that the basic physics of the universe is organized so that life as we know it can prosper, and that the magic number is related in some way to our special place in it. Perhaps the answer will be found when mathematicians finally succeed in uniting the basic forces of nature in a general unified theory (GUT). □ see COSMOLOGY, HUBBLE'S CONSTANT, SPEED OF LIGHT, GENERAL RELATIVITY, PROTON, ANTHROPIC PRINCIPLE, GENERAL UNIFIED THEORY

magnetic disk a disk which has a magnetizable surface on which computer data can be recorded □ see MAGNETIC TAPE

magnetic field the invisible influence round a magnet which causes one magnet to act on another. In particular, charged particles (eg electrons and protons) swerve as they pass through a magnetic field. Though we are not sure how it is produced, the Earth has a magnetic field stretching out into space which reacts with the solar wind to produce aurorae and the magnetosphere. Mercury, Jupiter, and the Sun also have magnetic fields of their own making. □ see MAGNETOSPHERE, AURORA, MERCURY, JUPITER, SOLAR WIND, SUN, EARTH

magnetic storms – see GEOMAGNETISM

magnetic tape a strip of material which has a magnetizable surface on which computer and other data can be recorded □ see MAGNETIC DISK

magnetism the attraction of magnetized iron, steel, some alloys, and of an electromagnet for other magnets and magnetizable substances. Magnetism is the basis of the operation of many devices including transformers, solenoids, relays, and motors used in control and sensing equipment. □ see MAGNETIC FIELD

magnetohydrodynamics (abbr **MHD**) the study of how electrically conducting fluids (eg ionized gases in the atmosphere of the Sun) behave in magnetic fields. The movement of the ionized fluids creates an electric field which interacts with the magnetic field and causes changes to the motion of the fluid. □ see ION, PLASMA, ANTIMATTER PROPULSION

magnetosphere a region surrounding the Earth where the Earth's magnetic field interacts with the flow of ionized particles in the solar wind. 60 000km out from the sunward side of the Earth, there is bowshock (a shock wave) which defines the beginnings of the magnetosphere where the solar wind first meets the Earth's magnetic field. Inside the bowshock is a turbulent region of ionization known as the magnetosheath. This boundary region is generally known as the magnetopause. On the opposite side of the Earth the magnetosphere is drawn out to a great distance in the wake of the solar wind. The magnetosphere includes the radiation belts (eg the Van Allen radiation belts) where solar particles are trapped in the Earth's magnetic field and cause the lights of the aurora borealis. The magnetosphere has been under investigation ever since the first rockets and satellites were sent above the Earth's atmosphere. Spacecraft have shown that the planets Mercury and Jupiter have magnetospheres, the result of their own magnetic fields interacting with the solar wind. □ see ION, PLASMA, SOLAR WIND, AURORA, ACTIVE MAGNETOSPHERIC PARTICLE EXPLORER

magnification – see MAGNIFYING POWER

magnify to increase the apparent size of different objects and so make them appear nearer ⟨*a telescope magnifies distant objects*⟩ □ see TELESCOPE, RESOLVING POWER

magnifying power *also* **magnification** the ratio of the angle subtended by the image of an object seen through a telescope (or microscope), and the angle subtended by the same object seen without the telescope □ see TELESCOPE, APERTURE

magnitude a measure of the relative brightness of stars and other celestial objects. The apparent magnitude (symbol m) is a measure of the brightness of the object as viewed from Earth, and its value depends on the intrinsic brightness of the star (ie its luminosity), on its distance from Earth, and on how much light is absorbed by interstellar gas and dust. The brightest star, Sirius, (Alpha Canis Major) has an apparent magnitude m = −1.5, and the nearest star (Alpha Centauri) has a value of m = +10.7: the brighter a star, the more negative its apparent magnitude; the dimmer it is, the more positive the value of m. Apparent magnitude gives no true indication of a star's luminosity, ie its intrinsic brightness. This is given by its absolute magnitude (symbol M) which is the apparent magnitude of a star if it were at a distance of 10 parsecs from the Earth. □ see INTERSTELLAR MEDIUM, PARSEC, CEPHEID VARIABLE

main stage the rocket engine which produces the greatest amount of thrust in a multistage rocket □ see MULTISTAGE ROCKET

malfunction a partial or total failure in the operation of an electronic device or system □ see BUG, TROUBLESHOOTING, ABORT

manipulator a device for the (remote) operation of equipment in a distant and/or hazardous environment □ see REMOTE MANIPULATOR SYSTEM, TELEPRESENCE

man-machine interface any hardware which allows a person to exchange information with a computer or other machine. At present, a keyboard is the most widely used man-machine interface in computer systems, but devices are being developed to allow users to talk to computers. □ see SPEECH SYNTHESIS, SPEECH RECOGNITION, TELEPRESENCE

manned of or being a spacecraft carrying astronauts or cosmonauts ⟨*a ~ mission to Mars*⟩ □ see APOLLO, EXPEDITION TO MARS

manned manoeuvring unit – see EXTRAVEHICULAR MOBILITY UNIT

Marecs [*Maritime Communications Satellite*] – see INMARSAT

Mariner a series of US interplanetary spacecraft that explored Venus, Mars, and Mercury during the 1960s and early 1970s. The 200kg Mariner 2 flew past Venus at a distance of 35 000km in 1962,

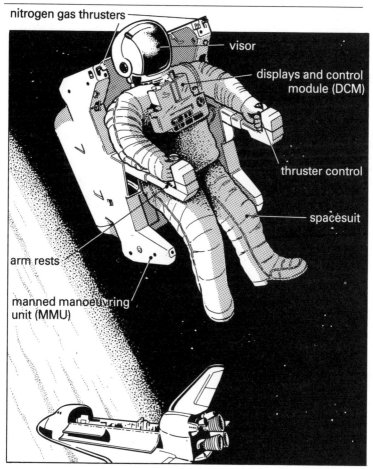

nitrogen gas thrusters

visor

displays and control
module (DCM)

thruster control

spacesuit

arm rests

manned manoeuvring
unit (MMU)

Manned manoeuvring unit

and measured Venus's temperature and detected the solar wind. The
260kg Mariner 4 was the first successful mission to Mars, flying past
this planet in 1965: the photographs it took showed Mars has a dry
and cratered surface – no sign of canals or life! The 250kg Mariner 5
passed close to Venus in 1967 and measured the variation of temperature with height in its atmosphere, and also its mass and diameter.
Two identical 400kg spacecraft, Mariners 6 and 7, passed close to
Mars in 1969 and photographed the planet's equatorial and polar
regions. The surface temperature, atmospheric pressure and composition, and mass of Mars were also measured. The 1000kg Mariner 9
became the first spacecraft to orbit Mars, in November 1971. Its elliptical orbit took close-up TV pictures of Mars's surface, and of its two

satellites, Phobos and Deimos. This mission could have been jeopardized by a planet-wide dust storm as Mariner 9 arrived to orbit Mars, but the dust cleared a few weeks later and detailed photographs showed giant canyons, and dry river beds that suggests water once flowed across planet Mars. The changing appearance of the polar caps with seasons was also recorded. The 500kg Mariner 10 reached Venus and Mercury in 1974, the first two-planet mission. After taking detailed photographs of Venus from 5700km in February 1974, which showed the structure of the dense cloud-cover of Venus in fine detail, the spacecraft used Venus's gravitational field to obtain a gravity-assist to Mercury. The Mercury fly-by occurred in March 1975. Subsequently, two more close-encounters of Mercury were made, in September 1974 and March 1975. The two magnetometers aboard Mariner 10 measured the interplanetary magnetic field and discovered that Mercury has a magnetic field. An infrared radiometer measured Mercury's surface temperature, and an ultraviolet spectrometer measured the absorption of solar ultraviolet radiation in Mercury's atmosphere. Mariner 1 (to Venus) and Mariner 8 (to Mars) suffered launch failure. Mariner 3 did not achieve its intended trajectory to Mars. □ see VENUS, MARS, MERCURY, CANALS, GRAVITY ASSIST, PHOBOS AND DEIMOS

Maritime Communications Satellite (abbr **Marecs**) – see INMARSAT

maroon to isolate in a remote place. To date, no astronaut has been marooned in space, though when the launch of Soyuz T-10 was aborted on 26 September 1983, three astronauts aboard the orbiting Salyut 7 space station were stranded in space for a while. Soyuz T-10 contained two astronauts who were to repair Salyut 7's solar panels and leave the fresher Soyuz T-10 for the resident astronauts on Salyut 7 to return in at the end of their stay.

Mars the fourth planet of the Solar System which has a diameter of about 6800 kilometres, orbits the Sun at about 228 million kilometres, and has two small moons, Phobos and Deimos. The modern era of Mars exploration began in 1965 when the cameras aboard the spacecraft Mariner 4 revealed an apparently dead world of huge extinct volcanoes and deep chasms, its reddish-brown surface pockmarked by countless craters caused by meteorites that had crashed into the planet. There was no indication of the canals and oases of earlier maps and the prospect of life on Mars dimmed considerably. Subsequent study by the flyby missions of Mariners 6 and 7 in 1969 confirmed this disappointing result as well as the fact that the atmosphere is a hundred times thinner than Earth's and composed mainly of carbon dioxide. The icecaps on Mars are composed of frozen carbon dioxide and recede and advance with the Martian seasons. As Mariner 9 approached the planet in 1971 it found the surface of Mars completely hidden from view by fierce windstorms. The latest probes were the 1976 Viking landers and orbiters which operated for several years and returned an enormous amount of data on the chemistry of the soil, the meteorological conditions of the landing sites, and local landforms. The valleys on Mars look as if they have been subjected to slow erosion by running water and indicate that liquid water and ice may exist at shallow depths below the surface. We shall expect first-hand information about the climate and geology of Mars when the - Russians land cosmonauts on Mars sometime towards the end of the century. □ see CANALS, VIKING, MARINER, SOLAR SYSTEM, VOYAGER, [1]KEPLER, EXPEDITION TO MARS, PHOBOS AND DEIMOS

maser [*m*icrowave *a*mplification by *s*timulated *e*mission of *r*adiation] a device used for amplifying very weak radio signals, esp those received from satellites and spacecraft. A maser increases the strength of microwaves by energizing atoms to a level where they give off radio energy of the desired frequency. An Earth station's dish aerial uses a maser to increase the strength of a weak radio signal over a million times – compare LASER □ see MICROWAVES, EARTH STATION, DISH AERIAL, RADIO ASTRONOMY

mass (symbol **m**) the quantity of matter in a body. Mass, which is measured in kilograms, gives a body two properties: its inertial mass (ie its resistance to a change in motion), and its gravitational mass which determines the gravitational field it produces. □ see WEIGHT, MACH'S PRINCIPLE, RELATIVITY, MASS-ENERGY EQUIVALENCE

mass-energy equivalence the mass, m, of a body is equivalent to an amount of energy, E, given by the equation $E = mc^2$ where c is the speed of light. This important equation was discovered by Einstein when developing his general theory of relativity. It explains how stars produce their energy (eg fusion reactions when hydrogen 'burns' to form helium in the Sun), and how energy is released in nuclear weapons (eg the splitting of plutonium atoms in atomic bombs). The energy obtained if a mass of 1kg is completely converted into energy is equal to about 10^{17} joules. □ see GENERAL RELATIVITY, REST MASS, PAIR PRODUCTION, ANNIHILATION

mass number the number of protons and neutrons in the nucleus of an atom. For example, helium has a mass number of 4, carbon a mass number of 12, and uranium-235 a mass number of 235. □ see ATOM, ISOTOPE

Materials Experiment Assembly (abbr **MEA**) – see MAUS

materials processing the manufacture of materials (in space) for use in new products on Earth. There is a big demand for new materials for many of the technical systems we take for granted. These include structural materials for aeroplanes and spacecraft, semiconductors for electronic devices, components for lasers, and glass fibres for optical communications systems. The stringent properties required of these materials can be met if they are made in space where the near absence of gravity (microgravity) in an orbiting spacecraft makes it possible to grow single crystals of high purity and strength. On Earth, single crystals of gallium arsenide or silicon are grown from the molten material held in a crucible, but the crucible contaminates the crystal, and convection of heat prevents crystals with uniform properties from forming. In microgravity, it is possible to suspend molten materials 'in space' without contact with a container, and convection does not occur. It is also possible to make pharmaceutical products for use in treating diseases. The sale of these new materials and products will help to finance the enormous cost of building and launching orbiting space stations and Moon bases. □ see GALLIUM ARSENIDE, MAUS, EURECA, SPACELAB, OFFICE OF SPACE AND TERRESTRIAL APPLICA-TIONS, MONODISPERSE LATEX REACTOR, MOON BASE, SPACE STATION, LIFE SCIENCES

mating the coming together of two large items of equipment, such as a Soyuz spacecraft with the orbiting Salyut space station □ see DOCKING

MAUS (acronym for the German words for 'autonomous materials processing in weightlessness') materials processing experiment packages provided by West Germany and designed to be accommodated in the payload bay of the Space Shuttle. The first MAUS was flown aboard Space Shuttle mission STS-5 in November 1982. It used X rays to study how the metals gallium and mercury react when heated and allowed to cool in zero-g. This experiment failed but was successfully repeated on the later STS-7 in June 1983 where it formed part of the OSTA-2 payload. OSTA-2 also carried a desk-sized US package called the Materials Experiment Assembly (MEA) which included an examination of the growth of semiconductor crystals from a vapour. A wide range of MAUS experiments has been booked on further flights of the Space Shuttle, some of which will be combined with MEA. These experiments are part of a joint effort between NASA's Marshall Space Flight Centre, the German Ministry for Research and Technology, and the German Aerospace Research Establishment which is aimed at an intensive investigation of materials processing in space. □ see MATERIALS PROCESSING, OFFICE OF SPACE AND TERRESTRIAL APPLICATIONS, GETAWAY SPECIAL, SPACELAB, GALLIUM ARSENIDE

MEA [*Materials Experiment Assembly*] – see MAUS

mean solar time time measured with reference to the passage of the Sun across the meridian. Thus the mean solar second is 1/86 400 of the mean solar day which is the 24-hour interval between two successive passages of the Sun across the meridian. However, mean solar time varies with the small changes in the rate of rotation of the Earth. □ see GREENWICH MEAN TIME, ATOMIC TIME

medium waves radio waves which have wavelengths between 200 and 700 metres, ie frequencies between about 1.5MHz and 5MHz □ see LONG WAVES, SHORT WAVES, MICROWAVES, FREQUENCY, ELECTROMAGNETIC RADIATION

megabit a quantity of binary data equal to 1 million (10^6) bits

megabyte (symbol **MB**) a quantity of computer data equal to 1 million (10^6) bytes

megahertz (symbol **MHz**) a frequency equal to 1 million (10^6) hertz. Radio waves at this frequency are known as medium waves. A microprocessor's clock may run at between 2 and 4 MHz. □ see FREQUENCY

megaparsec one million (10⁶) parsecs ⟨*the Virgo cluster of galaxies is at a distance of about 22 ~s*⟩ □ see PARSEC, CLUSTER 3

Mercury the innermost planet of the Solar System which has a diameter of about 4880km, orbits the Sun at an average distance of 58 million kilometres, and does not have any moons. The spacecraft Mariner 10 made three close approaches to Mercury in 1974 and 1975 and revealed a desolate, heavily cratered, and very hot world. The maximum daytime temperature is about 200C, but since Mercury has a very elliptical orbit, the temperature at perihelion (closest approach to the Sun) may reach 450C. Nighttime temperatures plunge to −180C largely because the atmosphere, consisting mainly of minute traces of helium, neon and argon, is very thin. Mercury has a high density suggesting it has a large nickel-iron core. □ see MARINER

Mercury project the pioneering US project to develop a system for launching a man into Earth orbit and to recover both man and spacecraft. The first sub-orbital 'hop' by an American was made by Alan Shepherd in a capsule, Mercury MR3, sent into the upper part of the atmosphere in May 1961 by a Redstone rocket. John Glenn was the first American to make an orbital flight in the capsule Mercury MA6, launched by an Atlas rocket in May 1962. The last capsule in the series, Mercury MA9, was launched in 1963. The Mercury project was followed by the two-man Gemini project. □ see GEMINI PROJECT, APOLLO

meson a class of elementary particles which includes the pi meson (pion), the mu meson (muon), K-meson, rho meson, etc. The pi meson belongs to a class of elementary particles known as hadrons. Three states of pi mesons exist: the neutral, positively charged, and negatively charged. The charge carried is equal to the charge on the electron. A charged pion decays into a charged mu meson and a muon-neutrino, whilst the unstable neutral pi meson decays into two gamma-ray photons. Pi mesons are produced in cosmic ray showers in the Earth's atmosphere. A pi meson is a combination of two quarks, an up quark and an anti down quark, and has a total charge of +e (the electronic charge). The neutral pion may be considered as a union of an up quark and an anti up quark (or a down quark and an anti down quark). □ see HADRON, NEUTRINO, QUARK, ELEMENTARY PARTICLES, COSMIC RAYS

mesosphere – see EARTH'S ATMOSPHERE

message any information which travels between a source and destination □ see TELEMETRY

metagalaxy the entire system of galaxies, including the Milky Way, within and beyond the visible universe, and the region of space it occupies □ see GALAXY, UNIVERSE

meteor – see MICROMETEOROID

meteorite a meteor which does not completely burn up in the Earth's atmosphere and reaches the ground □ see ASTEROID, MICROMETEOROID, COMET, TEKTITE

meteoroid

meteoroid – see MICROMETEOROID

meteorology the study of the Earth's atmosphere, esp weather and weather forecasting □ see WEATHER SATELLITE, EARTH'S ATMOSPHERE

¹meter an instrument for recording and measuring something ⟨*a flow rate* ~⟩ □ see PHOTOMETER, SPECTROMETER, ANEMOMETER, THERMOMETER, RADIOMETER

²meter 1 to measure with the aid of a meter **2** to supply something in measured quantities ⟨*to* ~ *an oxygen supply*⟩

metre (symbol **m**) the basic unit of distance in the SI system. It is the length equal to 1 650 763.73 wavelengths in vacuum of a particular orange line in the spectrum of the gas krypton-86. Structures in the universe range from the very small (eg the size of the nucleus of an atom which is about 10^{-14}m in diameter) to the size of the observable universe which is about 15 billion (15×10^9) light years or 10^{26}m. Note that the ratio of these two quantities, larger to smaller, is about 10^{-40}, and is regarded by some astrophysicists as a significant 'magic' number. □ see SI UNITS, KILOGRAM, SECOND, MAGIC NUMBER

MHD – see MAGNETOHYDRODYNAMICS

microcircuit a miniature circuit assembled on a tiny piece of pure silicon □ see SILICON CHIP, INTEGRATED CIRCUIT, MOORE'S LAW

microelectronics the production and use of silicon chips containing many thousands of components. Silicon chips are used in communications, control, and computer equipment without which space exploration would not be possible: radio communications systems enable us to keep in touch with astronauts and artificial satellites many millions of miles from Earth; computers aboard spacecraft automatically control the functions of the spacecraft as it swings by or lands on a planet in the Solar System; and a network of communications satellites in Earth-orbit form part of a worldwide radio and TV system. Applications of microelectronics in space exploration also 'spin off' products for commercial and domestic use and such products have the power to produce social change. These changes are already being felt in industry and commerce as robots replace people, and as more people become involved in generating, gathering, and processing information to satisfy the demand for knowledge. □ see SUPERCOMPUTER, SILICON CHIP, MOORE'S LAW, ROBOTICS, ARTIFICIAL INTELLIGENCE, EXPERT SYSTEM, GALLIUM ARSENIDE, OPTICAL COMMUNICATIONS, RADIATION HARDENING

microgravity almost zero gravity as an astronaut would experience in an Earth-orbiting spacecraft. Microgravity is the small mutual gravitational attraction between objects (eg between an astronaut and his/her toothbrush) as predicted by Newton's Law of Gravitation. There is considerable interest in using microgravity for making new alloys and very pure crystals. □ see MATERIALS PROCESSING, NEWTON'S LAW OF GRAVITATION, CONTINUOUS-FLOW ELECTROPHORESIS SYSTEM

micrometeorite – see MICROMETEOROID

micrometeoroid a dust particle in space which has a mass less than a millionth of a gram and a diameter less than a tenth of a millimetre. When micrometeoroids strike the Earth's atmosphere, they decelerate and float to the Earth's surface without burning up. They are then called micrometeorites. Meteoroids are objects with a diameter greater than 0.1mm and they sometimes burn up in the upper atmosphere and produce a streak of light known as a meteor or shooting star. If meteoroids do not burn up completely and reach the Earth's surface, they are called meteorites. There is a meteoroid cloud orbiting the Sun and many are found in the tails of comets. Thus spacecraft (eg Giotto) sent to investigate comets need protection against these small high-speed particles of cosmic dust. □ see SHOOTING STAR, COSMIC DUST, ¹GIOTTO, DUST SHIELD

micron a distance equal to one millionth (10^{-6}) of a metre. The micron is used by infrared astronomers to measure the wavelength of radiation from galactic objects such as quasars and new-born stars. □ see INFRARED ASTRONOMY, INFRARED SPACE OBSERVATORY

microsecond (symbol μs) a time interval equal to 1 millionth (10^{-6}) of a second. The time taken for light to travel a distance of 300 metres through space is about 1 microsecond.

microspaceship a very small and light (unmanned) spacecraft that is easy to launch, accelerates quickly to high speeds, is easy to store on a host spacecraft, and is more able to withstand the shock of launch and acceleration than a bigger craft. The design of a microspaceship has been studied by a team at America's Jet Propulsion Laboratory: it found that a spacecraft of about 25kg could carry out experiments in deep space. By comparison, the joint ESA/NASA Ulysses mission to study the Sun uses a 370kg spacecraft; the Voyager-2 interplanetary vehicle now has a mass of 788kg; and the Galileo mission launched to Jupiter in 1986 has a launch mass of about 2550kg. One way of reducing the size of a spacecraft is to use a very high frequency radio system for spacecraft communications. An antenna as small as $0.1mm^2$ using the K_a band (a frequency of 31GHz not presently used) could communicate with a spacecraft twice the distance away of the Sun from Earth. Power consumption on this microspacecraft would be reduced to about 30W (Voyager uses about 400W), and could be obtained from a much smaller solar panel. Of course, microelectronics has already reduced the size of electronics control, communications, computer, and measurement circuits. Thrust for interplanetary flight could be obtained using the pressure of sunlight (solar sailing), and attitude control from tiny pulsed plasma thrusters. Perhaps alien microspacecraft – maybe from a neutron star – are already exploring the Solar System, too small to be detected. However, small they may be, but their mass must be enormous! □ see SPACECRAFT, SOLAR SAILING, MICROELECTRONICS

microswitch a small mechanically operated switch usu needing only a small force to operate it. A microswitch is often incorporated into machinery (eg robotic devices) so that it is operated when some moving part comes into contact with a button or lever on the switch. Although generally of small size, some microswitches can easily switch several amperes of current. □ see ROBOTICS

microwatt (symbol μW) a unit of power equal to 1 millionth (10^{-6}) of a watt □ see JANSKY

microwave background radiation the remnant of the intensely hot radiation of the Big Bang from which the universe evolved. This radiation was discovered by spacecraft of the Explorer series in the early 1960s. It has a maximum intensity at a wavelength of about 2.5mm and represents a temperature of about 3K. In the vicinity of the Solar System, the microwave background radiation has about the same intensity in all directions. In intergalactic space, away from the radiation of stars, the microwave background radiation is dominant and has an energy density of about 4×10^{-14} joules per cubic metre (Jm^{-3}). □ see BIG BANG, MICROWAVES, EXPLORER 1

microwaves high frequency radio waves having wavelengths between 1 and 300 millimetres. Microwaves are used for straight line radio communications. □ see MICROWAVE BACKGROUND RADIATION, DISH AERIAL, POWERSAT

midcourse guidance – see GUIDANCE, COURSE CORRECTION

mid deck the portion of the Space Shuttle below the flight deck that contains the crew's sleeping quarters, waste management system, personal hygiene station, airlock to the payload bay, avionics, mission control station, payload control station, stowage, and equipment for emergencies. In the mid deck area, up to four seats may be occupied by astronauts during launch and landing, though only *Atlantis* is built for all four, the other three orbiters were built with three seats and a fourth can be added when brackets are fitted. Since four seats may also be occupied in the flight deck, eight crew members can be carried on Shuttle flights (to accommodate extra payload specialists when Spacelab flies in the payload bay). □ see SPACE SHUTTLE, FLIGHT DECK, PAYLOAD BAY, MISSION SPECIALIST, PAYLOAD SPECIALIST, SPACELAB

military carried on or supported by the armed forces ⟨*spy satellites are for ~ use*⟩ – compare CIVILIAN

Milky Way 1 *also* **Galaxy** the star system to which the Sun and the Solar System belong. The Galaxy is a flattened disc-shaped system containing prominent spiral arms which wind out from the central nucleus. It contains of the order of a hundred thousand million stars in a disc which is approximately 30 kiloparsecs in diameter and 4 kiloparsecs thick at the centre. The Sun is situated about 10 kiloparsecs from the nucleus in one of the spiral arms. Its shape is similar to another spiral galaxy, our nearest star system, the Andromeda galaxy. The entire Milky Way galaxy is rotating about an axis through the centre, completing one rotation in about 200 million years. □ see GALAXY **2** the diffuse band of light stretching in a great circle across the celestial sphere, and clearly visible on a cloudless moonless night. In 1750 the English instrument maker Thomas Wright published a remarkable book, *Original Theory or New Hypothesis of the Universe*, in which he suggested that the stars lie in a flat slab, a 'grindstone', of finite thickness but extending in all directions in the plane of the slab. Because the Solar System lies within the slab, we see more stars when we look along the plane of the slab than when we look in any other direction. This is what we see as the Milky Way. □ see GALAXY, ANDROMEDA GALAXY

milliwatt (symbol **mW**) a power equal to 1 thousandth (10^{-3}) of a watt

minor planet – see ASTEROID

mirror a polished or smooth surface (eg of polished metal or silvered glass) that forms images by reflection

mission a flight by a spacecraft to and/or from Earth to space ⟨*a manned ~ to Mars*⟩ □ see MISSION CONTROL, SPACE SHUTTLE, EXPEDITION TO MARS, [1]GIOTTO

Mission – see SPACE SHUTTLE, STS-1 – STS-9, MISSION 41B – MISSION 61C

Mission 41B (STS-10) **date:** 3–11 February 1984; **launch:** Kennedy Space Centre; **spacecraft:** *Challenger*; **crew:** Commander Vance Brand, Pilot Robert Gibson, Mission Specialists Ronald McNair, Robert Stewart, and Bruce McCandless; **orbit:** 306km × 322km inclined at 28.5° to the equator; **activities/findings:** two commercial communications satellites, Palapa-B2 and Westar-6, launched but failed to reach geostationary orbit owing to failure of Payload Assist Module (PAM) booster; extravehicular activity (EVA) by McCandeless and Stewart successfully tested manned manouevring unit; launch and retrieval of SPAS-01 were not carried out owing to failure of Remote Manipulator System; test of Shuttle radar using small balloon in preparation for rendezvous with Solar Max satellite in following April also cancelled as the balloon did not inflate; **problems:** most troublesome flight to that date; **landing:** first ever return to Kennedy Space Centre. □ see SPACE INSURANCE, PAYLOAD ASSIST MODULE, SHUTTLE PALLET SATELLITE, MANNED MANOEUVRING UNIT, REMOTE MANIPULATOR SYSTEM, SOLAR MAX

Mission 41C (STS-11) **date:** 1–13 April 1984; **launch:** Kennedy Space Centre; **spacecraft:** *Challenger*; **crew:** Commander Robert Crippen, Pilot Dick Scobee, Mission Specialists Ox Van Hoften, Terry Hart, and George Nelson; **orbit:** initially 218 × 498km inclined at 28.5° to the equator, then by several manoeuvres to 493km × 498km at the same inclination; **activities/findings:** successful capture and repair of Solar Max satellite launched in 1980; Nelson and Van Hoften set new record for extravehicular activity (EVA) remaining outside for 7hours 18minutes, just 19minutes short of record Apollo 17 lunar surface walk; Nelson used manned manoeuvring unit (MMU) for complex EVA operations with Solar Max; other main task of mission was release of Long Duration Exposure Facility (LDEF) which is a bus-sized 9700kg 12-sided cylinder acting as mounting platform for experiments requiring long-term exposure to space; LDEF was left in space to be recovered on a later flight; other payloads included student experiment involving 3300 honeybees to study how zero-g affects building of honeycombs; **problems:** Nelson failed to dock his MMU with Solar Max and it had to be captured with the Remote Manipulator System; **landing:** Edwards Air Force Base, California, because of bad weather at Kennedy Space Centre. □ see SOLAR MAX, REMOTE MANIPULATOR SYSTEM, MANNED MANOEUVRING UNIT, CRIPPEN

Mission 41D (STS-12) **date:** 30 August – 5 September 1984; **launch:** Kennedy Space Centre; **spacecraft:** *Discovery*; **crew:** Commander Henry Hartsfield, Pilot Michael Coats, Mission Specialists Richard Mullane, Steven Hawley, Michael Coats, and Judith Resnik; **orbit:** 297km × 314km inclined at 28.5° to the equator; **activities/findings:** first flight for the new lighter-weight *Discovery* and for Judith Resnik, the second US woman to go into space; spin-stabilized communications satellite SBS-4 deployed from payload bay, and successful firing of Payload Assist Module (PAM) booster raised it into geostationary orbit over the equator; also disc-shaped Syncom IV-2 released frisbee-style from payload bay, and under its own power achieved geostationary orbit; third satellite, Telstar-3 similar to SBS-4, also released; tests of solar panel array, OAST-1 carried out by Resnik; Walker carried out tests of continuous-flow electrophoresis system (CFES) designed for drug processing; **problems:** *Discovery* flight delayed from previous June when computer failure prevented correct firing sequence of engines; Shuttle toilet got blocked with ice and astronauts had to resort to alternative methods; **landing:** Edwards Air Force Base, California. □ see OAST-1 SOLAR ARRAY, SHUTTLE TOILET, CONTINUOUS-FLOW ELECTROPHORESIS SYSTEM, PAYLOAD ASSIST MODULE

Mission 41G (STS-13) **date:** 5–13 October 1984; **launch:** Kennedy Space Centre; **spacecraft:** *Challenger*; **crew:** Commander Bob Crippen, Pilot Jon McBride, Mission Specialists David Leestma, Sally Ride, and Kathryn Sullivan, Payload Specialists Marc Garneau (Canadian) and Paul Scully-Power (US Navy); **orbit:** 346km × 359km inclined at 57° to equator; **activities/findings:** largest crew to date and first time two woman astronauts together in space, Ride having previously been in orbit (STS-7); Sullivan became first US woman astronaut to make spacewalk with Leestma in simulation of refuelling of satellites; Earth surface observations made using Shuttle Imaging Radar (SIR-B); Measurement of Air Pollution from Satellites (MAPS); successful launch of Earth Radiation Budget Satellite (ERBS); OSTA-3 payload, and a number of Getaway Special (GAS) payloads also carried in the payload bay; **problems:** the SIR-B antenna did not readily deploy; ice clogged water spray boiler system resulting in uncomfortably high cabin temperatures for a time, but toilet operated well; **landing:** Kennedy Space Centre. □ see SULLIVAN, SHUTTLE IMAGING RADAR, OFFICE OF SPACE AND TERRESTRIAL APPLICATIONS, GETAWAY SPECIAL

Mission 51A (STS-14) **date:** 8–16 November 1984; **spacecraft:**

Discovery; **crew:** Commander Fred Hauk, Pilot David Walker, Mission Specialists Anna Fisher, Joseph Allen and Dale Gardner; **orbit:** 294km × 298km at an inclination of 28.5° to the equator, then raised to 353km×370km for satellite rendezvous; **activities/findings:** recovery of two satellites, Palapa-B2 and Westar-6, which were left in unusable orbits when their Payload Assist Module (PAM) boosters failed after launch on Mission 41G; successful launch of communications satellite Anik-D2 from payload bay, but the satellite was put into 'on-orbit storage' until its telephone and television communications channels were needed in 1986; also Syncom IV-1 communications satellite (identical to Syncom IV-2 released on Mission 41D) deployed frisbee-style from payload bay; **landing:** Kennedy Space Centre. □ see SPACE INSURANCE, ANIK

Mission 51B (STS-17) **date:** 29 April – 7 May 1985; **spacecraft:** *Challenger*; **crew:** Commander Robert Overmeyer, Pilot Frederick Gregory, Mission Specialists Don Lind, William Thornton, and Norman Thagard, Payload Specialists Taylor Wang, Lodewijk van den Berg; **orbit:** 346km × 355km inclined at 57° to the equator; **activities/findings:** payload bay carried Spacelab 3 (launched before Spacelab 2 – Mission 51F) which contained 15 experiments managed round the clock by splitting the crew into 'gold' and 'silver' teams; supervision included an 8-hour overlap period when all seven astronauts were on duty; one experiment called ATMOS (Atmospheric Trace Molecule Spectroscopy) used an interferometer to probe Earth's ozone layer for traces of manmade chemicals, such as nitric oxide and chlorine (from aerosol sprays); further flights will continue to monitor ozone layer for chemicals that could degrade this 20km-thick blanket of ozone, and which protects life on Earth from damaging ultraviolet radiation; Spacelab 3 also carried 24 rats and 2 squirrel monkeys which "demonstrated their suitability for research in orbit"!; **problems:** urine hose leaked during a medical experiment, and food and droppings escaped from the rat cages forcing astronauts to wear masks when cleaning out the animal enclosures; **landing:** Edwards Air Force Base. □ see SPACELAB, EARTH'S ATMOSPHERE

Mission 51C (STS-15) **date:** 24–27 January 1985; **spacecraft:** *Discovery*; **crew:** Commander Tom Mattingley, Pilot Loren Shriver, Mission Specialists James Buchli, Ellison Onizuka and Gary Payton (USAF astronaut); **orbit:** not given by NASA; **activities/findings:** very little information given, but a US Department of Defense satellite was launched using the inertial upper stage (IUS) rocket; the satellite was a signals intelligence (Sigint) satellite designed to operate in geostationary orbit over the USSR to listen into Soviet radio, microwave, telephone, teletype, and satellite communications through its two large deployable dish aerials; Ellison Onizuka, a native of Kona, Hawaii, provided fellow astronauts with a snack of macadamia nuts; **problems:** the launch was delayed a day since unusually cold weather in Florida caused ice to form on the spacecraft; **landing:** Kennedy Space Centre. □ see INERTIAL UPPER STAGE, STAR WARS

Mission 51D (STS-16) **date:** 12–19 April 1985; **spacecraft:** *Discovery;* **launch:** Kennedy Space Centre; **crew:** Commander Karel Bobko, Pilot Donald Williams, Mission Specialists Rhea Seddon, David Griggs and Jeffrey Hoffman, Payload Specialists Charles Walker and Jake Garn (US Senator); **orbit:** 315km × 460km inclined at 28.5° to the equator; **activities/findings:** marked 4th anniversary of first Shuttle flight (STS-1); carried $30 toys for astronauts to examine (so NASA said) effects of weightlessness on everyday objects; noises from Garn's stomach recorded to study the causes of space sickness – Garn was sick as he promised to be!; Canadian Anik C-1 satellite deployed successfully, but the electrical systems on-board Syncom IV-3 failed to be activated by a special arming lever which should have moved when the satellite rolled out frisbee-style from the payload bay (Syncom IV-1 and IV-2 successfully launched during Mission 41D and Mission 51A); with an improvised 'flyswatter' attached to the arm of the Remote Manipulator System by spacewalking astronauts Hoffman and Griggs, Rhea Seddon eventually managed to get the robot arm to tug the arming lever over; however, Syncom remained electrically dead, agreed to attempt a rescue mission during Mission 51I; **problems:** on landing, a strong crosswind forced Bobko to apply brakes strongly, and 2m before *Discovery* stopped, a brake seized and a tyre burst; **landing:** Kennedy Space Centre. ☐ see MISSION 51I

Mission 51F (STS-19) **date:** 29 July – 7 August 1985; **spacecraft:** *Challenger*; **crew:** Commander Gordon Fullerton, Pilot Roy Bridges, Mission Specialists Story Musgrave, Karl Henize, and Anthony England, Payload Specialists Loren Acton and John-David Bartoe; **orbit:** 311km × 319km inclined at 49.5° to the equator; **activities/findings:** payload bay carried Spacelab 2 which was new version of Spacelab built by European Space Agency; Spacelab 2 (launched after Spacelab 3 – Mission 51B) carried 13 experiments to study the Sun, galaxies, Earth's upper atmosphere, and cosmic ray nuclei; experiments included two British instruments, viz a small helium-cooled infrared telescope (CHASE – Coronal Helium Abundance Spacelab Experiment) built by Mullard Space Science Laboratory to find the ratio of hydrogen to helium in the outer regions of the Sun, and an X-ray telescope built by Brimingham University for examining the hot gases between galaxies; Pepsi Cola persuaded astronauts to try out a new-style 'spacecan' which provided thirsty astronauts with a squirt of Pepsi from a pressurized can – 'one giant sip for mankind'!; **problems:** during launch, *Challenger*'s middle engine shut down early (a computer false alarm) making final orbit 65km lower than intended; lower orbit presented difficulties for the solar and galactic observations which depended on an atmosphere-free view; **landing:** Edwards Air Force Base. ☐ see SPACELAB, HELIUM

Mission 51G (STS-18) **date:** 17–24 June 1985; **spacecraft:** *Discovery*; **crew:** Commander Daniel Brandenstein, Pilot John Creighton, Mission Specialists Shannon Lucid, Steven Nagel, and John Fabian, Payload Specialists Patrick Baudry (Frenchman) and Sultan Salman

Abdelazize Al-Saud (Saudi Arabian); **orbit:** 356km × 352km inclined at 28.5° to the equator; **activities/findings:** deployment of three communications satellites, Morelos-A (Mexican), Arabsat-1B (Arab League), and Telstar 3D (American Telephone and Telegraph Company), each of which was boosted to geostationary orbit by a Payload Assist Module; release and subsequent capture of Spartan-1, a free-flying astronomy platform which carried X-ray sensors to study the Perseus Cluster of galaxies and the central core of the Milky Way; test firing from Hawaii of Star Wars laser beam which successfully locked onto *Discovery*; Patrick Baudry carried out a series of biomedical experiments, including the French electrocardiograph experiment which supplemented data obtained by French cosmonaut aboard the Soviet Salyut 7 mission in 1982; materials processing and West German Getaway Special experiments carried in the payload bay; **problems:** initial failure of laser beam test due to wrong data in computer program – beam hit Shuttle on wrong side!; main landing gear sank 0.15m into not-so-dried up lake bed; **landing:** Edwards Air Force Base. ☐ see PAYLOAD ASSIST MODULE, CLUSTER 3, SPARTAN, STAR WARS, SALYUT

Mission 51I (STS-20) **date**: 27 August–3 September; **spacecraft**: *Discovery*; **crew**: Commander Jo Engle, Pilot Richard Covey, Mission Specialists James Ox Van Hoften and William Fischer, Payload Specialists John Lounge and George Nelson; **orbit**: 351 × 380km inclined at 28.55° to the equator; **activities/findings**: successful deployment of three communications satellites, Australia's Aussat-1, America's ASC-1 and Leasat-4; capture of faulty Syncom IV-3 satellite (deployed on Mission 51A) using the Remote Manipulator System operated by Fischer, and repair in the payload bay by Van Hoften and Nelson; **problems**: launch aborted at T-minus 5 minutes on 24 August owing to thunderclouds over launch site, and again on 25 August at T-minus 9 minutes when backup computer failed to synchronize with four primary computers; one of the payload bay sunshades temporarily became entangled with Aussat's aerial; **landing**: Edwards Air Force Base ☐ see MISSION 51D

Mission 51J (STS-21) **date**: 3–7 October 1985; **spacecraft**: *Atlantis*; **crew**: all military comprising Karel Bobco (USAF), Ronald Grabe (also USAF), David Hilmers (US Marine Corps), Robert Stewart (US Army), and William Pailes (USAF); **orbit**: 351 × 380km inclined at 28.5° to the equator; **activities/findings**: first flight test of *Atlantis*; military payload including the launch of two satellites and some onboard experiments; **landing**: Edwards Air Force Base.

Mission 51L (STS-25) **date**: 28 January 1986; **spacecraft**: *Challenger*; **crew**: Commander Francis Scobee, Pilot Michael Smith, Mission Specialists Ronald McNair, Ellison Onizuka, and Judith Resnik, Payload Specialists Gregory Jarvis and Mrs. Christa McAuliffe. *Challenger* destroyed and all crew killed when external tank exploded in ball of fire 72 seconds after launch. Major setback to Shuttle programme. Christa McAuliffe, a social studies teacher at Concord

High School, New Hampshire was to give her students lessons from space. *Challenger* was also due to deploy NASA's TDRS satellite and Spartan-Halley retrievable satellite.

Mission 61A (STS-22) **date**: 30 October–6 November 1985; **spacecraft**: *Challenger*; **crew**: Commander Henry Hartsfield, Pilot Steven Nagel, Mission Specialists Guion Bluford, James Buchli, and Bonnie Dunbar, Payload Specialists Reinhard Furrer, Ernst Messerschmid (West Germans), and Wubbo Ockels (Dutch); **orbit**: 322 × 333km inclined at 57° to the equator; **activities/findings**: flight of ESA's Spacelab-D1 devoted to materials processing and life sciences, planned and controlled by the Republic of West Germany; **landing**: Edwards Air Force Base. □ see SPACELAB, MATERIALS PROCESSING, LIFE SCIENCES

Mission 61B (STS-23) **date**: 26 November 1985; **spacecraft**: *Atlantis*; **crew**: Shaw, O'Connor, Cleave, Spring, Ross, Vela (Mexican), Walker; **activities**: deployed US Satcom Ku-2, Mexican Morelos-B, and Australian Aussat-2.

Mission 61C (STS-24) **date**: 18 December 1985; **spacecraft**: *Columbia*; **crew**: Gibson, Bolden, Chang-Diaz, Hawley, G. Nelson, Cenker, W. Nelson; **activities**: deployed Satcom Ku-1.

mission control any central computing and communications facility, esp one to monitor and control the flight of interplanetary spacecraft and manned missions into space through the use of Earth stations and communications satellites □ see JOHNSON SPACE CENTRE, JET PROPULSION LABORATORY, DEEP SPACE NETWORK

mission specialist a member of the crew of the Space Shuttle who coordinates all affairs connected with the Shuttle payload, including the allocation of payload space to various experiments on behalf of the users. He/she is assisted by a payload specialist. □ see SPACE SHUTTLE, RIDE

mission station a position located aft of the pilot's station on the right side of the Space Shuttle flight deck that has the displays and controls for the payloads carried aboard the Space Shuttle □ see CONSOLE, MID DECK, MISSION SPECIALIST

MLP – see MOBILE LAUNCH PLATFORM

MLR – see MONODISPERSE LATEX REACTOR

MMT – see MULTIPLE MIRROR TELESCOPE

MMU [*m*anned *m*anoeuvring *u*nit] – see EXTRAVEHICULAR MOBILITY UNIT

mobile launch platform (abbr **MLP**) the two-storey steel structure fitted with tracks for carrying the Space shuttle from the vehicle assembly building (VAB) to the launch pad, 5.5km away, at the Kennedy Space Centre, Florida. It is 7.6m high, 48.8m long, and 41.4m wide. Unloaded, it has a mass of about 3730 tonnes. When fully laden with a fuelled Space Shuttle, it has a mass of about 5700 tonnes, and can be moved by its four 2050kW diesel engines at 1.6kmh⁻¹. The diesel engines produce electrical power for the motors which drive four independently driven double-tracked crawlers spaced 27.4m apart at each corner of the MLP. □ see VEHICLE ASSEMBLY BUILDING, GANTRY, VANDENBERG AIR FORCE BASE

mock-up a full size replica of a spacecraft, or other device, made of wood, plastics, and other materials so that tests can be carried out before the actual device is built

modular consisting of one or more standard units ⟨*a space station is a ~ system*⟩ □ see SPACE STATION

modulate to change the amplitude, frequency, or phase of a wave in response to changes in some property of a message to be carried by the wave – compare DEMODULATE

module 1 any self-contained unit, esp an electronic unit, which forms part of a larger system ⟨*a communications ~*⟩ **2** a self-contained usu manned unit which is a building block in a larger system ⟨*a command ~*⟩ □ see APOLLO, SPACE SHUTTLE, SPACE STATION

molecule a group of atoms held together by electrical forces. It is the smallest unit of a substance which can exist and have all the chemical properties of that substance. Many different molecules have been discovered in the interstellar dust and gas of the Galaxy. Most of these molecules are organic, ie based on carbon, such as methane (CH_4). □ see ATOM, INTERSTELLAR MEDIUM

Molniya a series of Soviet communications satellites providing long-distance telephone, telegraph, and TV communications via the 'Orbit' receiving stations in the Soviet far north and far east, and central Asia. Electrical power is provided by a 'windmill' of six solar panels. Unlike the US Intelsat communications satellites, the Molniya are tracked daily as they pass overhead in highly elliptical orbits of approximately 500km by 40 000km. □ see GORIZONT

momentum the property of a moving body given by the product of its mass and velocity □ see LINEAR MOMENTUM, ANGULAR MOMENTUM

monitor to look at, listen into, or keep track of an operation or a system ⟨*to ~ radio signals from a spacecraft*⟩

monodisperse latex reactor (abbr **MLR**) an oven in which small uniform spheres are produced. 'Monodisperse' means that the spheres produced are all of the same size; 'latex' refers to the collection of small spheres produced; and 'reactor' refers to the oven in which they are made. The MLR package comprises a 45kg cylindrical housing which contains the oven and a 22kg electronics package. It was designed to fly mid-deck on Space Shuttle missions (eg STS-7), and the small spheres produced are the first space product to be sold commercially. Before launch, the MLR contains 'seed' polystyrene spheres two millionths of a metre (2μm) in diameter. During the flight, the MLR is activated by an astronaut, and a chemical reaction called polymerization makes the spheres grow to a size of about 30μm in the microgravity conditions inside the orbiting Space Shuttle. After the Space Shuttle lands, the MLR is taken to the Marshall Space Flight Centre and the spheres are separated ready to be sold in small phials containing 99.5% water and 0.5% microspheres. The microspheres are so uniform in size that they can be used to calibrate microscopes. They have a number of biomedical uses: modified with radioactive elements or dyes, they can be injected into the bloodstream of animals to follow blood flow, or they can be used to manufacture 'smart bombs' to carry drugs to cancer tumors where they would deliver a lethal dose to the tumour but leave the surrounding tissue unaffected. A multimillion dollar market is expected for these miniature plastic spheres. □ see SPACE SHUTTLE, MATERIALS PROCESSING

Moon the Earth's only natural satellite and visible in the Sun's reflected light. Its diameter is 3476km, and its average distance from the Earth is 384 400km. It completes one orbit of the Earth in 27.3 days and always keeps the same face turned towards us, ie it is in synchronous rotation about the Earth. The Moon has been studied in close-up by orbiting satellites, surface landers and rovers, and manned Apollo missions. It has an almost negligible atmosphere (traces of argon, neon, and helium have been found) so its surface is regularly exposed to the fierce radiation of the Sun and the severe cold of space during the 27-day lunar 'day'. Surface temperatures range from −180C to +110C and it is bombarded by micrometeorites. The origin of the Moon is in doubt: one view is that it was formed from one or more fragments which broke off from the Earth – the fission hypothesis – which is not generally supported; it may be a captured asteroid or other body formed elsewhere – the capture theory; it may have formed as part of a double planet with the Earth – a now popular theory; or it may have formed from the coalescence of a large number of minor planets. □ see APOLLO, LUNAR ORBITER PROBES, LUNIK PROBES, RANGER, SURVEYOR, ZOND, LUNOKHOD, SYNCHRONOUS ROTATION

Moon base *also* **lunar base** a permanently manned settlement on the Moon. No humans have been to the Moon since the Apollo 17 mission in December, 1972. But NASA's long-term goal is a return to the Moon during the first decade of the next century, and the Space Shuttle and the proposed US Space Station should provide the means to do it. The Space Shuttle will enable the US Space Station to be built in low Earth orbit by the 1990s, and by the year 2000 it is expected that the US Space Station will have two reusable vehicles to enable manned missions to be made as far as the Moon. One of these vehicles can be likened to a local taxi and the other to an intercontinental airline. The 'taxi' will be the orbital manoeuvring vehicle (OMV) and the 'airline' the orbital transfer vehicle (OTV). The OMV will be used to service satellites close to the Space Station and for other tasks, and the OTV will ferry payloads to and from geostationary orbit or launch spacecraft to the Moon and planets. Though the Apollo missions revealed no water, no organic matter, and no living organisms on the Moon, the main justification for a Moon base seems to be to mine its minerals. It not only has plenty of oxygen in its rocks, but also valuable minerals such as iron and titanium. However, there may be strong strategic reasons why America and Russia would want to establish a Moon base. □ see SPACE SHUTTLE, US SPACE STATION, ORBITAL MANOEUVRING VEHICLE, ORBITAL TRANSFER VEHICLE, LUNAR SLIDE LANDER, EHRICKE

moonquakes tremors inside the Moon detected by instruments (seismometers) set up during the Apollo landings. Moonquakes are much less intense than earthquakes and have a number of causes, such as the stress set up in the Moon by the gravitational attraction between the Earth and the Moon. The delay in the arrival of tremors occuring deep inside the Moon shows that the Moon has a 60km thick crust consisting of silicates of iron, calcium, and magnesium. Underneath the crust is a 1200km thick mantle, and a core of partially molten rock. □ see MOON, APOLLO, LUNIK PROBES, APOLLO LUNAR SURFACE EXPERIMENTS PACKAGE

Moore's Law the number of components that can be integrated onto a single silicon chip will double every year. This prediction was made by Gordon Moore, founder of the Intel Corporation, in the early 1960s when integrated circuits were made from a few tens of components (known as small-scale integration). To the present day Moore's prediction has been found to hold good. By the late 1960s, the annual doubling effect had led to several hundred components being integrated onto a silicon chip (known as medium-scale integration). During the 1970s, component counts per chip had reached several hundred thousand (known as large-scale integration). In the 1980s, microprocessor and memory chips now have close to a million components on a single chip, little larger than the early chips, and this is called very large-scale integration. □ see SILICON CHIP, FIFTH GENERATION COMPUTER

morning star the planet Venus (or sometimes Mercury) seen shining brightly in the eastern sky just before sunrise □ see VENUS

Mount Stromlo Observatory the site, near Canberra, of the Australian National University's 1.9m reflecting telescope. Mount Stromlo is at an altitude of 768m and has other telescopes on it.

Mount Wilson Observatory – see HALE OBSERVATORIES

MSS [*m*ultispectral *s*canner] – see LANDSAT

Mullard Radio Astronomy Observatory the site of Cambridge University's Five Kilometre radio telescope, and the One Mile radio telescope. The first comprises eight paraboloid radio dishes and has been in operation since 1971, and the second has one movable and two fixed dishes. The two telescopes are used as interferometers for the detailed study of cosmic radio sources. □ see RADIO ASTRONOMY, APERTURE SYNTHESIS, VLA

Multiple Mirror Telescope (abbr **MMT**) a revolutionary type of reflecting telescope comprising six identical mirrors, each 1.8m in diameter, arranged in a hexagonal array to provide the light-gathering power of a single 4.5m diameter mirror. It is the third largest telescope in the world but it costs considerably less than a single 4.5m mirror. The MMT is sited at the Smithsonian Astrophysical Observatory on Mount Hopkins near Tucson, Arizona, at an altitude of 2380m, and it can operate at visible and infrared wavelengths. The light from the six mirrors is reflected from a secondary mirror on the main axis of the mirrors. Laser beams reflected from the primary mirrors are monitored by detectors linked to a computer. The computer provides control signals which position each mirror so that it contributes light to a single image viewed by an eyepiece, photographic plate, or other sensing device. The world's largest optical telescope will be the Keck 10m Telescope which will cost $85M and be in operation on Mauna Kea, Hawaii, in 1992. Its mirror will be built from 36 separate mirrors each 1.8m across and only 75mm thick. These individual mirrors will be accurately positioned up to 300 times per second by a computer-controlled system. □ see REFLECTING TELESCOPE, OPTICAL ASTRONOMY, APERTURE SYNTHESIS, ALTAZIMUTH MOUNTING

multiplexing a method of making a single communications channel carry several messages simultaneously – compare DEMULTIPLEXING

multispectral scanner (abbr **MSS**) – see LANDSAT

multistage rocket a launch vehicle comprising two or more rockets mounted one on top of the other. Thus in a 3-stage rocket (eg Saturn 5) the third stage carries the payload (eg a spacecraft or satellite), and the rocket engine and its propellant. The third stage is the payload of the second stage which, in turn, is part of the payload of the first stage. At launch, the more powerful rocket engines of the first stage lift the vehicle off the launch pad and provide it with its initial velocity. It is discarded at relatively low altitudes when its fuel runs out. Then the second stage ignites and carries the payload to higher altitudes before it, too, is discarded. The third stage provides the payload with its required orbital velocity, or the escape velocity from the Earth. Thus multistage rockets enable more massive payloads to be launched than with single stage rockets since spent engines and fuel tanks are discarded at low altitudes. The 111m high Saturn 5 is the largest multistage rocket to be built so far. □ see LAUNCH VEHICLE, SATURN 5

Mutual Assured Destruction (abbr **MAD**) a belief that nuclear war cannot be fought, won, or even survived. Thus if either the US or the USSR initiates an attack on the other, enough nuclear weapons will remain in the armoury of the aggressor to destroy the other side. MAD rests on the fear of a counterstrike and provides defence by the threat of offence. □ see INTERCONTINENTAL BALLISTIC MISSILE, STAR WARS

MX missile also **'Peacekeeper'** a new US 3-stage solid-fuelled intercontinental ballistic missile carrying 10 independently-targetable reentry warheads. The development of the MX missile has been dogged by political controversy, but barring further political objections it should be fully operational in the late 1980s or early 1990s. The Soviet version of the MX missile is the SS-18. □ see INTERCONTINENTAL BALLISTIC MISSILE, MUTUAL ASSURED DESTRUCTION, STAR WARS

N

n the symbol for the prefix nano meaning one thousand millionth of (10^{-9}) □ see NANOMETRE

nadir the point on the celestial sphere that is directly opposite the zenith and vertically downwards from an observer. Both the zenith and the nadir lie on the observer's celestial meridian. – compare ZENITH □ see CELESTIAL COORDINATES

Nancy the site of a large 305m diameter radio telescope near Nancy, France, which is part of the Observatory of Paris □ see RADIO ASTRONOMY, APERTURE, DISH

nanometre (symbol **nm**) a distance equal to 1 thousand millionth (10^{-9}) of a metre. The nanometre is often used to express the wavelength of electromagnetic radiation (eg the wavelength of red light is about 700nm). □ see ANGSTROM

nanosecond (symbol **ns**) a time interval equal to 1 thousand millionth (10^{-9}) of a second. The time interval for sunlight to travel a distance of 1 metre is about 3ns.

NASA [*N*ational *A*eronautics and *S*pace *A*dministration] an American agency founded in 1958 to develop nonmilitary uses of space. It successfully developed the Mercury and Gemini projects of the 1960s which put the first American into space. This achievement led to the Apollo landings on the Moon, the manned space station Skylab, and many artificial satellites and interplanetary spacecraft. NASA is now cooperating with the European Space Agency in several projects including the US Space Station and the Hubble Space Telescope. □ see EUROPEAN SPACE AGENCY, BRITISH AEROSPACE, NATIONAL SPACE DEVELOPMENT AGENCY, INSTITUTE OF SPACE AND ASTRONAUTICAL SCIENCE, HUBBLE SPACE TELESCOPE, DEEP SPACE NETWORK

NASDA – see NATIONAL SPACE DEVELOPMENT AGENCY

National Aeronautics and Space Administration – see NASA

National Oceanic and Atmospheric Administration (abbr **NOAA**) a US organization that replaced the Environmental Science Services Administration (ESSA) of the US Department of Commerce in 1970. ESSA was responsible for launching the first generation TIROS (Television Infrared Orbital Satellite) polar-orbiting series of weather satellites, beginning with TIROS-1 in 1960. An advanced (second generation) TIROS was launched on 1 April 1966 and provided more data than its predecessor. The TIROS satellites were updated with the third generation ITOS (Improved TIROS Operational System); ITOS-1 was launched on 23 January 1970 and could obtain data day and night in real time as well as store data for later transmission. When NOAA took over from ESSA, the ITOS satellites were replaced by the fourth generation NOAA series beginning with

NOAA-1 launched on 11 December 1970. The first of the TIROS-N series of NOAA satellites (NOAA-6) was launched into polar orbit on 27 June 1979, followed by NOAA-7 on 23 June 1981, NOAA-8 in 1983 and NOAA-9 in 1984. □ see NOAA-9, WEATHER SATELLITE, REMOTE SENSING, LANDSAT

National Space Development Agency (abbr **NASDA**) a Japanese organization which funds space projects. Suborbital experiments on materials processing are at present carried out using the TT-500-A two-stage rocket that provides about 7 minutes of freefall at one-thousandth g, and it is intending to put a materials processing package aboard the Spacelab J mission scheduled for 1988, accompanied by a Japanese payload specialist. □ see INSTITUTE OF SPACE AND ASTRONAUTICAL SCIENCE, MATERIALS PROCESSING, SPACELAB

navigation any method by which the position, course, and distance travelled by a spacecraft is measured during a journey so that course corrections can be made if necessary. Navigation usu implies a person is aboard the spacecraft, or that the spacecraft is capable of being controlled remotely (by a navigator); if navigation is automatic, it is called guidance. □ see GUIDANCE

navigator a person who remotely controls a spacecraft from an Earth station ⟨~s *at the Jet Propulsion Laboratory*⟩ □ see GUIDANCE, JET PROPULSION LABORATORY, JOHNSON SPACE CENTRE

Navstar – see GLOBAL POSITIONING SYSTEM

nebula (plural **nebulae/nebulas**) a cloud of interstellar gas and dust that may appear as a luminous patch (an emission nebula), or as a dark hole against a brighter background (a dark nebula). The dust in an emission nebula (eg the Great Nebula in Orion) may be caused by ionization of the gas and dust by ultraviolet radiation coming from nearby hot stars or by reflection of starlight. If there are no nearby stars, a nebula (eg the Horsehead nebula, also in Orion) may obscure light from stars behind it and appear as a dark untidy blob. □ see INTERSTELLAR MEDIUM, COSMIC DUST

negative 1 charged with an excess of electrons **2** *of a part of a circuit* having a low electrical potential ⟨*the ~ terminal of a battery*⟩ **3** *of a number* numerically smaller than zero – compare POSITIVE

Neptune the eighth planet of the Solar System which has a diameter of about 4500km, orbits the Sun at about 4500 million kilometres, and has two moons, Triton and Nereid. Triton is orbiting so close to Neptune that it appears to be in danger of being torn apart by Neptune's gravity. Methane is the main atmospheric gas and, as with all of these outer (Jovian) planets, Neptune appears to be composed mainly of hydrogen and helium which change from gas to liquid to solid at increasing depths in the atmosphere, and surround an iron-silicate core. Earth-based observations, using specially sensitive charge-coupled devices at the focal plane of telescopes, are revealing more about this distant planet of the Solar System. But the best views of Neptune this century should be obtained by Voyager 2 spacecraft. After a close encounter with Uranus in 1986, Voyager 2 will fly within

7500km of Neptune's north pole in August 1989. Five hours later it should sweep within 44 000km of Triton. Observations should indicate whether Neptune, like Jupiter, Saturn, and Uranus, has rings round it. At this distance, data from Voyager 2 will take about 4 hours to reach Earth. □ see JUPITER, SATURN, URANUS, VOYAGER, CHARGE-COUPLED DEVICE

neutrino (symbol γ) a massless electrically neutral elementary particle which, according to the special theory of relativity, travels at the speed of light. At least two types of neutrino are known: electron-types and those produced in association with muons (mu mesons). The neutrino has an antiparticle, the antineutrino ($\bar{\gamma}$). Neutrinos are a by-product of the energy-producing nuclear reactions inside stars whereby hydrogen is converted into helium. Since these neutrinos react very weakly with matter, they pass out of the Sun and through the Earth, too, so they are difficult to detect. However, an idea of how many neutrinos are emitted would help astrophysicists to test the theory of the processes going on inside stars. Experiments have been carried out to detect neutrinos from the Sun using large quantities of perchloroethylene (dry-cleaning fluid) inside tanks deep underground. A very small number of neutrinos should react with the chlorine atoms to produce radioactive argon gas. But much less argon is produced than required by theory. Perhaps the conditions inside the sun are not as we think, and further experiments to detect lower energy neutrinos are underway. Neutrinos are certainly important particles. It is thought that the titanic eruption of a supernova, when a shell of element-rich gas is ejected from the exploding star, is the result of the pressure exerted by neutrinos generated inside the superdense core of the star just prior to the explosion. We owe our existence to the fact that the neutrinos have only a weak interaction with atoms. If the weak interaction were much weaker, they would not be able to exert so much pressure on the outer atmosphere of the star, so there would be no supernova and no distribution of the very elements (eg carbon and iron) on which life on Earth is based. And if the weak interaction were much stronger, the neutrinos would be trapped inside the core of the star. This is another example of the apparent 'fine tuning' of the physical processes in the universe which renders life as we know it possible. □ see ELEMENTARY PARTICLES, ANTIPARTICLE, MESON, NEUTRON, QUARK, WEAK INTERACTION, BIG BANG, GENERAL UNIFIED THEORY, ANTHROPIC PRINCIPLE, SUPERNOVA

neutron (symbol **n**) the uncharged particle found with protons in ordinary atomic nuclei. Its mass is almost equal to that of the proton (about 1.67×10^{-27}kg). Inside the nucleus the neutron is stable but, unlike the proton, it cannot exist by itself and rapidly decays into an electron, a proton, and an antineutrino ($\bar{\gamma}$). The neutron has an antiparticle, the antineutron (ñ), but both types of neutron are now regarded as being made up of the combination of more fundamental particles called quarks. For example, the neutron is regarded as being made up of one 'up' quark and two 'down' quarks. □ see ELEMEN-

TARY PARTICLES, QUARK, PROTON, HADRON, DEUTERIUM, ANTIMATTER

neutron star a star that has contracted under its own gravity to such a degree that most of its material has been compressed into neutrons. Neutron stars are probably formed when a star becomes a supernova. The body which remains has a diameter of between 20 and 30km, a density of 10^{17}kgm^{-3}, and produces an extremely powerful magnetic field. □ see NEUTRON, SUPERNOVA, PULSAR, BLACK HOLE

newton (symbol **N**) the SI unit of force defined as the force required to accelerate a mass of 1 kilogram by 1 metre per second per second ($1N = 1kgms^{-2}$). The maximum force (thrust) generated by the Space Shuttle main engines is about 28 million newtons. □ see SI UNITS, JOULE, WATT, THRUST

Newtonian telescope – see REFLECTING TELESCOPE

Newton, Sir Isaac (1642–1727) English scientist and mathematician born in the year Galileo died. In 1665, Newton graduated from Cambridge University without particular distinction. When London was in the grip of the plague, he returned to the safety of his mother's farm in Lincolnshire, and there (so the story goes) watched an apple fall to the ground and began to wonder whether the same force that attracted the apple also held the Moon in its grip, ie that the Moon, too, is falling towards the Earth (though never getting closer since it is in orbit about it). At that point he put aside his theoretical considerations about gravity in favour of optics; but in 1684 he was able to declare to Hooke (a lifelong academic adversary) and Wren that the force of attraction between two bodies weakened as the square of the distance between them increased. He showed that this 'inverse square law' would make the planets revolve about the Sun in elliptical orbits – the ellipses that Kepler had shown earlier from his observations on the orbits of Mars's and Jupiter's satellites. In 1687 Newton's book *Philosophiae Naturalis Principia Mathematica* was published in Latin (in English in 1725). This is considered to be the greatest work ever written. In it he expounded three laws of motion, and the universal law of gravitation. Newton's theories have stood the test of time and are the cornerstone for calculations of spacecraft trajectories. They provide an overall scheme for understanding the mechanics of the universe (though later modified by Einstein for high relative speeds and strong gravitational fields). Newton's experiments with light and colour led him to conclude that light was a stream of particles, not waves as others supposed. In this he was not entirely correct for light can behave as either particles (quanta) or waves. He thought that it was not possible to get rid of the false colours surrounding the images produced by refracting telescopes and designed a reflecting telescope (the Newtonian design) which did not show this (chromatic) aberration. Musing on his intellectual achievements, Newton said: "I do not know what I may appear to the world, but to myself I seem to have been only like a boy playing on the seashore, and diverting myself now and then finding a smoother pebble or a prettier shell than ordi-

nary, whilst the great ocean of truth lay all undiscovered before me".
□ see COPERNICUS, [2]KEPLER, [2]GALILEO, NEWTON'S LAWS OF MOTION, NEWTON'S LAW OF GRAVITATION, REFLECTING TELESCOPE, QUANTUM THEORY, EINSTEIN

Newton's Law of Gravitation the force of attraction, F, between two bodies is proportional to the product of their masses, m_1 and m_2, and inversely proportional to the square of their distance apart (d), ie $F = Gm_1m_2/d^2$. The constant of proportionality, G, is known as the Gravitational Constant. This law is the basis of all calculations about the orbiting planets in the Solar System; it was used successfully to predict the existence of planets (eg Pluto) invisible to the naked eye, and it is the basis for calculating the trajectories of spacecraft sent to explore the Solar System. Though quite adequate for all these calculations, Newton's Law of (Universal) Gravitation was challenged by Einstein's General Theory of Relativity which was published in 1916. □ see GRAVITATIONAL CONSTANT, RELATIVITY, KEPLER'S LAWS

Newton's Laws of Motion the three basic laws of motion which Newton published in *Principia* in 1687, which are the foundation of celestial mechanics and govern the operation of rockets. The laws are: **1** a body at rest remains at rest and a body in motion remains in motion at constant velocity as long as outside forces are not involved (ie a body has inertia); **2** the rate of change of linear momentum is proportional to the applied force; **3** to every action there is an equal and opposite reaction. □ see NEWTON, NEWTON'S LAW OF GRAVITA-TION, MOMENTUM, INERTIA

NOAA – see NATIONAL OCEANOGRAPHIC AND ATMOSPHERIC ADMINISTRATION

NOAA-9 a US weather satellite launched from the Vandenberg Air base, California, on 1 December 1984 and placed in a 870km high polar orbit. From this vantage point, NOAA-9 transmits weather pictures and other meteorological data to users round the world to help in local weather forecasting (eg the development and path of hurricanes), and to help local agriculture, fishing, and forestry industries. Sophisticated instruments measure cloud and snow cover, the temperature of the Earth's surface and atmosphere at different altitudes, and the total solar energy arriving at the Earth. As part of its mission, NOAA-9 also receives, processes, and retransmits data from free-floating balloons and remote weather stations round the world. NOOA-9 also carries special instruments which are part of an international life-saving programme which makes use of satellites to rescue people from crashed aircraft and ships in distress. This project is used by Canada, France, the Soviet Union, and the United States and, to date, has saved over 400 lives. NOAA-8, which was launched in 1983, performed satisfactorily until June 1984 when difficulties arose with its master clock. This affected its attitude control system and it began tumbling in orbit unable to relay its signals effectively to ground stations. Unexpectedly, the clock began to function again in May 1985

and the satellite was quickly stabilized and brought back into service.
□ see NATIONAL OCEANIC AND ATMOSPHERIC ADMINISTRATION,
WEATHER SATELLITE, TUMBLING, ATTITUDE CONTROL, RADIOMETER

NORAD – see NORTH AMERICAN AEROSPACE DEFENSE COMMAND

North American Aerospace Defense Command (abbr
NORAD) a 1.8 hectare complex of computers and communications
systems deep inside Cheyenne Mountain, Colorado, USA, with which
the United States Air Force monitors military activities from space.
The computers speedily process information from a worldwide network of radar installations and satellites that keep watch on possibly
hostile activities in space. NORAD has hot-line links with the Pen-
tagon and the White House to help coordinate any retaliation
necessary to protect its aerospace resources from attack. The 900 people working in this underground refuge are protected from nuclear
attack by 25 tonne 1m-thick doors. The buildings rest on steel springs
which act as shock absorbers. Possibly harmful radioactive, chemical,
and biological substances are filtered from incoming air, and internal
communications are protected against possible damage from the
electromagnetic pulse (EMP) produced in a nuclear attack. □ see
ELECTROMAGNETIC PULSE, STAR WARS

northbound node – see ASCENDING NODE

nose cone the cone-shaped shield covering the front end of a
rocket to give it a streamlined shape and to protect it against overheating when moving through planetary atmospheres □ see ABLATING
MATERIAL, STREAMLINE FLOW

nova (plural **novae**) a star that suddenly and unpredictably emits
much more light than normal. Ten to fifteen novae occur in the
Galaxy each year, and they are classified as 'fast' or 'slow' depending
on the time taken for their brightness to decline. The brightness of fast
novae (eg Nova Cygni 1975) declines over a period of months, whilst
slow novae (eg Nova Herculis 1934) continue to emit copious radiation over a period of years. Novae are thought to be members of close
binary systems: a white dwarf (the nova itself) and a close cool red
star. Material from the red star streams onto the white dwarf where it
forms a disc-shaped orbit, and somehow precipitates a sudden nova
explosion – probably the result of the runaway burning of hydrogen.
In any case, spectroscopic observation shows that a nova ejects an
expanding shell of hydrogen-rich gas accompanied by a great deal of
heat and light. A nova is named after the constellation in which it
occurred and the year when it was observed. □ see STAR, BINARY
STAR, SUPERNOVA, HYDROGEN BURNING

nozzle 1 the carefully shaped aft section of a rocket chamber that
controls the expansion of the exhaust gases so that the heat energy
produced in the combustion chamber is efficiently converted into kinetic energy **2** the tapered part of a pump or turbine which directs fluid
into or leads it away from the impeller or turbine wheel □ see
THRUST, ROCKET ENGINE, SOLID ROCKET BOOSTER

nuclear relating to the nucleus of an atom, atomic power, or the atomic bomb □ see NUCLEUS

nuclear energy the energy liberated by a nuclear reaction (fission or fusion) or by radioactive decay □ see RADIOISOTOPE THERMOELECTRIC GENERATOR, NUCLEAR FUSION, NUCLEAR FISSION

nuclear fission the splitting up of an atomic nucleus with the emission of radiation (eg neutrons and gamma rays) and the release of energy. Radioactivity is caused by spontaneous fission of uranium, plutonium, thorium, etc. Nuclear reactors are designed to control the release of energy from fission of uranium and plutonium, and are being considered as the power source for future rockets – compare NUCLEAR FUSION

nuclear fuel material (eg plutonium) that can be fissioned to produce energy in a nuclear reactor

nuclear fusion the joining together of two light nuclei (eg hydrogen or deuterium) at high temperatures to form a heavier nucleus with the production of immense amounts of energy. The energy of all the stars, including the Sun, is produced by fusion in what is an example of a thermonuclear reaction. Iron and lighter nuclei produce energy in fusion reactions, but heavier nuclei must be split to produce energy – compare NUCLEAR FISSION □ see STAR, HYDROGEN BURNING

nuclear radiation the emission of particles (eg neutrons) and/or radiation from an atomic nucleus as a result of nuclear fission or nuclear fusion □ see RADIOACTIVITY

nuclear reaction a reaction that involves changes to the nucleus of an atom. These changes may be the result of nuclear fission, nuclear fusion, neutron capture, or radioactive decay

nuclear reactor the part of a nuclear power station or nuclear engine that produces heat from the fission of atoms of uranium or plutonium □ see NUCLEAR FISSION

nuclear weapons a collective term for weapons that obtain their destructive power from the detonation of atomic bombs or hydrogen bombs □ see ATOMIC BOMB, NUCLEAR FUSION, NUCLEAR FISSION, THERMONUCLEAR, CRUISE MISSILE, STAR WARS

nucleon the general name for the proton or the neutron in the nucleus of an atom □ see NUCLEUS

nucleonics the physics and technology of nuclear energy and its applications

nucleosynthesis the creation of the elements (eg iron) by nuclear reactions in stars. Nucleosynthesis began about a hundred seconds after the Big Bang, the event that began the universe some 20 billion (20×10^9) years ago, when deuterium and helium formed from the very hot 'soup' of elementary particles which is all the universe consisted of at that time. Once the universe had expanded and cooled sufficiently, and galaxies of stars had formed, heavier elements were created in the stars themselves. This is a continuing process. As a star evolves, a contracting superdense core of helium is produced from the conversion of hydrogen nuclei into helium nuclei. Eventually, the high temperature and pressure inside the star produces carbon from helium. If the star is more than twice the Sun's mass, a sequence of nuclear reactions then produces heavier elements such as oxygen, magnesium, potassium, and iron. The fusing of nuclei stops with the production of iron, nickel, and cobalt which have a mass number of 56. If there is a supply of free neutrons in a star, the synthesis of nuclei heavier than iron occurs by neutron capture. In this way, elements such as californium (mass number 254) and uranium (mass number 238) can be produced. The heaviest elements require intense neutron fluxes and these are produced in supernovae which catastrophically spread the elements formed in earlier life throughout the surrounding space. This material contributes to the dust and gas out of which planets condense and revolve round a newly formed star. The life on these planets depends on elements (eg oxygen, magnesium, and iron) created in the hearts of other stars and in the explosions of supernovae millions of years ago. □ see BIG BANG, COSMIC ABUNDANCE, STAR, SUPERNOVA, HYDROGEN, HELIUM, DEUTERIUM, NUCLEAR FUSION, INTERSTELLAR MEDIUM CARBON, ANTHROPIC PRINCIPLE

nucleus (plural **nuclei**) **1** the central and relatively small part of a comet. It is often described as a 'dirty snowball' since it is made up of about 75% ice (a mixture of water ice, carbon dioxide ice, ammonia ice, and other ices) and about 25% dust. No one has seen the nucleus of a comet which is why Halley's comet is now under such close scrutiny by visiting spacecraft. This comet may have a nucleus about 35km in diameter which the solar wind reduces by about 1m every time it reaches perihelion. The material which evaporates forms the coma and the comet's tail. □ see COMET, HALLEY'S COMET **2** the central and relatively small part of an atom which is made up of protons and neutrons bound together by powerful nuclear forces. The number of protons in the nucleus of an atom is called its atomic number, and the total number of protons and neutrons is its mass number. In a neutral atom, the positive charge carried by the protons is exactly balanced by the negative charge carried by the electrons surrounding the nucleus. □ see ATOM, ISOTOPE, NUCLEAR FISSION, NUCLEAR FUSION

OAO – see ORBITING ASTRONOMICAL OBSERVATORY

OAST-1 Solar Array a large, ultralight, 31m-long wing-shaped solar array which astronaut Judith Resnik tested while flying in the Space Shuttle *Discovery* during Mission 41D on 1 September 1984. OAST-1 was folded up accordion-like in a flat box only 0.18m high, and its 84 plastic panels were deployed to test the stability of the array when shaken by thruster firings. On later missions of the Space Shuttle, OAST will provide extra electrical power for Shuttle experiments and housekeeping functions. □ see SOLAR CELL, SPACE SHUTTLE, MISSION 41D

objective *also* **object lens** that part of a telescope or binoculars that faces the object. The objective produces the image viewed by the eyelens. □ see EYEPIECE, APERTURE, TELESCOPE

object lens – see OBJECTIVE

oblate flattened at the poles ⟨*Jupiter has an ~ shape*⟩

observatory a building, institution, or spacecraft that is used for astronomical observation. Though Earth-based observatories were once equipped only with optical telescopes, nowadays they usually include infrared and/or radio telescopes. These telescopes observe and record what they 'see' with a range of instruments, including photographic film, spectrometers, photometers, and, of course, the human eye. Early civilizations (in India, Egypt, and Britain, for example) built observatories in easily accessible places. But modern observatories are sited in remote dry mountainous regions where the atmosphere is so thin that starlight is not affected by clouds or water vapour. The site is usually chosen well away from troublesome lights of cities. Even radio telescopes need to be built far from man made electrical interference. The best place for observatories is in space, well above the disturbing effects of the Earth's atmosphere and man made light and electrical signals. □ see HALE OBSERVATORIES, GAMMA-RAY OBSERVATORY, HUBBLE SPACE TELESCOPE, SEEING

occultation the eclipsing of one celestial body by another, usu larger, body ⟨*the ~ of Venus by the Moon*⟩ □ see ECLIPSE

offgassing *also* **outgassing** the release of gas from a material, esp material which is exposed to a vacuum

Office of Space and Terrestrial Applications (abbr **OSTA**) an organization which flies experiment packages in the payload bay of the Space Shuttle. OSTA-1 was the first payload flown on STS-2 in November 1982 and carried instruments for measuring air pollution, and radar for high-resolution imaging of the Earth's surface; in June 1983, OSTA-2 carried materials processing experiments; OSTA-3 went up on Mission 41G in October 1984 and continued with the work of OSTA-1. The OSTA payloads will continue to be carried on Shuttle flights. □ see MAUS, SHUTTLE IMAGING RADAR, SHUTTLE PALLET SERVICE, SPACE SHUTTLE

Office of Technology Assessment (abbr **OTA**) a (supposedly) non-partisan department of the US Congress which makes objective appraisals of government spending on technological projects. Its most recent study of proposals for the US Space Station (which, in January 1984, President Reagan had directed NASA to develop) said that the United States does not have clearly defined long-range goals for the use of outer space. Furthermore, the OTA says that there should be a national (US) debate on the utilization of space, and it suggests that a less ambitious orbiting space station could be used to launch voyages to the Moon, carry out manufacture of new products, and service satellites. Such a 'Salyut type' space station could be assembled using existing hardware, eg Spacelab modules and a service module based on the European Retrievable Carrier (Eureca) satellite, and Leasecraft, a commercial orbiter being developed by Fairchild. As expected, NASA officials were less than pleased with OTA's comparison with the Soviet Salyut space station and with the suggestion that they have failed to define long-term goals. That successive space projects appear not to build on accumulated experience and hardware to keep down costs has been evident in the immediate post-Apollo period; perhaps the same mistake is being made with regard to the US Space Station in that insufficient use is being made of current Shuttle-related hardware in its design. □ see US SPACE STATION, EURECA, MOON BASE

Olympus – see DIRECT BROADCAST SATELLITE, BRITISH AEROSPACE

OMS – see ORBITAL MANOEUVRING SYSTEM

OMV – see ORBITAL MANOEUVRING VEHICLE

Oort cloud a reservoir of perhaps a thousand million comets which is thought to exist far beyond the orbit of Pluto. From time to time, a few icy objects 'leak' from this vast congregation into the Solar System and become visible comets in the night sky. They are nudged from their orbits by the gravitational effects of a 'Planet X' which is believed to orbit beyond Pluto, or they may be perturbed by passage through clouds of interstellar dust. Perhaps this dust also 'tops up' the Oort cloud which may have started as the left-over debris from the time the Solar System was formed. □ see COMET, INTERSTELLAR MEDIUM, HALLEY'S COMET, PLANET X

open cluster – see CLUSTER 2

open-loop *of a control system* having no feedback for comparison and corrective adjustment and so giving a preset output – compare CLOSED-LOOP

opposition the alignment of a superior planet (eg Jupiter) with the Sun so that they appear in opposite parts of the sky, ie the elongation is 180°. When the superior planets are at opposition it is the best time to study them since they are closest to the Earth and observable throughout the night. □ see ELONGATION, CONJUNCTION, ECLIPSE

optical of or using light ⟨~ *astronomy*⟩

optical astronomy the use of telescopes to study planets of the Solar System, stars, galaxies, and other celestial objects, from the light they emit. Observations from Earth-based observatories are limited by absorption and distortion of light in the atmosphere, although the use of electronic cameras and computers can vastly improve the quality of the images obtained. Earth-orbiting observatories provide much better 'seeing' and, coupled with the latest electronic sensing, control, and communications systems, provide more accurate observations over a wider electromagnetic spectrum than studies from the Earth's surface. □ see ELECTROMAGNETIC SPECTRUM, WINDOW, HUBBLE SPACE TELESCOPE, RADIO ASTRONOMY, SEEING, TELESCOPE

optical axis the imaginary line that passes through the optical centre of a lens or mirror, and on which the image of a star or some other distant object, is produced □ see TELESCOPE

optical communications *also* **fibre-optics communications** the use of long thin glass fibres for sending messages by the light of a laser. Optical communications is fast becoming an attractive alternative to communication by wire for several reasons: strong electric and magnetic fields generated by lightning and electrical machinery do not interfere with the message carried by the light beam; there is no interference (crosstalk) between signals in neighbouring fibres; broken fibres are not a fire hazard since the escaping light is harmless; glass and plastics fibres are cheaper and lighter than copper wires; and, most importantly, by using laser light switched on and off rapidly in response to the message being transmitted, a single fibre can carry considerably more information than a copper wire. These reasons make optical communications very attractive in spacecraft and space stations. For example, the immunity of the optical signals to electrical fields means that a spacecraft's internal communications system would be protected from the strong electrical behaviour of Jupiter's atmosphere. For long distance communication, the main problem at the moment is the production of transparent materials of sufficiently high purity to reduce the number of boosters required to regenerate the weakening signals. Nevertheless, the first submarine cable has been laid between Britain and Holland, and a trans-Atlantic cable will follow. □ see BANDWIDTH, ELECTROMAGNETIC PULSE

optical fibre a thin glass or plastic thread through which light can travel without escaping from its sides. Coloured light spilling out from the ends of a bundle of plastic fibres is often used in decorative lamps. Doctors use an optical fibre to illuminate and examine internal organs of the body without causing too much discomfort. But, without doubt, optical fibres are destined to have their most far-reaching effects in communications systems. □ see OPTICAL COMMUNICATIONS

optical pulsar a pulsar (eg the pulsar in the Crab nebula) that produces flashes of visible light □ see PULSAR

orbit the path followed by a natural or artificial body around a centre of gravity ⟨*Halley's comet travels in an ~ round the Sun*⟩ □ see ARTIFICIAL SATELLITE, KEPLER'S LAWS, TRAJECTORY, ORBITAL ELEMENTS

orbital elements six pieces of data that uniquely specify the position and path of a celestial body in its orbit. The shape and size of an elliptical orbit is described by its eccentricity and semimajor axis. The orientation of the orbit in space is described by the inclination of its orbital plane to (usu) the ecliptic and by the longitude of its ascending node. The orientation of the orbit in the orbital plane is usually specified by the angular distance between the periapsis and the ascending node. The sixth orbital element, the time of periapsis, determines the position of the body in its orbit. □ see APSIDES

Orbital Manoeuvring System (abbr **OMS**) two rocket engines located on pods on each side of the aft fuselage of the Space Shuttle to provide (small) changes to the speed of the spacecraft whilst in orbit. These rockets are able to provide a change of speed of about $300ms^{-1}$. A small proportion of this change is used to insert the Space Shuttle into orbit at the end of its ascent. The OMS is also used for orbital transfer, rendezvous with satellites, and deorbit operations. The OMS uses a hypergolic propellant, such as nitrogen tetroxide and monomethyl hydrazine in the normal oxidizer-to-fuel mixture of 1.65. □ see HYPERGOLIC FUEL, SPACE SHUTTLE

orbital manoeuvring vehicle (abbr **OMV**) *also* **space tug** an unmanned remote-controlled rocket-powered vehicle designed to move satellites around in orbit. The OMV is being considered by NASA for launch by the Space Shuttle in about 1990. It would have a size of about 5m by 1m and be sent out from the payload bay of the Space Shuttle to retrieve satellites that have either failed or need refuelling; it might also be used to boost failing satellites into fresh orbits. The OMV is justified since the Space Shuttle uses too much fuel in repairing and recovering satellites. □ see ORBITAL TRANSFER VEHICLE, REMOTE MANIPULATOR SYSTEM

orbital period the time taken for one celestial body (eg the Moon) to make one complete orbit round another (eg the Earth). For a body which is in an elliptical orbit of semimajor axis, a, about a much more massive body of radius R, the orbital period, P, is given by the equation $P = 2\pi a^{3/2}/\sqrt{[gR^2]}$ where g is the gravitational acceleration at the Earth's surface. If an artificial satellite orbits the Earth in a circular path of radius r = 42 300km, its orbital period is equal to 24 hours and this is known as a geostationary orbit ⟨*the ~ of the Space Shuttle in a circular orbit 300km above the Earth (eg Mission 51G in January 1985) is 90min*⟩. □ see ORBITAL VELOCITY, ORBITAL ELEMENTS, ECCENTRICITY, KEPLER'S LAWS

orbital transfer vehicle (abbr **OTV**) a reusable rocket for boosting satellites into a higher geostationary orbit from the orbiting US Space Station. The Space Shuttle will take satellites and propellants to the Space Station to be assembled as cargo and propellant for the OTV which will then boost them into geostationary orbit. Its job done, the OTV will return to the Space Station. The OTV will use liquid oxygen and hydrogen as propellant. The oxygen may be manufactured from Moonrock. □ see US SPACE STATION, PAYLOAD ASSIST MODULE, INERTIAL UPPER STAGE, CENTAUR, MOON BASE

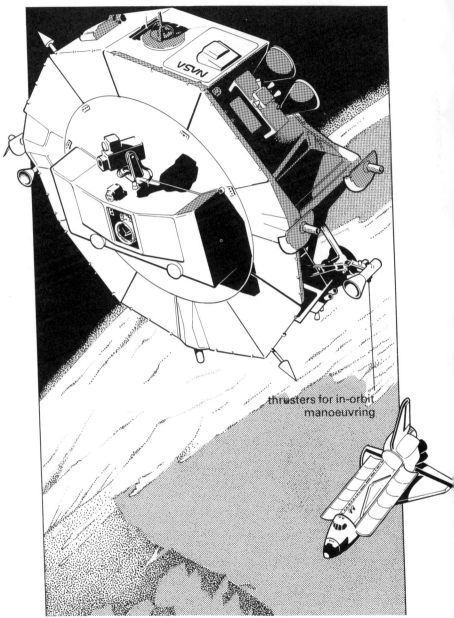

Orbital manoeuvring vehicle

orbital velocity the speed at which one celestial body (eg a communications satellite) orbits another body (eg the Earth). The orbital velocity, v, at any point on the path of an artificial satellite orbiting the Earth in an elliptical orbit is given by $v = \sqrt{[gR^2(2a-r)/ra]}$, where R is the radius of the Earth, r is the distance of the satellite from the centre of the Earth, a is the semimajor axis of the elliptical orbit, and g is the acceleration due to gravity at the Earth's surface. For an elliptical orbit (eg the path of Halley's comet) the orbital velocity is highest when r is smallest, ie when the comet is closest to the Sun (at perihelion) and least when the comet is farthest from the Sun (at aphelion). For a circular orbit, a=r and $v = \sqrt{[gR^2/r]}$. Thus the orbital velocity of the Space Shuttle in a circular orbit 300km above the Earth is about 7.8kms^{-1}. For a satellite to escape from the pull of the Earth, a = infinity, and the escape velocity from an orbit of radius r is given by $v = \sqrt{[2gR^2/r]}$. □ see ORBITAL PERIOD, ORBIT, ESCAPE VELOCITY, KEPLER'S LAWS

¹Orbiter that part of the US's Space Transportation System (STS) that orbits the Earth. Though less often used by the general public, the term is synonymous with Space Shuttle. Four Orbiters have so far been into Earth orbit; *Columbia, Challenger, Discovery,* and *Atlantis*. □ see SPACE SHUTTLE

²orbiter a spacecraft designed to orbit a planet or satellite without landing on its surface. The spacecraft Viking, Galileo, and Veneras 9 and 10 comprised two parts, an orbiter and a lander. The orbiter section normally relays signals from the lander to Earth. □ see LANDER, VIKING, GALILEO, VENERA

Orbiting Astronomical Observatory (abbr **OAO**) either of two ultraviolet astronomy satellites, OAO-2 which was launched in 1968 and OAO-3 in 1972. (OAO-4 failed to go into orbit). OAO-2 surveyed about one sixth of the sky in the ultraviolet wavelength range 115 to 320nm. OAO-3 (named Copernicus) carried a UV telescope with a 0.9m aperture, a high-resolution ultraviolet spectrometer, as well as X-ray equipment. □ see ULTRAVIOLET ASTRONOMY

ORS [*O*rbiting *R*efuelling *S*ystem] – see SULLIVAN

oscillating universe – see PULSATING UNIVERSE

oscillation a flow of electricity, property of a wave, or the movement of an object that periodically changes □ see YAW, FREQUENCY

oscillator a circuit or device for producing a periodic waveform of specific properties (sinusoidal, triangular, square, etc). Oscillators are widely used building blocks in analogue and digital electronic systems, esp in digital clocks and watches, and computer systems. □ see ATOMIC TIME

oscilloscope – see CATHODE-RAY OSCILLOSCOPE

OSO – see ORBITING SOLAR OBSERVATORY

OSO [*O*rbiting *S*olar *O*bservatory] any of NASA's seven successful satellites launched between 1962 and 1975 to study the Sun. They observed solar flares, scanned the solar disc, measured fluctuations in the corona, and noted changes in radiation intensities. □ see SUN

OSTA – see OFFICE OF SPACE AND TERRESTRIAL APPLICATIONS

OTA – see OFFICE OF TECHNOLOGY ASSESSMENT

OTV – see ORBITAL TRANSFER VEHICLE

outer planets planets of the Solar System which lie further from the Sun than the Earth. Thus Mars, Jupiter, Saturn, Uranus, Neptune, and Pluto are the outer planets of the Solar System – compare INNER PLANETS □ SEE GIANT PLANETS

outgassing – see OFFGASSING

¹output 1 the act or process of delivering information **2** the terminal of a device (eg an amplifier) from which information is delivered **3** a signal which represents information ⟨*the ~ displayed on a cathode-ray oscilloscope*⟩ – compare ¹INPUT

²output to send information from one device to another ⟨*to ~ data from a computer to a printer*⟩ – compare ²INPUT

overload to make a device or system exceed its operating capacity ⟨*to ~ a solar panel*⟩

oxidizer a material that supplies oxygen for the combustion of a solid propellant or a liquid fuel □ see CRYOGENIC ENGINE, STORABLE PROPELLANT, HYPERGOLIC FUEL

Ozma Project – see SEARCH FOR EXTRATERRESTRIAL INTELLIGENCE

pad – see LAUNCH PAD

paddle a solar panel □ see SOLAR CELL

pair production the simultaneous creation from a gamma-ray photon of an electron and a positron in the field of an atomic nucleus – compare ANNIHILATION □ see COSMIC RAYS, BIG BANG, ANTIMATTER

Palapa – see SPACE INSURANCE

pallet a platform that supports equipment and instruments in the cargo bay of the Space Shuttle □ see SPACELAB

¹PAM [*pulse amplitude modulation*] – see PULSE CODE MODULATION

²PAM – see PAYLOAD ASSIST MODULE

panspermia the interstellar transportation of spores which might have started life on Earth. Though many astronomers do not believe that spores could survive the rigours of lengthy space travel, recent experiments with the bacterium *Bacillus subtilis* indicate they may be wrong. The bacterium were found to be much more resistant to ultraviolet light (which is present in starlight), than expected. Moreover, when the spores were covered with a thin mantle of frozen water, methane, ammonia, and carbon dioxide, all of which are found in intersteller dust clouds, their chances of survival were much improved. It has been calculated that spores protected in this way could survive for between 4.5 and 45 million years in space. Fred Hoyle, a strong supporter in the steady-state model of cosmology, also believes that life 'blew' in from space early in the Earth's history. An ageless universe would give spores plenty of time to don their icy 'spacesuits' and reach the Earth from a nearby solar system. □ see HOYLE, ASTROBIOLOGY

parabola – see CONIC SECTION

parabolic shaped like a parabola ⟨*a ~ orbit*⟩ □ see REFLECTING TELESCOPE, ORBITAL ELEMENTS

parabolic orbit an orbit shaped like a parabola, ie that has an eccentricity of 1. A body (eg a comet) in a parabolic orbit escapes from the body it orbits. □ see ESCAPE VELOCITY, ORBITAL ELEMENTS

paraboloid a curved surface formed by the rotation of a parabola about its axis. Most mirrors of reflecting optical and radio telescopes have this type of surface since a beam of radiation parallel to its axis is reflected to a focus on the axis – compare ELLIPSOID, HYPERBOLOID

parachute – see DRAG PARACHUTE

paraffin – see KEROSENE

parallax the apparent displacement of a (celestial) object when seen from two widely displaced points. Thus it is the (usu small) angle measured in arc seconds that a baseline connecting the two points would subtend at the object. Knowing the parallax of a star enables astronomers to calculate the distance of that star from the Earth as the reciprocal of parallax is equal to distance in parsecs. ☐ see PARSEC

parallax second – see PARSEC

parameter a factor or quantity that influences the behaviour or properties of something ⟨*glide path* ~s⟩

park to put a satellite in a particular orbit ⟨*communications satellites are generally* ~ed *in a geostationary orbit*⟩

parking orbit a temporary orbit round a celestial body in which a spacecraft or satellite waits before being injected into a final orbit or trajectory. The Apollo missions used parking orbits round the Earth and Moon as part of the procedure for a Moon landing. ☐ see APOLLO, EXPEDITION TO MARS

parsec [*par*allax *sec*ond] (symbol **pc**) a unit used for measuring astronomical distances and equal to 30.86×10^{12}km or 3.26 light-years. It is defined as the distance at which the mean radius of the Earth's orbit (1 astronomical unit) would subtend an angle of 1 arc second. ☐ see MEGAPARSEC, PARALLAX, LIGHT-YEAR, ASTRONOMI-CAL UNIT, GALAXY

particle any very small piece of matter less than about 1 millionth (10^{-6}) of a metre in diameter ☐ see COSMIC DUST, MICROMETEOROID, ELEMENTARY PARTICLES

particle accelerator – see SYNCHROTRON

pascal (symbol **Pa**) the SI unit of pressure equal to 1 newton per square metre (Nm^{-2}). The pressure of the Earth's atmosphere at sea level is about 10^{5}Pa.

pass the passage overhead of an aircraft or satellite when it is within range of a ground station for telemetry signals ☐ see TELEMETRY

path 1 a connecting route for signals in a circuit **2** a route between two points in a communications channel **3** a sequence of instructions in a computer program

Pauli Exclusion Principle the principle that no two elementary particles of the same type can have the same set of quantum numbers (eg spin and charge). Roughly speaking, the principle means that there is a sort of force trying to keep two electrons, for example, apart, quite independently of the electrical repulsive force between them. This principle determines the way elementary particles share the available energy states in an atom, and it also has important conse-quences for the evolution of the universe. ☐ see ELEMENTARY PARTICLES, DEUTERIUM, PROTON, STAR, SPIN, ANTHROPIC PRINCIPLE

payload the part (eg astronauts, passengers, cargo, and mail) of the total mass of a launch vehicle over and above that which is necessary for its operation. The payload is usually a very small part of the total mass of the space vehicle at lift off. ☐ see PAYLOAD BAY, PAYLOAD SPECIALIST

Payload Assist Module (abbr **PAM**) a solid-fuelled rocket engine designed as the upper stage of a launch vehicle to send a satellite into geostationary orbit. This engine was used to launch the two com-munications satellites, Westar-4 and Palapa-B2, from the payload bay of the Space Shuttle Challenger on 3/4 February 1984. However, these

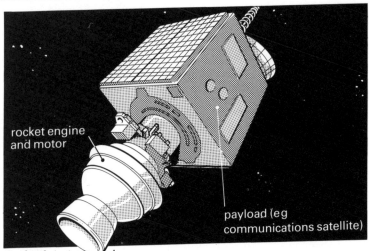

rocket engine
and motor

payload (eg
communications satellite)

Payload Assist Module

PAMs failed to work properly on this occasion because their nozzles contained faulty material causing them to burn through far short of the planned 85 second burn time. A new more powerful version of PAM has been designed to launch the rest of the Navstar Global Positioning satellites from the Space Shuttle by the end of the 1980s. □ see SPACE INSURANCE, GLOBAL POSITIONING SYSTEM, INERTIAL UPPER STAGE, CENTAUR

payload bay *also* **cargo bay** the unpressurized part of the Space Shuttle fuselage designed for carrying satellites to be released into orbit, and other equipment. Its maximum usable dimensions are: length 18.3m, and width 4.6m. Hinged doors extend the full length of the payload bay which is large enough to hold Spacelab, the Hubble Space Telescope, or components for the US Space Station. □ see SPACELAB, HUBBLE SPACE TELESCOPE, CENTAUR, EXTRAVEHICULAR MOBILITY UNIT

payload specialist an astronaut who is in charge of, and has expert knowledge of, the experiments aboard the Space Shuttle. Payload specialists may come from industry, university, government agencies, or even be self-employed. In addition to operating experiments, the payload specialist is also knowledgeable about food, hygiene, life support, caution and warning, and other basic systems on board the Space Shuttle. The payload specialist is responsible to the mission specialist □ see MISSION SPECIALIST, MID DECK

PCA – see PNEUMATIC CONTROL ASSEMBLY

PCB – see PRINTED CIRCUIT BOARD

PCM – see PULSE CODE MODULATION

Peacekeeper – see MX MISSILE

perigalacticon the point in a star's orbit (eg the Sun's orbit) that is nearest to the galactic centre. The furthest point is called the apogogalacticon. □ see APSIDES, MILKY WAY

perigee the point in the orbit of an artificial satellite or the Moon that is nearest the Earth's centre, and at which its speed is highest – compare APOGEE □ see APSIDES, ORBIT

perihelion the point in the orbit of an artificial satellite, planet, or comet in solar orbit when it is nearest to the Sun ⟨*Halley's comet is at ~ on Feb 9 1986*⟩ – compare APHELION □ see APSIDES, KEPLER'S LAWS

perilune a point in the orbit of an artificial satellite or spacecraft about the Moon that is nearest to the centre of the Moon – compare APOLUNE □ SEE ORBIT

period the time that elapses before a cyclic event (eg the flashes of light from a pulsar) repeats itself □ see FREQUENCY, ORBITAL PERIOD

periodic recurring at regular intervals ⟨*Halley's comet is a ~ comet*⟩ □ see APPARITION

perturbation any small disturbance to the regular motion of a celestial body (eg the perturbation of the orbit of a satellite by the solar wind) □ see SOLAR WIND, COURSE CORRECTION

Phobos and Deimos the two moons of Mars, both of which were discovered in 1877. The Mariner 9 spacecraft took the first close-up pictures of these moons whilst in orbit about Mars during 1971–1972. They are irregular potato-shaped objects, Deimos measuring about $11 \times 12 \times 15$km, and the more heavily cratered Phobos about $19 \times 21 \times 27$km. These moons are widely thought by astronomers to be asteroids which were later captured by Mars. But in 1960 a really imaginative theory of their origin was made by the Soviet astrophysicist, Shklovskii. Observations had shown that Phobos's orbit is decaying, ie it is losing height. Shklovskii calculated that for its orbit to decay at the measured rate, Phobos must have a density one-thousandth that of water. Thus, considering its size, Phobos must be hollow. So Shklovskii concluded that Phobos, and possibly Deimos, were 'spaceships' and had been launched from Mars by a long extinct Martian civilization! We now know that Phobos and Deimos do have a low density, but about twice that of water. A clearer picture of these two moons should emerge during the two Soviet missions to Mars beginning in 1988. Each mission will carry three French experiments designed to study the surface properties of Mars and Phobos. While the Soviet spacecraft hover 50m above the surface of Phobos, instruments will fire a beam of ions to disturb the soil and then measure its composition remotely. Phobos (Fear) and Deimos (Terror) are named after the two horses that pulled the chariot of the god of war, ie Mars, in Greek mythology. □ see MARS, MARINER, EXPEDITION TO MARS, VIKING, VOYAGER, ¹KEPLER, CANALS

photoelectric cell a device whose electrical properties are changed by light. Two commonly used photoelectric cells are the solar cell and the light-dependent resistor. □ see SOLAR CELL

photometer an instrument for measuring the brightness of a source of light (eg a star) or the relative brightness of a pair of lights. If the photometer measures the brightness of the source at particular wavelengths, it is called a spectrophotometer. The eye is now rarely used as the light sensor in photometers, and photocells and photomultipliers are used instead. The spectrophotometer is an important instrument aboard interplanetary spacecraft. □ see VOYAGER, MAGNITUDE

photometry the measurement of the intensity of light □ see MAGNITUDE

photomultiplier a device which increases the brightness of an image or the intensity of an electrical signal by increasing the number of electrons produced by the action of light on a metal in a vacuum □ see IMAGE INTENSIFIER

photon the smallest 'packet' or quantum of light energy. Light sensors such as photodides absorb photons and produce electrons which enable the light to be detected. □ see QUANTUM, PLANCK'S CONSTANT

photosensitive responding to light ⟨*a solar cell is a ~ device*⟩

photosphere the visible face of the Sun □ see SUN, SUNSPOTS

photovoltaic of or being a device (eg a solar cell) which produces electrical energy when it absorbs light □ see SOLAR CELL

picosecond a time interval equal to 1 million millionth (10^{-12}) of a second. Light takes about 1 picosecond to travel 0.3mm.

piezoelectricity the electricity that certain crystals (eg quartz) produce when squeezed. The effect is put to good use in some types of gas lighters, in most hi fi pickups, and in very stable electronic oscillators.

pilot the career astronaut who helps to fly the Space Shuttle, or other spacecraft, under the authority of its commander. On the Shuttle, he/she operates the Remote Manipulator System (RMS) to deploy or capture satellites, is proficient in operating the systems aboard Spacelab, and is the second crewperson for spacewalks. □ see COMMANDER, REMOTE MANIPULATOR SYSTEM, FLIGHT DECK, SPACEWALK

Pioneers 10 and 11 two spacecraft which, in 1973 and 1974 respectively, made the first flybys of Jupiter. They became the first spacecraft to negotiate the asteroid belt and, subsequently, the first manmade objects to leave the Solar System. Pioneer 10 returned information about Jupiter's Great Red Spot and its radiation belts. Pioneer 11 went on to make a successful flyby of Saturn in September 1979. The signal that Pioneer 10 had crossed Neptune's orbit was sent by its 8W transmitter over a distance of 4500 million kilometres, and took 4 hours to reach Earth. Like the later Voyager spacecraft, the Pioneers carry information about life on Earth should any intelligent life intercept them on their way to the stars. Pioneer 10's closest approach to a star will occur in 3200 years time, but we shall have lost track of it by then. NASA hopes to maintain communication with the spacecraft until about 1993 and obtain data about the boundary between the heliosphere and the interstellar medium, a region known as the heliopause. □ see VOYAGER, SEARCH FOR EXTRATERRESTRIAL

INTELLIGENCE, PIONEER VENUS, HELIOPAUSE

Pioneer Venus a spacecraft launched in 1978 and put in orbit round the planet Venus to make detailed observations of its atmosphere using an on-board ultraviolet spectrometer. To enable this instrument to examine the constituents of Halley's comet, NASA mission control tilted Pioneer Venus in orbit using the seven small thrusters aboard the spacecraft. Since Pioneer Venus spins in orbit, tilting the spacecraft a little at a time enabled a complete image of the entire comet to be built up and transmitted to Earth. □ see INTERNATIONAL COMETRY EXPLORER, VENUS, ULTRAVIOLET SPECTROMETER

pip – see BLIP

pitch the up and down movement of the front end of an aerospace vehicle in relation to its rear end – compare YAW

Planck's constant (symbol h) the factor by which the frequency of electromagnetic radiation must be multiplied to obtain the energy of a quantum of the radiation. Planck's constant is equal to $6.62 \times 10^{-34} Ws^2$. □ see QUANTUM, QUANTUM THEORY

planet a body that orbits the Sun or another star and that is seen by reflected light. All the planets of the Solar System probably share a common origin with the Sun some 4600 million years ago. The existence of planets round distant stars can be inferred from any disturbances of the star's motion. Also, infrared sensitive telescopes (eg IRAS) have found evidence of 'dust' bands round some stars. Improvements in resolution and sensitivity of space-based telescopes, especially the Hubble Space Telescope, should give us some better clues as to whether the distant stars have planetary systems of their own. □ see SOLAR SYSTEM, INFRARED ASTRONOMY SATELLITE, HUBBLE SPACE TELESCOPE, ASTROMETRY

Planet-A (renamed Suisei ('Comet') after launch) a Japanese interplanetary probe launched in August 1985 to make a close study of Halley's comet in March 1986. Planet-A was the first interplanetary spacecraft to be launched outside the USA and the USSR. It is identical with Tansei-5 which was launched early in 1985 to test the engineering and communications systems for the encounter with the comet. Like Tansei-5, Planet-A was boosted direct into heliocentric orbit using the Japanese Mu-3S 11 solid-propellant rocket. It had a launch mass of 140kg which included 12kg of scientific payload and 10kg of hydrazine propellant for its three thrusters used for trajectory corrections and attitude control. It is a spin-stabilized cylindrical spacecraft which is covered with about 2000 solar cells to provide between 67W and 104W of electrical power while it orbits the Sun at distances ranging from 1AU to 0.68AU. It communicates via its parabolic high-gain antenna with a new 64m radio antenna at Usuda, 170km north-west of Tokyo. Planet-A carries two instruments on a cylindrical platform: an ultraviolet TV camera which makes use of an image-detecting CCD (charge-coupled device) to study the emission of ultraviolet light from hydrogen gas in the coma of the comet; and a

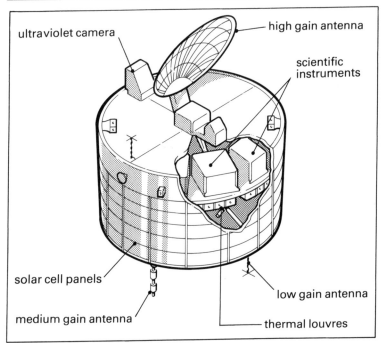

Planet-A

solar wind particle analyser to find out how ions in the solar wind are distributed within 30° of the ecliptic. Unlike the European Giotto spacecraft which was also in the vicinity of Halley's comet in March 1986, Planet-A did not carry any protection against collision with cometary debris. Thus its closest flyby distance was limited to about 10 000km. □ see HALLEY'S COMET, [1]GIOTTO, VEGA, DUST SHIELD

planetary system a group of bodies including planets, comets, and asteroids that orbits a star. Planetary systems may be common for the Sun is only one of thousands of millions of stars in the Galaxy. □ see SOLAR SYSTEM, BETA PICTORIS, INFRARED ASTRONOMY SATELLITE

planetesimals the dust, boulders, and larger bodies that are thought to have formed the planets of the Solar System by a process of accretion □ see ACCRETION, ASTEROID, SOLAR SYSTEM

planetoid – see ASTEROID

Planet X a hypothetical tenth planet beyond the orbit of the outer-most known planet, Pluto. It is the barely detectable and unaccountable perturbations (small movements) in the orbits of the outer planets which makes astronomers propose the existence of this other body. (It was the perturbations in the orbit of Neptune which led to the prediction and discovery, in 1930, of Pluto itself, also called

Planet X at the time). Theorists suggest that Planet X has a mass between one and five times the mass of the Earth and it orbits the Sun at between 50 and 100 astronomical units. Whether a Planet X exists is hotly argued among astronomers, for there are other explanations for the small descrepancies in the motions of the outer planets. A whole clutch of comets beyond the orbit of Neptune has been proposed to explain why there are so many short-period comets in the Solar System. These comets might be part of the Oort Cloud, a hypothetical reservoir of comets beyond the planets. Perhaps both Planet X and a comet belt exist and Planet X influences the paths of these comets so that they leak into the Solar System from time to time. The precipitation of a whole shower of comets in this way could also be an explanation of why the fossil record indicates that life on Earth has experienced mass extinctions every 28 million years. □ see PLUTO, COMET, OORT CLOUD, TEKTITE

plasma a highly ionized gas containing approximately equal numbers of positively charged ions and negatively charged electrons. Plasma accounts for over 99% of the universe: the Sun and stars are made of plasma. The thin upper atmosphere of the Earth is also a plasma, and will be investigated by one of the Spacelab missions, called Space Plasma Lab 1, scheduled for June 1988. Two pallets in the Spacelab will be packed with instruments to study the plasma in the Earth's upper atmosphere: one of these experiments is called WISP (Waves in Space Plasma) and will have an antenna 300m long! It will transmit waves in the frequency range 300Hz to 30MHz, and the way these waves are transmitted and reflected from the plasma will enable the density of the electrons to be found. Another experiment will fire a beam of electrons, ions, and neutral gas into the atmosphere to study the artificial aurora created by the beam. □ see SPACELAB, EARTH'S ATMOSPHERE, ACTIVE MAGNETOSPHERIC PARTICLE EXPLORER, MAGNETOSPHERE, ELECTRIC PROPULSION, MAGNETOHYDRODYNAMICS

plasma rocket a rocket engine that obtains thrust by the acceleration of a plasma in a magnetic field. Lengthy interstellar journeys could be made using plasma rockets since they would be capable of high speeds over long periods of time. □ see PLASMA, ANTIMATTER PROPULSION, ELECTRIC PROPULSION

Pluto the ninth and probably the most distant planet of the Solar System. It has a diameter of about 6400km and orbits the Sun in a highly elliptical orbit at an average distance of about 6000 million kilometres. Since Pluto orbits the Sun once in every 248 years, it has yet to complete an orbit following its discovery in February 1930. Very little is known about Pluto but its surface is probably covered by a layer of methane ice. In 1985 a rare alignment of the Earth with Pluto and its only known satellite, Charon (discovered in 1978), gave astronomers a new approach to the study of this distant planet. Charon passes in front of Pluto once every 124 years causing the brightness of Pluto to vary by about 4%. This variation is superimposed on a 30% brightness change every 6.4 days because one hemisphere of Pluto is 30% brighter than the other. Measurements of the times, durations, and changes of brightness are allowing the masses, diameters, and densities of Pluto and Charon to be calculated with better precision. Pluto is not on the itinerary of Voyager 2 which will reach Neptune in 1989. A ballistic flight (ie one not using continuous rocket thrust or gravity-assist) would take about 30 years to reach the planet. However, plans have been put forward to launch a spacecraft to Pluto in about 1989. By using the favourable position of Jupiter for gravity assist to increase the spacecraft's speed during the latter part of the journey, this trip would take only 10 years. The spacecraft would be provided with on-board data storage facilities. The data (eg pictures of Pluto and Charon) would be transmitted to Earth at the leisurely rate enforced by Pluto's great distance from us. □ see VOYAGER, NEPTUNE, GRAVITY ASSIST, PLANET X

plutonium (chemical symbol **Pu**) a heavy metallic manmade radioactive element of atomic number 94. Its most important isotope is Pu-239 (mass number 239) which is used as a nuclear fuel and in weapons. □ see ISOTOPE, NUCLEAR FISSION

pneumatic moved or worked by air (or other gas) pressure □ see PNEUMATIC CONTROL ASSEMBLY

pneumatic control assembly (abbr **PCA**) a pressurized gas system that operates when the main engine of the Space Shuttle is being prepared for launch. The PCA controls the liquid oxygen and hydrogen bleed valves, and provides emergency shutdown of the flow of propellant to the engines if there is an electrical failure to the main engine. □ see BLEED, SPACE SHUTTLE

pn junction the boundary between a p-type and an n-type semiconductor and the basis for rectifier diodes, light-emitting diodes, transistors, integrated circuits, and solar cells. Its basic function is to allow current to flow easily from the p-type to the n-type material but to stop current flowing through in the opposite direction. In the solar cell, a voltage is produced across the pn junction when it receives light energy, and this is a source of electrical energy for artificial satellites and spacecraft. □ see SOLAR CELL, MICROELECTRONICS

pogo to vibrate up and down (as on a pogo stick). The term is used to describe the vertical vibrations of a rocket at launch caused by the interaction of the forces generated by the burning fuel with the mechanical properties of the rocket's structure. □ see RUMBLE

polar 1 rotating about an axis passing through the poles ⟨~ *axis of a telescope*⟩ □ see EQUATORIAL MOUNTING **2** passing over the North and South poles of the Earth ⟨*a satellite in* ~ *orbit*⟩ □ see POLAR ORBIT

polar orbit the orbit of an Earth satellite (or an artificial satellite of another planet) that passes over the poles – compare EQUATORIAL ORBIT □ see WEATHER SATELLITE

pole star the star closest to the North or South celestial pole. At the moment, the North pole star is Alpha Ursa Minor which can be found from the two 'pointers', Dubhe and Merak, in the constellation, the

Plough. The pole star is the primary of a multiple star system and is a creamy yellow supergiant. The gravitational pull of the Sun on the Earth's equatorial bulge causes the Earth's axis to precess (ie wobble) with a period of precession of 25 800 years. Thus the pole star changes: in 3000BC it was Alpha Draconis, and in AD 14000 it will be Vega. □ see PRECESSION, CONSTELLATION, GIANT STAR

pollution contamination with manmade waste. Since the dawn of space exploration in the late 1950s, there has been a steady increase in the amount of left-over debris from activities in space. There is so much uncharted material in Earth orbit that there is a good chance that it will collide with satellites and space stations with possibly disastrous results. Spent rockets and various discarded waste materials tumble in Earth-orbit as useless artificial satellites. The Apollo landings on the Moon left behind a large quantity of spent scientific experimental equipment and containers, and spacecraft which reach the surface of planets (eg Venera and Viking) are, perhaps, the start of pollution of the Solar System. In the 1970s, the US Air Force was criticised, esp by radio astronomers, for its West Ford Project. This was a passive communications system involving the release of 400 million hair-thin dipoles (to reflect radio waves) which produced a ring 3200km above the Earth. Fortunately, the dipoles dispersed without causing any danger to other satellites or bother to radio astronomers. On the positive side, many satellites have been designed to keep watch on pollution of the Earth's atmosphere, its rivers, and oceans. □ see LANDSAT, UPPER ATMOSPHERE RESEARCH SATELLITE

populations of stars the classes into which stars (and other celestial objects) can be divided according to age, location in space, and metal content. In 1944, the astronomer W Baade classified stars into two populations. Population I stars, such as our Sun, are associated with gas and dust in the spiral arms of galaxies, ie they are located in the galactic plane. These stars are very luminous and their spectra show they are rich in metals. Population II stars are old stars located at the centre of a galaxy and in the surrounding galactic halo. Spectra of these stars show they are deficient in metals, and were formed from hydrogen and helium in the original gas cloud. Population I stars become progressively enriched with metals and other heavy elements produced in supernovae explosions. We now know that stars have a range of populations between Baade's two classes. □ see STAR, GALAXY, NUCLEOSYNTHESIS, SUPERNOVA

position 1 the place occupied by a celestial body ⟨*a star's ~*⟩ □ see ASTROMETRY **2** a crew member's station aboard a spacecraft □ see WORKSTATION

positive 1 having a deficit of electrons **2** *of a part of a circuit* having a relatively high electrical potential ⟨*the ~ terminal of a battery*⟩ **3** *of a number* numerically greater than zero – compare NEGATIVE

positron (symbol β^+) *also* **antielectron** the positively charged antiparticle of an electron □ see ELECTRON, ANTIPARTICLE, PAIR PRODUCTION, ANNIHILATION

potential energy the energy a body has by virtue of its position or the arrangement of its parts. An artificial satellite gains potential energy (and kinetic energy) from the chemical energy of the rocket fuel in the launching rocket. If a weight W (newtons) is raised vertically a distance h (metres) in a gravitational field, its potential energy is equal to W × h joules. Here W = m × g where m is the mass of the body and g the acceleration due to gravity (9.81ms^{-2} at the Earth's surface) – compare KINETIC ENERGY □ see ESCAPE VELOCITY, TETHERED SATELLITE

power 1 the number of times that a number is to be multiplied by itself □ see EXPONENT **2** (abbr **PWR**) the rate at which energy is absorbed, radiated, or delivered by a device ⟨*the electrical ~ delivered by a solar panel*⟩ **3** the reciprocal of the focal length, in metres, of a lens or a mirror. A converging lens or mirror has a positive power; a diverging lens or mirror has a negative power. The power of a lens is a measure of its focusing or defocusing ability ⟨*a thick lens has a high ~*⟩. □ see REFRACTING TELESCOPE

power dissipation the power generated as heat in a device owing to the current flowing through it. Every electronic device has a maximum rating of power dissipation and to exceed this maximum may damage the device. The heat generated by electronic devices on spacecraft needs to be got rid of otherwise the temperature in instrument and control packages may rise too high and damage the circuits.

powerful 1 *of a rocket engine* able to provide a large thrust **2** *of a communications system* able to carry lots of information **3** *of a computer* able to carry out many instructions at high speed

power pack – see POWER SUPPLY

power plant an assembly of engines and accessories (eg fuel supply, ignition, and cooling systems) that generates power for propulsion ⟨*a spaceship's ~*⟩ □ see POWERSAT

Powersat *also* **solar power satellite** a large array of solar panels in Earth orbit to convert solar energy into electrical energy for use on Earth. This 'space power station' is still on the drawing board, but the US plans to build the first of them (perhaps using facilities provided by the US Space Station) in the early part of the next century. The satellite would have to be large; a flat structure is planned, 10km long and 5km wide, and with a mass of about 36 thousand tonnes. A high-power beam antenna 1km in diameter would be attached to it, and to this would be assembled microwave power amplifiers and waveguides. On Earth, a vast array of dipole aerials would be built, 10km by 13km, to receive and convert the microwave radiation into a form which could be fed into the power system. The microwaves would be generated at a frequency of 2.45 gigahertz. The radiation level on Earth would be very low and harmless to humans, animals, birds, and aircraft – so it is said! □ see OAST-1 SOLAR ARRAY, MICROWAVES, DIPOLE AERIAL, WAVEGUIDE, FREQUENCY

power supply *also* **power pack** a unit that is usu portable (eg built into a spacesuit) and provides electrical power ⟨*a DC ~*⟩

preamplifier *also* **preamp** an amplifier that boosts the power or voltage of weak signals (eg from a radio aerial) before they are fed to a receiver or a further stage of amplification □ see MASER

precession the slow change in the direction of the axis of rotation of a body in response to a turning force (a torque). The two ends of the axis of a precessing body trace out circles. In the case of the Earth, the poles complete one circle every 25 800 years. As the axis precesses, the celestial equator moves relative to the ecliptic so that stellar coordinates change slowly with time. The axis of spin-stabilized satellites may also precess if gravitational or other forces produce a torque on the satellite. □ see TORQUE, POLE STAR

precision the degree of refinement in a measurement or operation ⟨*the ~ with which Voyager performs its mission*⟩ – compare ACCURACY □ see RESOLUTION

pressurize 1 to maintain near-normal air pressure in the cabin of a spacecraft ⟨*the Space Shuttle cabin is ~ed*⟩ **2** to apply pressure above that of the pressure outside a chamber ⟨*to ~ the gas in the liquid propellant tanks of a rocket*⟩

primary the celestial body that is nearest to the centre of mass of a system of orbiting bodies. Though all members of a system orbit round its centre of mass, the primary appears to be the centre of the system ⟨*the Sun is the ~ of the Solar System*⟩. □ see BINARY STAR, CENTRE OF MASS, KEPLER'S LAWS

primary mirror a circular disc with a concave reflecting surface which collects and focusses light at the prime focus in a reflecting optical telescope. The disc is usually made of a glass/ceramic compound, fused quartz, or silica, which expand very little with temperature change. Once the reflecting surface has been ground to the precise shape, it is (usu) aluminized to obtain a highly reflecting surface. The ability of a telescope to pick out faint celestial objects (ie its 'light grasp') is directly related to the area of its primary mirror. The Hale 5m diameter reflecting telescope on Mount Wilson in California, is made of pyrex and has a mass of 13 tonnes, though even this is not the largest, for the Soviet Academy of Sciences has a 6m telescope in use. The Hubble Space Telescope uses a 2.4m diameter mirror. □ see OBJECTIVE, TELESCOPE, HUBBLE SPACE TELESCOPE, ABERRATION 1, APERTURE

primeval *also* **primaeval** existing in the earliest stages of the universe ⟨*the Big Bang began from a ~ atom*⟩ □ see BIG BANG

printed circuit board (abbr **PCB**) a thin board made of electrically insulating material (usu glass fibre) on which a network of thin copper tracks is formed to provide connections between components soldered to the tracks. Electronic equipment (eg computers and instruments) aboard a spacecraft is usu assembled on one or more printed circuit boards. □ see MICROELECTRONICS

printout computer data, graphics, telemetry, etc recorded on paper by a printer

probe any device for investigating something, usu remotely ⟨*Voyager is an interplanetary* ~⟩ □ see SPACECRAFT, REMOTE SENSING

Prognoz a series of Russian satellites launched in the 1970s to study the Sun, and the effects of the solar wind on the Earth. For example, Prognoz-6 followed a highly elliptical Earth orbit and investigated the effect of solar activity on the interplanetary medium and the Earth's magnetosphere. □ see INTERSTELLAR MEDIUM, MAGNETOSPHERE, EARTH'S ATMOSPHERE

program a list of instructions written in a language that can be translated into something a computer can understand. A program is usu entered from a keyboard and stored in a computer's random access memory, or it is loaded into it from an external program store (eg a magnetic tape or disk). □ see SUPERCOMPUTER

Progress the unmanned 'cargo ships' which are launched to dock automatically with the Soviet orbiting Salyut space station. Once the resident astronauts have unloaded provisions from a Progress cargo ship, it separates and is guided to a destructive burn-up in the Earth's atmosphere. For example, two weeks after Kizim, Solovyov, and Atkov arrived at Salyut 7 aboard Soyuz T-10 on 8 February 1984, Progress 19 arrived and docked with Salyut's rear unit. It brought 2 tonnes of propellants, scientific equipment, consumables, and mail. On 31 March, Progress 19 undocked and was destroyed in the atmosphere. On 20 April, Progress 20 ferried more provisions and equipment to Salyut 7. □ see SALYUT, SOYUZ, ORBITAL TRANSFER VEHICLE

project a specific undertaking which may involve research and development, design, construction, and evaluation of a system and/or hardware to accomplish a scientific or technological objective ⟨*the Apollo* ~⟩

propagate to send information through something ⟨*radio waves can be* ~d *through space*⟩ □ see SPEED OF LIGHT

propel to drive forward ⟨*a rocket engine* ~s *a spacecraft*⟩ □ see PROPELLANT, THRUST, BOOSTER ROCKET

propellant the mixture of fuel (eg hydrogen) and oxidizer which burns to give a rocket thrust. Propellants are broadly divided into solids and liquids. □ see STORABLE PROPELLANT, CRYOGENIC ENGINE, OXIDIZER, HYPERGOLIC FUEL, THRUST, BOOSTER ROCKET, SOLID ROCKET BOOSTER

proportional counter a gas-filled sensor for counting and measuring the relative strength of ionizing radiations (eg X rays). The gas is at a higher pressure than in a Geiger tube so that radiation entering the tube produces pulses of current proportional to the energy of the radiation. A proportional counter is an important instrument aboard X-ray astronomy satellites. □ see GEIGER COUNTER, X-RAY ASTRONOMY

proton (symbol **p**) the charged particle found with neutrons in ordinary atomic nuclei. Its mass is almost equal to that of the neutron

(about 1.67 × 10⁻²⁷kg). The number of protons in the nucleus of an
atom determines its chemical properties as an element. An isolated
proton is stable (unlike the neutron) and it has an antiparticle called
the antiproton. It is interesting that when protons are produced in the
laboratory by collisions between atomic nuclei, every one is accompanied by an antiproton. So presumably, in the Big Bang, equal
numbers of protons and antiprotons were created, but as the temperature fell, they rapidly annihilated each other to produce gamma ray
photons. Similarly, electrons and positrons produced in the Big Bang
disappeared by mutual annihilation. However, a few (about one in a
billion) protons and electrons survived annihilation to produce the
galaxies of stars and other matter with which we are familiar. But the
mystery is, where are all the positrons (antielectrons) and antiprotons
which also escaped this annihilation? Perhaps galaxies of antimatter
do exist. But since galaxies occasionally collide, there should be a
stronger background of gamma rays than there is. It suggests that the
universe is asymmetric, and most of it is made up of the small residue
of matter (as opposed to antimatter) which was left over after the Big
Bang. □ see NUCLEUS, NEUTRON, PROTON, ELEMENTARY PARTICLES,
QUARK, DEUTERIUM, BIG BANG, ANTIMATTER, UNIVERSE, NUCLEO-
SYNTHESIS, HADRON, ANTHROPIC PRINCIPLE

Proton a powerful Soviet rocket used to launch heavy payloads (eg
the Salyut spacestation) into Earth orbit. It can be fitted with strap-on
boosters to improve performance. One version of the rocket, the
Proton SL-12, launched the Vega spacecraft towards Venus in December 1984. □ see BOOSTER ROCKET, TITAN 2, SATURN 5, VEGA

protoplanet a planet in the making. It is the stage when the planet
has emerged by accretion □ see ACCRETION, SOLAR SYSTEM,
ASTEROID, INFRARED ASTRONOMY SATELLITE

protostar a star in the making. It is the stage in a star's life after it
has separated from a cloud of gas and dust but before it has become
hot enough for nuclear reactions to begin. □ see INFRARED ASTRON-
OMY SATELLITE, STAR

prototype the first working device or system on which later designs
are based ⟨a ~ space suit⟩ □ see PROJECT

psychomotor relating to muscular action following directly from a
mental process. The psychomotor response of an astronaut is important in determining the effectiveness with which he or she carries out
the coordinated manipulation of controls aboard a spacecraft. □ see
SPACE MEDICINE

pulsar a rapidly spinning neutron star which produces regular pulses
of light and radio waves. The neutron star is a giant atomic nucleus of
immense atomic weight which is the greatly contracted remnant of an
awesome supernova explosion. It spins very rapidly and has a powerful magnetic field associated with it. Electrons caught up in this
magnetic field emit light and radio waves (and in some pulsars X rays
and gamma rays) in a narrow rotating beam. It behaves as a cosmic
'lighthouse'. The frequency of the pulses produced by pulsars range

from 30 times per second (30Hz) (eg the one in the centre of the Crab
nebula) to 0.25Hz. Over 300 pulsars are now known and it is calculated that in the Galaxy alone there are over a million. The rate of
formation of pulsars is calculated to be about 25 per century, though
this number conflicts with the rate of formation of supernovae, so the
theory of the formation of pulsars may have to be revised. The frequency of all pulsars is slowly decreasing as the neutron star loses
energy. Occasionally a pulsar may speed up briefly (eg the Vela
pulsar). These glitches or 'spin ups' are thought to be due to a rear-
rangement of the crust or core of the neutron star. All light- and
radio-emitting pulsars are single stars, though X-ray pulsars occur in
close binary systems. In these the X rays are produced when the hot
gases from the companion star are channelled onto the pulsar at its
magnetic poles. X rays are produced in these hot gases every time the
pulsar rotates. □ see SUPERNOVA, NEUTRON STAR, BLACK HOLE,
QUASAR

pulsating universe *also* **oscillating universe** a model of our
universe that repeatedly undergoes contraction and expansion □ see
BIG BANG, STEADY STATE, COSMOLOGICAL MODEL

pulse a short-lived variation of a signal ⟨*the ~s produced by a
Geiger counter*⟩ □ see BURST, PULSAR, PULSE CODE MODULATION

pulse amplifier an amplifier designed for use in radiation detectors
(eg Geiger counters) where it amplifies and shapes pulses generated
by the radiation before passing them on to counting circuits □ see
DISCRIMINATOR, GEIGER COUNTER, PROPORTIONAL COUNTER

pulse amplitude modulation (abbr **PAM**) – see PULSE CODE
MODULATION

pulse code modulation (abbr **PCM**) a way of changing an
analogue signal into a digital signal for transmission along telephone
lines or by radio waves. The analogue signal (eg a voice from a
microphone) is sampled to determine its amplitude at very short intervals of time, a process known as pulse amplitude modulation (PAM).
At least 8000 samples need to be taken per second to represent speech
patterns, and 40 000 to represent music. Each of the samples is converted into a binary code (usu 8 bits) and this code is the information
transmitted. At the receiving end, the digital code is changed back
into an analogue signal. Pulse code modulation is used in preference
to analogue transmission since it is possible to receive the original signal more accurately at the receiving end, even though it may be
immersed in a lot of background noise. A form of pulse code modulation is used to transmit pictures from spacecraft to Earth. □ see
TELEMETRY, TRACKING AND DATA RELAY SATELLITE

pump a machine that exerts a force on a fluid flowing through it so
as to lift the fluid, increase its rate of flow, or increase its pressure ⟨*a
liquid oxygen ~*⟩ □ see PNEUMATIC CONTROL ASSEMBLY

purge to get rid of residual fluid (fuel and oxidizer) in the tanks or
pipe lines of a rocket after it has been fired

pyrometer an electronic instrument for measuring temperatures

above 600C using, for example, a thermocouple as a sensor □ see
THERMOCOUPLE

QSO [*quasi-stellar object*] – see QUASAR

quantum (plural **quanta**) the smallest 'packet' by which certain properties (eg energy, gravity, and momentum) can change or be transmitted. A quantum of light energy is known as a photon and the (proposed) quantum of gravitational field strength is called a graviton. The energy of a quantum is proportional to the frequency of the radiation and is given by the simple equation: quantum energy = Planck's constant × frequency. The idea of quanta is the basis of the quantum theory. □ see PLANCK'S CONSTANT, QUANTUM THEORY, RELATIVITY, GRAVITON

quantum theory a theory of physics which is able to explain very successfully many of the complex ways energy and matter interact by assuming that electromagnetic radiation (eg light) comprises packets of energy called quanta. The quantum theory and the theory of relativity are two of the most important theories put to work in trying to explain the complex and often violent behaviour of the universe. □ see QUANTUM

quark a hypothetical elementary particle of which all hadrons (eg protons) are supposed to be composed. Theory predicts that there are several types of quark differing from each other by their charges and other innate properties. For example, each proton is supposed to contain two distinct 'flavours' of quarks: two so-called 'up' or u quarks, each with an electric charge of 2e/3, and a 'down' or d quark with a charge of –e/3, where e is the electronic charge. The combination has an overall positive charge of +e as required. Note that the labels up and down have nothing to do with the orientation of quarks. The neutron is a combination of one u and two d quarks giving an overall charge of zero. There are four other known quarks: charmed (c) and strange (s) which have charges of +2e/3 and –e/3, respectively; and top (t) and bottom (b) which have charges of +2e/3 and –e/3, respectively. These six quarks each come in three different colours (for want of a better name) making 18 quarks in all. The force that 'glues' quarks together in the proton must be very strong and possibly increases with distance; certainly high-energy collision experiments in particle accelerators have not succeeded in breaking protons apart. The strong interquark forces are supposed to be due to the interchange of massless particles called 'gluons' of which there are eight different flavours. Like other elementary particles, quarks have their antiparticle counterparts, eg the anti d quark. It has been suggested that neutron stars consist of a 'soup' of free quarks. □ see ELEMENTARY PARTICLES, NEUTRON, PROTON, NEUTRON STAR, BIG BANG

quartz a crystalline form of silicon oxide that is piezoelectric, ie it can be used for the generation of precise timing signals in clocks and watches ⟨*a ~ crystal watch*⟩ □ see PIEZOELECTRICITY

quasar [*quasi*-stell*ar* object] the most distant, the most energetic, and the most curious of all cosmic objects. The brightest quasars look like stars but emit as much energy as hundreds of galaxies and have a volume far smaller than the Milky Way. We know from the large red-shift in their spectra that the farthest quasars are rushing away from us at more than 90 per cent of the speed of light; the light from them has taken a thousand million light years to reach us. It is difficult for astrophysicists to explain the nature of quasars, though much information about them has been gathered by space-based telescopes. Recent evidence (eg that from the Ultraviolet Explorer Satellite) suggests that a quasar is a galaxy which has a massive black hole at its centre. It is therefore proposed that the immense and concentrated power of a quasar is due to gas spiralling inwards under the influence of the black hole's enormous gravity and, in the process, becoming hot and emitting radiation. The first quasar to be discovered was 3C-273 in the Virgo cluster of galaxies. □ see VIRGO A, BLACK HOLE, GALAXY, SEYFERT GALAXY, RED SHIFT, PULSAR, APERTURE SYNTHESIS, QUASAT, SYNCHROTRON RADIATION, RADIO ASTRONOMY

quasar satellite – see QUASAT

Quasat [*qua*sar *sat*ellite] a joint NASA/ESA project to place a 15m diameter radio telescope in Earth orbit to make a detailed study of the structure of quasars. Quasat would follow an elliptical orbit, swinging in as close as 5700km and out as far as 12 500km. It would be an 'outrigger' station for a radio astronomy technique called very long baseline interferometry (VLBI). VLBI uses two or more radio telescopes to synthesize the effect of a radio telescope having a larger resolving power than that possible with an individual dish. The technique allows astronomers to pry deep into the hearts of quasars in an attempt to understand what produces their enormous energies. Acting in association with radio telescopes on Earth, Quasat would transmit to Earth the characteristics of the radio waves it receives from quasars. Synchronized by an atomic clock, radio signals from the same quasar would be received simultaneously by Earth-based radio telescopes. Tape recordings of the radio waves would be transported to a central processing facility where radio maps of the quasars would be made. Quasat's radio dish must be compact enough to fit into the payload bay of the Space Shuttle or of the Ariane launcher. It must be rigid when deployed in space, and its surface must be accurate to around 1mm so that it will not distort the reflected radio signal when working at radio wavelengths as short as 13.5mm – which is considered to be a tall order for a 15m radio telescope which has to survive the rigours of a rocket launch. But, at this wavelength, Quasat should be able to see detail as small as the distance between the Earth and the Sun from the centre of the Galaxy, thirty thousand light-years away, where recent radio maps have indicated that matter might be falling on to a central object, perhaps our own Galaxy's extinct quasar core. Quasat could increase our understanding of the most extraordinary object imaginable – the supermassive black hole which could be the 'engine' at the heart of a quasar. □ see QUASAR, RADIO ASTRONOMY, APERTURE SYNTHESIS, VERY LARGE ARRAY

quasi-stellar object (abbr QSO) – see QUASAR

quench to stop the electrical discharge in a radiation detector (eg a Geiger tube) immediately after a particle or photon has produced ionization in it. The discharge must be quenched otherwise particles which follow each other in quick succession cannot be detected by their individual ionization of the gas in the detector. In a Geiger tube, the ionization is quenched by introducing a small quantity of a gas (eg xenon) into the tube.

quiescent Sun the Sun when relatively inactive, as occurs round the middle of the sunspot cycle □ see SOLAR FLARE, SUNSPOTS, SUN

rad [*r*adiation *a*bsorbed *d*ose] a unit of absorbed dose of ionizing radiation (eg X rays) as opposed to the actual radiation present. The distinction is necessary since different materials absorb ionizing radiations to different degrees. A dose of one rad means the absorption of one hundredth of a joule per kilogram (10^{-2}Jkg^{-1}) of the absorbing material – compare REM □ see X RAYS

radar [*ra*dio *d*etection *a*nd *r*anging] an electronic system which sends out high frequency radio waves to locate distant objects by measuring the time taken to receive an echo. A radar altimeter measures the height of an aerospace vehicle above the ground. Radar has been used by astronomers to obtain images of the surface of cloud-covered Venus. □ see VENUS RADAR MAPPER, VENERA, SHUTTLE IMAGING RADAR

Radarsat a satellite system being designed and built by Spar Aerospace Company of Canada, and scheduled for launch in 1990. From its polar orbit, Radarsat will survey Canada's mineral, forest, and agricultural resources, as well as ice packs in northern waters. The satellite will use synthetic aperture radar (an imaging system which is not adversely affected by poor weather). □ see SHUTTLE IMAGING RADAR, TELEPRESENCE, POLAR ORBIT

radial velocity the movement of an object along the observer's line of sight

radian a unit of angle equal to the angle subtended at the centre of a circle by an arc equal in length to the radius of the circle. Thus 2π radians equals 360°. □ see ARC SECOND, RESOLVING POWER

radiate to send out energy as light, heat, radio waves, or as any other form of electromagnetic radiation ⟨*a quasar ~s radio waves*⟩ □ see ELECTROMAGNETIC RADIATION, PROPAGATE, SYNCHROTRON RADIATION

radiation electromagnetic waves (eg light) or particles (eg protons) which carry energy through matter or space □ see ELECTROMAGNETIC RADIATION, NUCLEAR RADIATION

radiation damage the harmful effects of radiation on matter. Instruments aboard spacecraft need protection from radiation (eg cosmic rays). □ see RADIATION HARDENING, ¹GALILEO

radiation dose the amount of radiation absorbed by a substance, system, or living tissue □ see RAD, REM, SPACE MEDICINE, RADIATION HARDENING

radiation hardening the process of producing microelectronic devices that are resistant to the damaging effects of radiation. Radiation hardening is necessary if spacecraft are to operate in the radiation belts of the Earth and Jupiter. The electronic circuits on board the spacecraft Galileo, which will begin to orbit Jupiter in 1987, will have to be able to withstand the harsh radiation belts surrounding the planet. The chips will be able to withstand a radiation dose of 150 000 rad (a chest X-ray gives about 0.025 rad and a whole body dose of 1000 rad is normally fatal to humans). □ see RADIATION, RAD, ¹GALILEO

radiation illness ill-health caused by excessive exposure to ionizing radiation (eg gamma rays). The symptoms are nausea, vomiting, diarrhoea, blood cell changes, and, in the later stages of the illness, by haemorrhaging and loss of hair. □ see SPACE MEDICINE

radiation medicine the branch of medicine dealing with the prevention or cure of illness caused by the body's exposure to high-energy ionizing radiation such as gamma rays, X rays, and energetic particles □ see SPACE MEDICINE

¹radio to send information by radio waves

²radio communication at a distance using electromagnetic waves having frequencies in the range 15kHz to 900MHz □ see LONG WAVES, MEDIUM WAVES, SHORT WAVES, MICROWAVES

radioactive cloud *also* **atomic cloud** an enormous cloud of radioactive vapour and dust carried into the upper atmosphere from a nuclear explosion near the Earth's surface □ see NUCLEAR WEAPONS, ELECTROMAGNETIC PULSE

radioactive dating a method of measuring the age of rocks (eg lunar rocks) by finding out how much of a long-lived radioactive isotope and its stable decay product remain in a sample of the rocks. The technique depends on knowing the half-life of the isotope. For estimating the age of rocks, the following pairs of isotopes are used: potassium-40 which decays to argon-40 with a half-life of 1250 million years, and rubidium-87 which decays to strontium-87 with a half-life of 4880 million years. □ see RADIOACTIVITY

radioactive decay – see RADIOACTIVITY

radioactive isotope – see RADIOISOTOPE

radioactivity *also* **radioactive decay** the spontaneous breakdown of unstable atomic nuclei into other nuclei with the emission of energy. The energy is carried by alpha particles, beta particles (ie electrons), or positrons. Gamma-rays sometimes accompany the particles. The elements that release energy in this way are called radioisotopes. □ see DECAY, ISOTOPE, HALF-LIFE

radio astronomy the use of radio receiver equipment and specially designed aerials to study the behaviour of the Sun, stars, and galaxies. Radio astronomy, using those radio waves that can penetrate the Earth's atmosphere (through radio 'windows'), and optical astronomy are the two main ways by which electromagnetic waves from distant objects in the universe are received. Both Earth-based observations and space-based radio telescopes, using the latest advances in electronics, have greatly expanded our knowledge about the structure and origins of the universe. □ see ELECTROMAGNETIC RADIATION, DISH AERIAL, VERY LARGE ARRAY, QUASAT, RADIO GALAXY, CYGNUS A, CASSIOPEIA A, OPTICAL ASTRONOMY

radio frequency (abbr **RF**) a frequency in the range 15kHz to

900MHz used for communication □ see FREQUENCY, DISH AERIAL, MICROWAVES, MASER, X-BAND, ELECTROMAGNETIC RADIATION

radio galaxy a galaxy which is a powerful source of radio waves and usu identified with an optical galaxy. These galaxies often show a binary structure (eg Cygnus-A) and emit more power at radio wavelengths than normal galaxies (eg the Andromeda galaxy). Some radio galaxies are enormous: for example the galaxy 3C 236 is a double radio source and emits radio waves spanning almost six megaparsecs of space. The enormous power output of a radio galaxy is thought to be due to a supermassive black hole at its centre, as with the smaller, but often equally powerful quasars. □ see RADIO ASTRONOMY, QUASAR, CYGNUS-A, VIRGO, RED SHIFT, SYNCHROTRON RADIATION

radioisotope *also* **radioactive isotope** a (manmade) radioactive substance (eg plutonium) which emits radiation and changes into another substance □ see ISOTOPE, RADIOACTIVITY, RADIOISOTOPE THERMOELECTRIC GENERATOR

radioisotope thermoelectric generator (abbr **RTG**) an electrical power source for spacecraft which uses the heat generated in a quantity of radioactive isotope to produce electricity using a bank of thermocouples. An RTG is used instead of, or in addition to, solar panels, especially for spacecraft (eg Pioneers 10 and 11, and Voyagers 1 and 2) exploring the outer regions of the Solar System where sunlight is weak. The RTG is sometimes fitted at the end of a boom away from the rest of the spacecraft to ensure that any radiation leakage does not damage sensitive electronic circuits and instruments. The radioisotope in an RTG is not capable of sustaining a chain reaction. However, the Soviet Cosmos 954 which crashed in Canada in 1978 did carry a nuclear energy source which was a true nuclear reactor. □ see THERMOCOUPLE, ISOTOPE, NUCLEAR FISSION

radiometer an instrument that measures the total energy radiated from a body, esp electromagnetic radiation at infrared and radio wavelengths □ see WEATHER SATELLITE

radiosonde a balloon-borne instrument package that transmits data about weather conditions as it is carried upwards into the atmosphere. Inside the radiosonde, a meteorograph continuously records temperature, pressure, humidity, etc, and this is recovered for analysis after the balloon lands. □ see EARTH'S ATMOSPHERE, WEATHER SATELLITE, NATIONAL OCEANIC AND ATMOSPHERIC ADMINISTRATION

radio telescope an instrument for receiving and analysing radio waves from celestial objects. A radio telescope has an aerial, comprising one or more dishes or dipoles, connected by feeders to recording and display equipment. □ see RADIO ASTRONOMY, JODRELL BANK, FEEDER, QUASAT

radome [*rad*ar d*ome*] a housing sheltering a radio antenna, esp a radar antenna on an aircraft □ see DISH AERIAL

range 1 the difference between the upper and lower values of a quantity ⟨*the ~ of temperatures on Mars*⟩ □ see SPECTRUM, BANDWIDTH

2 an area in and over which rockets are tested ⟨*the Eastern Test* Range⟩ □ see WOOMERA TEST RANGE

Ranger the first American space programme intended to provide close-up photographs and chemical data of the surface of the Moon using flyby and lander spacecraft. No soft-landings were achieved by the early probes, and in 1963 the programme was changed to reconnaisance only. Rangers 1 to 5 were launched between August 1961 and October 1962, but the first Ranger to be effective was Ranger 7 which crashed in the crater Tycho on 31 August 1964, transmitting 4306 photographs until impact. These photographs, and those obtained by Rangers 8 and 9 launched in 1965, greatly exceeded the resolving power of the Earth's most powerful telescopes. The Ranger programme made lasting contributions to our knowledge of the Moon. □ see LUNAR ORBITER PROBES, SURVEYOR, ZOND

rate the number of events per second, minute, or hour ⟨*data ~ from a spacecraft*⟩ □ see DATA RATE

ray 1a any of the lines of light that appear to radiate from a bright object ⟨*~s of starlight*⟩ **1b** a narrow beam of radiant energy ⟨*X*-rays⟩ **1c** a stream of (radioactive) particles travelling in straight line ⟨*alpha ~s*⟩ **2** relatively bright lines of dust radiating from impact craters on the Moon's surface

reaction time the time interval between a physiological stimulus (eg an audible warning signal) and the response to it (eg operating a cut-off valve) ⟨*an astronaut's ~ must be small*⟩ □ see PSYCHOMOTOR

readout the display of a measurement on a meter or paper printout ⟨*a temperature ~*⟩

real-time being or involving the almost immediate processing, presentation, or use of data ⟨*the ~ transmission of TV pictures from the Voyager spacecraft*⟩

real-time data data from an event available in a usable form at essentially the same time as the event occurs ⟨*~ for controlling the launch of a rocket*⟩

real-time programming any computer-controlled activity where the program makes immediate decisions on events as they arise ⟨*~ of a Martian buggy*⟩ □ see ARTIFICIAL INTELLIGENCE, SUPERCOMPUTER

receiver 1 that part of a telephone or radio intercom containing the mouthpiece and earpiece **2** that part of a communications system which receives, decodes, and makes available to a user the information transmitted to it or received by it ⟨*a radio astronomy ~*⟩ – compare TRANSMITTER □ see RADIO TELESCOPE, COMMUNICATIONS SYSTEM

reconnaissance a preliminary survey to obtain information about an enemy's territory and military resources ⟨*the US satellite Big Bird is used for ~*⟩ □ see BIG BIRD, COSMOS, STAR WARS

record to store information on magnetic tape or disk or another storage medium for future use ⟨*to ~ telemetry signals from a satellite*⟩ □ see VENUS RADAR MAPPER

recover to bring something into view again, esp a comet after its orbit has taken it to some distant part of the Solar System ⟨*Halley's comet was* ~ed⟩

recovery the process of picking up (eg from the sea) a satellite, a satellite instrumentation package, or some other part of a rocket vehicle ⟨*the* ~ *of the Space Shuttle's solid booster rockets after launch*⟩ □ see SOLID ROCKET BOOSTER, DROGUE PARACHUTE

red giant a giant star with a relatively cool surface temperature of between 2000 and 3000K (which makes it glow red) and a diameter of 10 to 100 times that of the Sun. ⟨*Arcturus, in the constellation Bootes, is a* ~⟩ □ see GIANT STAR

red shift a displacement towards longer wavelengths of lines in the spectrum of a celestial object. If the light is in the visible spectrum, the lines are shifted towards the red end of the spectrum. The red shift in the spectra of light from stars was first noticed in 1868 by Sir William Huggins who correctly interpreted it as a Doppler Effect on electromagnetic waves from a star moving away from the Earth (a star moving towards the Earth has a blue shift). For example, the wavelength of a line in the spectrum of the star Capella (α Auriga) is longer than the corresponding line in the spectrum of the Sun by 0.01%; this shift to the red indicates that Capella is receding from us at 0.01% of the speed of light, or 30kms^{-1}. In the 1920s, Edwin Hubble found red shifts in the spectra of light from distant galaxies using the Mount Palomar 5m telescope. He found the more distant the galaxy, the greater the red shift. Some quasars show enormous red shifts in their spectra; one in particular seems to be receding from us at the enormous speed of over 90% of the speed of light. Hubble's observation that the galaxies are receding from each other has been interpreted as part of the evidence that the universe is expanding. This idea is consistent with the view that the universe has evolved from a 'Big Bang'. Other evidence, notably microwave background radiation, also points to an enormous convulsion within the last 20 thousand million years. That the red shift exists is well proven: that it is caused by an expansion of the universe is questioned. Perhaps the red shift is caused by strong gravitational fields in intergalactic space, or by some physical process which has not yet been discovered. Only further detailed measurements using telescopes in space, especially the Hubble Space Telescope, will help us to understand better the structure of the universe and its origins. □ see DOPPLER EFFECT, MICROWAVE BACKGROUND RADIATION, BIG BANG, GALAXY, QUASAR, HUBBLE'S LAW

redundancy the building-in of duplicated components in a system, such as hardware in a computer, for backup purposes to prevent failure of the entire system (eg a spacecraft) in the event of failure of some one function □ see SPACEWALK RADIO, BACKUP, FLIGHT DATA SUBSYSTEM

reentry the return to the Earth's atmosphere of an aerospace vehicle ⟨*the Soviet 'cargo ship', Progress, burns up on* ~⟩

reentry window the area in the Earth's upper atmosphere through which a returning spacecraft can pass to accomplish a successful reentry

reflectance – see ALBEDO

reflecting telescope *also* **reflector** an instrument which uses a mirror to bring light of distant objects to a focus. This image is looked at with an eyepiece, is photographed, or is otherwise analysed. There are several different designs of reflecting telescopes. The first and most well-known is the Newtonian design which was introduced by Newton in 1670 to overcome the false colouring (chromatic aberration) of the images produced by the earlier Galilean and Keplerian refracting telescopes. The mirrors of these early reflecting surfaces were made of polished speculum-metal. In 1783 William Herschel built a reflecting telescope which had a 1.2m diameter metal mirror, and the largest telescope of this type was made by William Parsons in 1845. In the latter part of the 19th century, refracting telescopes made use of the achromatic objective lens which reduced chromatic aberration and gave better images than reflectors. But, in the early years of this century, it became possible to produce highly reflective silver surfaces on glass mirrors, and a new era began in the design of large reflecting telescopes. Since a lens has to be supported round its edge, it sags under its own weight and distorts the image, but mirrors can be supported all over their rear face, so giant reflecting telescopes, having a big 'light grasp', began to be made. One of the first of these was the 5m reflector of the Hale Observatories – compare REFRACTING TELESCOPE □ see HALE OBSERVATORIES, OPTICAL ASTRONOMY, HUBBLE SPACE TELESCOPE, MULTIPLE MIRROR TELESCOPE, APERTURE

reflection the turning back by a reflecting surface of some of the radiation incident upon it – compare REFRACTION

reflector 1 a (polished) surface for reflecting radiation which strikes it □ see LASER **2** – see REFLECTING TELESCOPE

refracting telescope *also* **refractor** an instrument which uses a lens to bring light of distant objects to a focus. This image is looked at with an eyepiece, is photographed, or is otherwise analysed. The first refracting telescopes (eg the Galilean and Keplerian designs) used a single-lens objective. The image they produced suffered from false colouring (chromatic aberration) and so reflecting telescopes were used because they did not have this defect. The first achromatic lenses were made by John Dolland in 1756 but they were only of small diameter. However, by the early 1800s Joseph Fraunhofer succeeded in making large uniform discs of crown and flint glass, and several refractors were constructed including a 0.32m refractor for the Royal Greenwich Observatory. By the end of the 1800s, Alvan Clarke made large refractors for the American observatories: a 0.91m for the Lick Observatory and a 1.02m for the Yerkes Observatory. Then, in the early 1900s, it became possible to put a highly reflective silvered surface onto glass mirrors and hence to construct large reflecting telescopes. However, the older refractors did have an advantage over reflectors as the image they produced was steadier: their optical system could be enclosed in a tube and this reduced the air currents that caused an unsteady image – compare REFLECTING TELESCOPE ☐ see ABERRATION 1, OPTICAL ASTRONOMY

refraction the change in direction of a radiation as it passes from one (optical) substance to another – compare REFLECTION

refractor – see REFRACTING TELESCOPE

refurbish to repair and/or renew ⟨*to ~ a communications satellite*⟩ ☐ see SPACE INSURANCE, SOLAR MAX

regulator a device that controls the flow of something (eg the rate at which liquid propellant is fed to a rocket engine) ☐ see BLEED

relativity two theories that were developed by Albert Einstein in the first quarter of this century: the special theory of relativity (1905) explains how the everyday laws of physics do not apply in systems that are moving rapidly relative to us; and the general theory of relativity (1916) explains how a strong gravitational field curves the space it occupies ☐ see SPECIAL RELATIVITY, GENERAL RELATIVITY, EINSTEIN, NEWTON'S LAWS OF MOTION

relay to receive a signal and send it to another destination ☐ see TRANSPONDER, REPEATER, DIRECT BROADCAST SATELLITE

relay station a system of aerials and electronic equipment on Earth or in space, for receiving radio and TV signals from one place and beaming them to another. Relay stations are used for outside broadcasts on a TV network, for transmission of telephone signals across country, and for global communications using artificial satellites. ☐ see TRANSPONDER, EARTH STATION, DISH AERIAL, DIRECT BROADCAST SATELLITE, MICROWAVES

reliability the probability that an electronic device, circuit, or system will operate without failure ☐ see BACKUP, SPACEWALK RADIO, SPACE INSURANCE

relief valve a pressure release valve in a pipe line which operates automatically when pressure rises to a certain value ☐ see BLEED

rem [*r*oentgen *e*quivalent *m*an] the quantity of ionizing radiation (eg gamma rays) that produces the same biological effect per gram of living tissue as one rad of 200 to 250kV X rays ☐ see RAD, SIEVERT

remote control command over an operation (eg space exploration) by means of signals sent from a distance. The signals may be sent by wires, sound waves, light (esp infrared as in TV remote control), or as radio waves when controlling spacecraft at great distances from the Earth. ☐ see ¹GALILEO, PIONEER VENUS, VOYAGER, FLIGHT DATA SUBSYSTEM, BUGGY, TELEPRESENCE

Remote Manipulator System (abbr RMS) a 15m-long articulated arm that is operated by remote control from the short-sleeve environment of the aft flight deck of the Space Shuttle. The elbow and wrist movements of the RMS enable payloads, esp satellites, to be lifted into and out of the Shuttle's payload bay, or allow repairs or adjustments, such as moving TV cameras, to be made. A TV camera and lights at the end of the RMS enable its operator to see what his/her hands are doing. The Spar Aerospace Company of Canada is developing a computerized space vision system to be connected to the TV cameras. It will enable more accurate readings of a satellite's range, velocity, attitude, and position to be measured. □ see SPACE SHUTTLE, SHUTTLE PALLET SATELLITE, RIDE, SOLAR MAX, TELEPRESENCE

remote sensing the use of instruments (eg radiometers, spectrometers, and radar) to obtain information (eg temperature, weather conditions, and surface features) about distant objects (eg planets). India is a large country with largely unexplored natural wealth. Encouraged by Landsat's success in identifying and monitoring global resources, the Indian Space Research Organization (ISRO) has used its home-built SLV-3 launcher to place the Rohini series of satellites, equipped with CCD cameras, into orbit. Their home-built Bhaskhara 1 satellite was launched from the Soviet Union in July 1979 and carried two TV cameras and a microwave radiometer. In 1986, the Soviet Union is to launch the 800kg India Remote Sensing (IRS) satellite and place it in a polar Sun-synchronous orbit where it can map out a particular stretch of the Earth's surface every 21 days.
□ see LANDSAT, NATIONAL OCEANIC AND ATMOSPHERIC ADMINISTRATION, RADIOMETER, SPECTROMETER, RADAR, SHUTTLE IMAGING RADAR, VENUS RADAR MAPPER, INDIAN SPACE RESEARCH ORGANIZATION

rendezvous to meet at an appointed place and time □ see FLYBY, GRAVITY ASSIST

repeatability the capability of device (eg an electronic component) or a system (eg a computer) to operate the same way every time it is called upon to do so

repeater 1 a device fitted into a long distance communications system (eg a submarine cable) to strengthen a weakening signal **2** a station in a microwave communications system that receives, strengthens, and retransmits radio signals □ see TRANSPONDER, EARTH STATION

rescue ball an 860mm diameter ball in which an astronaut crouches while he/she is pulled from one Space Shuttle to another during a 'ship-to-ship' rescue mission in space. The Rescue Ball would be used by astronauts who do not have spacesuits, as it has its own short-term air supply and radio communications. □ see SPACEWALK, SPACEWALK RADIO, AIRLOCK, EXTRAVEHICULAR MOBILITY UNIT

resolution 1 – see RESOLVING POWER **2** a measure of the detail present in the image of an object. The better the resolving power of a telescope, the greater the detail which can be expected in the final image.

resolving power *also* **resolution** the ability of a telescope to distinguish between two closely separated objects. Resolving power is usu measured in angular units of radians, and for a radio or optical telescope it depends on the diameter of the dish or antenna used to collect the radiation from the object under observation: doubling the diameter of the aperture, halves the angle between two objects which may just be resolved (ie distinguished as separate) and the smaller the angle, the greater the resolving power. Large-aperture telescopes are not only sensitive, ie able to 'see' a faint distant object, but they are also able to show fine detail in the image of that object. For an optical telescope, the eyepiece magnifies the image produced by the objective lens or mirror. This magnification need only be sufficient to enlarge the detail already present in the image so that the eye, photographic plate, or other sensor, can see the detail present. Any further magnification is useless (sometimes called empty) for it does not reveal any further information. A perfectly made 5m diameter mirror in a reflecting telescope should be able to see detail on a human face at 1000km (600 miles), but most optical telescopes on the Earth's surface cannot operate at their optimum resolving power because the Earth's atmosphere distorts the light passing through it. However, a telescope (eg the Hubble Space Telescope) in Earth orbit is free from this distorting effect and can operate much nearer its theoretical resolving power. The resolving power of a radio telescope with a dish antenna also improves with its diameter, but since the wavelength of radio waves is much longer than light waves, the resolution is poor. However, by combining a number of individual radio telescopes, it is possible to synthesize a telescope of much larger aperture. Aperture synthesis can also be used in optical astronomy. □ see APERTURE, AIRY DISC, HUBBLE SPACE TELESCOPE, APERTURE SYNTHESIS, VERY LARGE ARRAY, MULTIPLE MIRROR TELESCOPE

resonance the build-up of large amplitude vibrations in an electrical or mechanical system by stimulating the system with a vibration close to or nearly equal to the natural frequency of vibration of the system. Resonance is the operating principle of tuned circuits (eg in a radio receiver) and of lasers and masers. □ see LASER, MASER

response time the time taken for a cause to have an effect (eg the time taken for a signal to be received on command from a spacecraft)

restart the reignition of a rocket engine after an unpowered phase in a mission ⟨*the rocket was* ~ed *in the parking orbit*⟩ □ see PAYLOAD ASSIST MODULE, PARKING ORBIT

rest mass (symbol m_a) the mass that a body has when it is at rest relative to an observer. Einstein's special theory of relativity predicts that the mass, m, of a body increases with its speed, v, according to the equation: $m = m_a/(1-v^2/c^2)$ where c is the speed of light. □ see SPECIAL RELATIVITY, GRAVITATIONAL CONSTANT

retrieval the process of capturing an Earth-orbiting satellite for stowage or repair ⟨~ *of a communications satellite using the Shuttle's remote manipulator arm*⟩ □ see REMOTE MANIPULATOR SYSTEM, ROBOTICS, TELEPRESENCE

retrofire to ignite a retrorocket

retrofit the provision of new parts or equipment not available at the time of manufacture ⟨*the* ~ *of a Space Shuttle*⟩ – compare REFURBISH

retrograde *of a planet* **1** orbiting round the Sun or revolving on an axis in the opposite direction to other celestial bodies ⟨*Uranus and Venus have* ~ *motion on their axes*⟩ **2** moving westwards among the

stars as seen from Earth ☐ see HALLEY'S COMET, URANUS
retroreflector – see SATELLITE LASER-RANGING
retrorocket a small rocket motor on a spacecraft, artificial satellite,
or aerospace vehicle for decelerating it or changing its orbit ⟨*the
Venera landers used a ~ to make a soft landing on Venus*⟩ – compare
BOOSTER ROCKET ☐ see RETROFIRE, THRUSTERS, REVERSE THRUST
reusable intended to be reclaimed for reuse ⟨*the Space Shuttle is a
~ spacecraft*⟩ – compare EXPENDABLE
reverse thrust a force applied (by rocket engines) to a spacecraft
to slow it down ☐ see RETROROCKET
revetment a wall of concrete, earth, or sandbags for protecting
against the blast of exploding fuel in the event of rocket failure ☐ see
BLOCKHOUSE
revolution one complete orbit of a planet about the Sun, a satellite
about a planet, or an artificial satellite about the Earth or other celestial body ☐ see ORBIT
revolve to move in a curved path about a centre ⟨*the planets ~
around the Sun*⟩ – compare ROTATE ☐ see ORBIT
RF – see RADIO FREQUENCY
RGO – see ROYAL GREENWICH OBSERVATORY
rich *of a fuel mixture* high in the combustible component relative to
the oxidizer (eg excess liquid hydrogen in a LH_2/LO_2 cryogenic
propellant)
Ride, Dr K Sally (*b*1951) astronaut and first American woman to fly
into space aboard the Space Shuttle, *Challenger*, during its five-day
mission (STS-7) in June 1983. *Challenger* was making its second flight
into space and was the most ambitious Shuttle flight at that time: in
addition to launching two commercial communications satellites, the
flight proved that it was possible to manoeuvre close to an orbiting
satellite, capture it, place it in the Shuttle's payload bay, and bring it
back to Earth. Ride, who was accompanied by four male crew members, was selected as a Mission Specialist for the flight primarily
because of her expertise in operating the Remote Manipulator System
which makes possible the capture and release of satellites. Having a
PhD in experimental general relativity and X-ray astrophysics, Ride
also operated many of the scientific experiments carried aboard
Challenger. ☐ see SPACE SHUTTLE, REMOTE MANIPULATOR SYSTEM,
MISSION SPECIALIST, SULLIVAN
right ascension (symbol α) the angular distance of a celestial body
measured eastwards along the celestial equator from the vernal
equinox to the intersection of the hour circle through the body. The
right ascension is usually expressed in hours (h), minutes (m), and
seconds (s) from 0h to 24h ⟨*the nearest star, Alpha Centauri, has a ~
of 14h 28m*⟩. ☐ see CELESTIAL COORDINATES, CELESTIAL EQUATOR
rising the daily appearance above an observer's horizon of a particular celestial body. On the equator, the stars appear to rise (and set) at
right angles to the horizon; at the poles they never rise (or set) but
move parallel to the horizon. Stars rise in the same position relative to

other stars about four minutes earlier each day, but the Sun, Moon, and the planets rise at different points on successive days. □ see
EVENING STAR

RMS – see REMOTE MANIPULATOR SYSTEM

robot 1 an artificial satellite or an interplanetary spacecraft that is designed to carry out measurements of the local environment and transmit its findings to Earth **2** a programmable controller (eg a computer-controlled machine for assembling a space station) that accurately performs and monitors machine operations **3** an automated system for handling goods and materials on a production line **4** a (fictional) humanoid machine that walks and talks □ see BUGGY, ARTIFICIAL INTELLIGENCE, EXPERT SYSTEM, ROBOTICS, SUPERCOMPUTER, TELEPRESENCE

robot arm a jointed mechanical device driven by stepping motors, solenoids, or other electromechanical devices, which is able to manipulate objects under manual or automatic control. Robot arms are important parts of spacecraft, esp those sent to explore the surface of planets. For example, a robot arm was fitted to the Viking so that it could sample the soil on the surface of Mars, and the Space shuttle uses a robot arm to capture satellites in Earth orbit. □ see ROBOTICS, REMOTE MANIPULATOR SYSTEM, TELEPRESENCE

robotics the use of computer-controlled machines to carry out a predetermined set of operations (eg the analysis of the gases in Jupiter's atmosphere by the spacecraft Galileo) □ see ROBOT, ARTIFICIAL INTELLIGENCE, EXPERT SYSTEM, CYBERNETICS

robotize 1 to control a device or system using robots ⟨*the assembly of the space station was ~d*⟩ **2** to equip with robots □ see ROBOT, ROBOTICS, BUGGY

Roche limit the minimum distance at which a satellite can orbit a primary without being torn apart by gravitational forces between it and the primary. Assuming that the satellite is held together solely by the mutual attraction between its parts, the Roche limit is about 2.4 times the radius of the primary. The rings of Saturn may have been formed after the break-up of a satellite that strayed inside the Roche limit. □ see ASTEROID, SATURN

rocket – see LAUNCH VEHICLE

rocket booster – see BOOSTER ROCKET

rocket engine a device in which propellants react together to produce heat that is converted into kinetic energy to propel a vehicle to which it is attached. Strictly, a rocket engine is defined as one in which liquid propellants (eg liquid hydrogen and oxygen) are brought together, and includes a complex system of tanks, pipes, valves, etc. On the other hand, rocket motor refers to a relatively simpler device that burns solid propellants, and requires, basically, a fuel in a case, an igniter, and a nozzle. □ see CRYOGENIC ENGINE, PAYLOAD ASSIST MODULE, CENTAUR, SOLID ROCKET BOOSTER, ELECTRIC PROPULSION, ANTIMATTER PROPULSION, SOLAR SAILING

rocket motor – see ROCKET ENGINE

roll to turn an aerospace vehicle about its longitudinal (lengthwise) axis – compare YAW, PITCH

rollout the last part of the flight of an aerospace vehicle following touchdown

rookie somebody on his/her first space flight; a novice astronaut ⟨*Judith Resnik was a ~ on Space Shuttle flight 41D on 30 August 1984*⟩ – compare VETERAN □ see PAYLOAD SPECIALIST

Rosat [*Ro*tgen*sa*tellit] a German satellite carrying a German 0.8m diameter X-ray telescope and a UK Wide Field Camera (WFC), to be launched by the Space Shuttle in 1987. The German instrument will survey the whole sky at X-ray wavelengths, and the WFC is to study the extreme ultraviolet (EUV) and soft X-ray region of the electromagnetic spectrum. These two instruments are expected to have a great impact on astrophysics. They will provide accurate maps of X-ray sources and their spectra will help scientists understand what is happening at high temperatures in the atmospheres of stars and quasars, and in intense gravitational fields in supernova remnants. □ see X-RAY ASTRONOMY, ULTRAVIOLET ASTRONOMY, SUPERNOVA REMNANT, GENERAL RELATIVITY

rotate to turn about an internal axis ⟨*to ~ a spacecraft in preparation for reentry*⟩ □ see SPIN STABILIZED

rover – see BUGGY

Royal Greenwich Observatory (abbr **RGO**) an observatory built at Greenwich, London, at the behest of Charles II, to provide accurate star charts to improve navigation at sea. The meridian of zero longitude passes through the grounds of the observatory. John Flamstead was appointed first Astronomer Royal of the RGO in 1676. Atmospheric pollution and city lights of an expanding London has forced the RGO to be resited at Herstmonceux Castle in Sussex. Its principal instrument was the 2.5m Isaac Newton telescope, but this has been moved to the Northern Hemisphere Observatory on the Canary Islands where viewing is much better than in England. □ see GREENWICH MEAN TIME, REFLECTING TELESCOPE

RTG – see RADIOISOTOPE THERMOELECTRIC GENERATOR

rumble a low-pitched rumbling noise generated by inefficient combustion in a liquid-propellant engine □ see POGO

S

Sagan, Dr Carl (b1934) Professor of Astronomy and Director of the Laboratory for Planetary Studies at Cornell University. He is particularly interested in planetary surfaces and atmospheres: for example, he worked out the greenhouse model for Venus' atmosphere and he played a leading role in the Mariner, Viking, and Voyager missions to the planets. For 12 years he was Editor-in-Chief of *Icarus*, the leading professional journal devoted to planetary research. He is author of more than a dozen books (eg *Cosmos*), many of which discuss the probability of life on the planets and the origin of life on our own. In the 1960s he was a member of the group that mimicked the physical conditions on Earth millions of years ago in an attempt to produce the building blocks of DNA (deoxyribonucleic acid), the chemical responsible for transmitting genetic information. One of the substances produced in these experiments was ATP (adenosine triphosphate), the primary energy store of living tissue. Sagan and his team suggested that life on Earth began in the oceans when some molecules began to store energy from the Sun. These molecules then became the building blocks for complex nucleic acids and proteins which are the basis for life. □ see ASTROBIOLOGY, SEARCH FOR EXTRATERRESTRIAL INTELLIGENCE, SOLAR SYSTEM

Salyut a series of Soviet orbiting manned space stations equipped to carry out Earth, solar, and astronomical observations and materials processing experiments. The prototype 18-tonne Salyut space station was launched on 19 April 1971 from the Baikonur cosmodrome. Soyuz 11 docked with Salyut 1 on 23 April and its three crew worked in the space station for a record 570.4 hours. All three were killed on reentry. Salyut 6 was launched on 29 September 1979 and was extensively used for continuous periods of 96.4 days and 139 days by cosmonauts who travelled to and from the space station in Soyuz spacecraft. Astronauts aboard the most recent Salyut 7 space station, which was launched on 19 April 1982, broke the endurance record for a stay in space when they returned to Earth on 2 October 1984 after a 237 day flight. Occupants of Salyut 7 have a varied diet, including fresh fruit and vegetables, and are able to take hot showers thanks to regular supply missions by the specially adapted Progress 'ferries' that dock automatically with the space station. Unlike the 'one-off' Skylab space station, Salyut 7 can be expanded as required by adding further modules (eg the Soyuz spacecraft which carries visiting cosmonauts to and from Salyut 7). This modular design of space stations is favoured by Soviet engineers and by the designers of the future US Space Station. □ see SOYUZ, PROGRESS, COSMOS, SKYLAB, US SPACE STATION, KOMAROV, SAVITSKAYA

sample a small quantity of something 〈~s *of Moon rocks*〉

sample-and-return mission any project in which an unmanned spacecraft lands on a celestial body and brings back a sample of its surface rocks

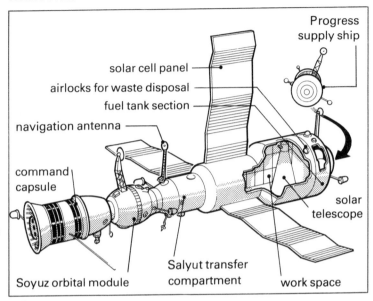

Salyut

SAS-1 [*Small Astronomy Satellite-1*] *also* **Uhuru** one of a series of three US X-ray and gamma-ray astronomy satellites launched between 1970 and 1973 from the San Marco platform in the Indian Ocean off the coast of Kenya. SAS-1 was one of the Explorer series of scientific satellites – Explorer 42, and it detected a black hole in the constellation Cygnus. □ see X-RAY ASTRONOMY, HIGH ENERGY ASTRONOMICAL OBSERVATORY

satellite 1 a natural celestial body that orbits another more massive body ⟨*the Moon is a ~ of the Earth*⟩ □ see MOON, GALILEAN SATELLITES **2** – see ARTIFICIAL SATELLITE

satellite laser-ranging accurate measurement of the movement of the Earth's crust and variations in the strength of its gravitational field, by reflecting laser beams from satellites. The UK has the most advanced facility in the world for these measurements using equipment developed by the University of Hull and the Royal Greenwich Observatory. The technique involves directing a laser beam at a satellite specially equipped with reflectors, and measuring the time interval for its return. Since the speed of the laser beam is known, the distance of the satellite can be calculated. If two stations send laser beams to a satellite in a known orbit, the distance between the two stations can be worked out. Repeated measurements then reveal any relative motion between two sections of the Earth's crust on which the stations stand, as well as small discrepancies in the Earth's gravitational field. The UK facility makes use of two of the eleven satellites fitted with laser reflectors: these are LAGEOS (Laser Geodynamics Satellite) launched by NASA and orbiting at an altitude of 6000km, and STARLETT orbiting at 1000km. LAGEOS is 600mm in diameter and covered with 400 corner cube reflectors. A neodymium laser sends out 150 picosecond pulses of laser light and the time for between 1 and 10 photons to return is measured to an accuracy of 100 picoseconds using a caesium atomic clock. In this way it is possible to measure distances accurate to 50mm. Just in case an aircraft strays into the path of the beam, putting its pilot in danger of suffering eye damage, anti-aircraft radar is used to automatically disable the laser beam within 10 milliseconds, only to cut back in when the aircraft leaves the radar beam. Better prediction of earthquakes is an obvious long-term goal of laser-ranging. □ see GEOPHYSICS, ATOMIC TIME, GLOBAL POSITIONING SYSTEM

Saturn the sixth planet of the Solar System which has a diameter of about 120 000km (second largest planet), orbits the Sun at about 1400 million kilometres, and has at least ten moons. Like Jupiter, Neptune, and other Jovian planets, Saturn is composed mainly of hydrogen and helium; the planet is so light it would float on water if there were an ocean big enough to support it. Saturn is surrounded by at least a thousand thin rings composed of countless bits of ice and frosted rock orbiting individually around its equator. The rings are separated by a number of gaps (eg the Cassini and Encke divisions). Dramatic and beautiful pictures of these rings have been captured by Voyagers 1 and 2 in their flyby missions in 1980 and 1981, respectively: their cameras revealed intricate hierachies of rings, wave patterns, spokes, and even braids. The sharpest resolution of the ring structure was obtained by an instrument called a photopolarimeter subsystem (PPS). The PPS was pointed at the bright star Delta Orionis, shortly before its closest encounter with Saturn. The star appeared to slide behind the rings and the PPS measured the variation in brightness of the star as light was blocked by varying amounts of ring material. Readings of this variation were taken 100 times per second so that the PPS obtained a resolution of about 100m. So much data has been sent back to Earth from these Voyager flybys that theoreticians are still trying to work out how material comes to be collected in such narrow tight rings. Voyager 2 also detected an unusual radio signal coming from Saturn which resembled static produced by thunderstorms. Analysis of the measurements indicated that this static does indeed come from massive thunderstorms in the atmosphere of Saturn. Thus Saturn, like the Earth, Venus, and Jupiter, becomes the fourth planet to show thunderstorm activity. Efforts are underway to detect these thunderstorms using Earth-based radio telescopes in Mexico, France, and Chile. □ see VOYAGER, JUPITER, RADIO ASTRONOMY, CASSINI

Saturn 5 the giant multi-stage rocket that was developed to launch the Apollo missions to the Moon in the late 1960s and early 1970s.

Saturn 5 had three sections, known as stages, each with its own engines. The first stage burned a mixture of kerosene (paraffin) and liquid oxygen which was fed to the five engines at 15 tonnes per second. After reaching a height of 61km and a speed of 8530kmh^{-1} in 2.5 minutes, the first stage was released and dropped into the sea. The second stage engines, burning liquid oxygen and hydrogen, then fired for 6 minutes to raise the vehicle to a height of 183.5km and a speed of 24 625kmh^{-1}. Then that stage was also released. The third stage, also burning liquid hydrogen and oxygen, fired for 2.75 minutes to place the Apollo spacecraft into a parking orbit at 28 000kmh^{-1}. After checking for 'all systems go', the lunar astronauts fired the third stage engine again to increase the speed to 39 270kmh^{-1} to speed the Apollo spacecraft to the Moon. Soon after, the Apollo spacecraft separated, turned round, and docked with the other end of the third stage, and drew out the lunar module ready for the Moon landing. Later, the third stage was fired again to make it crash into the Moon. The Apollo spacecraft itself contained a fourth stage rocket for manoeuvres in lunar orbit. □ see APOLLO, CRYOGENIC ENGINE, MOON, ^{2}TITAN, ARIANE

Savitskaya, Svetlana the first woman to 'walk' in space. She left the Russian space station, Salyut 7, on 25 July 1984 to carry out a 3 hour 35 minute EVA (extravehicular activity), the main aim of which was to test a new Universal Manual Tool (UR1 in Russian abbreviation). This electrically-operated tool produced an electron beam with which she burned a hole through a 0.5mm thick titanium plate: she also welded two metal samples together, and sprayed a silver coating on two black aluminium discs, all accomplished successfully in the vacuum of space outside Salyut 7. The next day, Svetlana's spacewalk was shown on Russian TV. Her spacesuit was specially equipped for female use and she had previously flown aboard Salyut 7 in 1982. □ see SALYUT, PROGRESS, TERESHKOVA, RIDE, SULLIVAN, SPACEWALK, SOYUZ

Schmidt camera – see SCHMIDT TELESCOPE

Schmidt telescope *also* **Schmidt camera** a telescope specially designed for photographing stars over a large region of the sky, ie it is a wide-field telescopic camera. Instead of using a mirror with a paraboloidal surface (as does the Newtonian telescope), a Schmidt telescope has a spherical surface. The spherical surface produces images suffering from spherical aberration, but it does not produce coma. (A paraboloidal mirror produces coma for a field of view greater than half a degree, so it cannot be used for wide-field photography.) In the Schmidt telescope, spherical aberration is corrected by putting a shaped transparent glass correcting plate in the path of the light rays entering the telescope before they reach the spherical mirror. An aberration-free image is then formed on a convex curved surface which is covered by the film. In this way, very sharp star images are formed over a field of about 10°. The 1.2m UK Schmidt telescope at Siding Spring Observatory, Australia, has been used to produce photographic plates for the Southern Sky Survey, which has mapped the positions of stars in the Southern Hemisphere. This survey has benefitted from the Palomar Sky Survey carried out in the Northern Hemisphere by the 1.2m Schmidt telescope at the Mount Wilson Observatory, California. □ see ABERRATION 1, HALE OBSERVATORIES, SIDING SPRING OBSERVATORY, REFLECTING TELESCOPE

Schwarzchild radius the radius of a particular black hole that must be exceeded if light is to escape from its surface. For a body of mass M, the Schwarzchild radius, R, is given by $R = 2GM/c^2$. For any radius less than this, the escape velocity is equal to the speed of light and the body becomes a black hole. The Earth would have to shrink to a diameter of about 10mm for it to become a black hole and to become invisible from the outside. □ see EVENT HORIZON, BLACK HOLE, ESCAPE VELOCITY

scintillation the flash or flashes of light produced by certain phosphors or crystals which have absorbed a single electron or proton or which have been struck by a quantum of gamma or X-ray radiation. Scintillation is responsible for the light emitted from the screen of a cathode-ray tube in a TV or VDU. □ see SCINTILLATION COUNTER, CATHODE-RAY OSCILLOSCOPE

scintillation counter an electronic instrument for measuring the amount of radiation given off by a radioactive source or the intensity of cosmic rays. Scintillation counters are used in X-ray and gamma-ray telescopes aboard spacecraft. □ see X-RAY ASTRONOMY, GAMMA-RAY ASTRONOMY, GAMMA-RAY OBSERVATORY

scope – see CATHODE-RAY OSCILLOSCOPE

Scorpius X-1 (abbr Sco X-1) the first cosmic X-ray source to be discovered, by instruments in a sounding rocket launched from White Sands, New Mexico, in June 1962. Its distance from Earth is not certain, but it is probably in the range 300 to 600 parsecs. This source has been identified with an optical binary system. □ see BINARY STAR, X-RAY ASTRONOMY

Search for Extraterrestrial Artefacts (abbr SETA) a proposed programme to look for objects that would indicate the prescence of alien spacecraft in the Solar System. These objects might be beacons, robots, or discarded debris from an alien search party, which were left on a planet or moon, or in orbit, and which, if discovered, would be of immense interest. But what would such objects look like, and where might they be? □ see SEARCH FOR EXTRATERRESTRIAL INTELLIGENCE

Search for extraterrestrial intelligence (abbr SETI) the search for intelligent life in the universe at large by attempts to detect the electromagnetic radiations (eg radio waves) produced by (advanced) technological civilizations. The number of life forms that, at this moment, might be sufficiently advanced technologically to be using radio communication is very uncertain: some astronomers believe that hundreds of thousands of advanced civilizations exist in the Galaxy. Others put the number much lower. Intelligent messages

second

at wavelengths ranging from ultraviolet to radio frequencies have been looked for from nearby stars. The Ozma Project of 1960 began pioneering studies using a 26m diameter radio telescope at Green Bank, West Virginia. Frank Drake and co-workers searched, fruitlessly, for intelligent signals (at the 21cm radio frequency hydrogen line emission) for about 150 hours without success. Their targets were the two Sun-like stars, Tau Ceti and Epsilon Eridani. In 1971, there was a proposal to build an immense circular array of perhaps 1000 radio dishes to search for signals from extraterrestrial civilizations. This giant radio 'eye', dubbed 'Cyclops', was not funded, but it would have been useful for radio astronomy and for tracking spacecraft. If intelligent life exists, we would expect to discover alien spaceprobes (perhaps similar to our own Viking spacecraft which searched for life on Mars), or the presence in the Solar System of interplanetary probes such as Pioneers and Voyagers. However, the volume of space which has been studied so far is extremely small, and the distances between planet-bearing stars is so large, that it is quite possible that even a huge artificial alien habitat in the asteroid belt of our own Solar System would be visually indistinguishable from asteroids to terrestrial observers. And for what reason would such a more advanced civilization want to talk with human beings? □ see ASTROBIOLOGY, PIONEERS 10 AND 11, UFO, SEARCH FOR EXTRA-TERRESTRIAL ARTEFACTS, CARBON, ANTHROPIC PRINCIPLE

second (symbol **s**) the fundamental unit of time in all systems of units. It is defined as the duration of 9 192 631 770 periods of the radiation corresponding to the transition between the two hyperfine levels of the ground state of the caesium atom. The second is used to express the characteristic lifetime of a system or the shortest time taken for a system to transmit appreciable information. Thus the lifetime of the universe is about 15 billion (15×10^9) years, ie about 10^{17}s), and the time for light to travel across a nucleus is about 10^{-23}s. Note that the ratio of these two quantities, larger to smaller, is about 10^{40}, and is regarded by some astrophysicists as a significant 'magic' number. □ see SI UNITS, KILOGRAM, METRE, ATOMIC TIME, ARC SECOND

Second Law of Thermodynamics a law stating that whenever any irreversible change takes place in a closed system the entropy (a measure of the system's degree of 'disorder') increases. For example, the conversion of chemical energy (eg from food) into information in our memories is always associated with the dissipation of some waste body heat. Thus, chemical energy has been irreversibly changed to heat which, as a random motion of molecules, represents an increase in disorder or chaos. The firing of a rocket engine produces a similar result: chemical energy from its fuel gives it and its payload potential and kinetic energy, but always with a proportion of wasted heat. If the universe is regarded as a closed system, the inference is that the total chaos of the universe is going up. Once all ordered forms of energy have been used up, the temperature throughout the universe evens

204

out and no life or useful work is possible. This conclusion is called the 'heat death' of the universe. However, by the time this theoretical state of affairs has been reached, the universe may have collapsed in on itself again, or intelligent life might have learned how to cope with it. □ see TIME, THERMODYNAMICS, BIG BANG

second of arc – see ARC SECOND

section part of an aerospace vehicle (eg the centre section or aft section) □ see SKIRT, MID DECK, MULTISTAGE ROCKET

seeing the clarity with which telescopic observations can be made at a particular time. Seeing varies with the amount of turbulence in the Earth's atmosphere. Poor seeing results in erratic movements of the image of a star; good seeing produces a steady disc-shaped optical image. No Earth-based optical telescope is able to operate at its theoretical maximum resolving power because of poor seeing, but a telescope above the Earth's atmosphere produces an image determined only by its resolving power. □ see RESOLVING POWER, HUBBLE SPACE TELESCOPE, ATTITUDE CONTROL, AIRY DISC

seismology the study of seismic (sound) waves which pass through the material of the Earth, or other solid body such as the Moon. On Earth, seismology helps to predict earthquakes and volcanic eruptions. When seismic waves are generated artificially, they are analysed to provide information about coal and oil resources, and to obtain information about the internal structure of the Earth. □ see SEISMOMETER, MOONQUAKES, APOLLO LUNAR SURFACE EXPERIMENTS PACKAGE

seismometer an (electronic) instrument for detecting and measuring (usu as a continuous graph) the strength of Earth tremors. Seismometers were taken to the Moon by the Apollo astronauts and detected 'moonquakes', and a lunar module was deliberately crashed into the surface to provide a source of seismic waves to gain an understanding of the structure of the Moon. □ see SEISMOLOGY, MOONQUAKES, APOLLO

selenography the study of the physical features of the Moon □ see MOON

selenoid a spacecraft in lunar orbit

selenology a branch of astronomy dealing with the Moon □ see MOON, APOLLO LUNAR SURFACE EXPERIMENTS PACKAGE, APOLLO

semimajor axis one half of the longest dimension of an ellipse □ see ELLIPSE, ORBIT

sensitivity the ability of a piece of electronic equipment (eg a radio receiver) to amplify a weak signal □ see CHARGE-COUPLED DEVICE, HUBBLE SPACE TELESCOPE

sensor a device which detects changes (eg variations in atmospheric pressure) in its surroundings and outputs an electrical or other signal. Many types of sensor are used in measurement, control, and communications systems aboard manned and unmanned spacecraft. □ see REMOTE SENSING, THERMOCOUPLE, STRAIN GAUGE, [1]METER

separation 1 the casting off of a spacecraft or satellite from its

launch vehicle or parent craft **2** (symbol ρ) the angular distance
between two bodies making a visual binary or double star. Its value is
expressed in arc seconds. The resolving power of an optical telescope
is often tested by seeing if it will 'separate' close binary stars. □ see
BINARY STAR, ARC SECOND, RESOLVING POWER, ASTROMETRY
separation velocity the speed of a spacecraft or satellite at the
time it separates from a launch vehicle or parent craft
service module – see APOLLO
servomechanism *also* **actuator** an (electromechanical) device that
supplies and transmits energy to control the operation of other
devices. A servomechanism uses sensors to monitor its movement and
to provide control signals. Servomechanisms are widely used in
spacecraft (eg Voyager) to position cameras, aerials, and other devices
for photographic and communications purposes.
servosystem a complete electromechanical system for precisely
controlling the position of something through the use of sensors which
monitor its movement and provide control signals. Servosystems are
used extensively in space hardware and especially in devices which
have to operate automatically in space. The Earth-orbiting Hubble
Space Telescope, the planet-touring Voyager spacecraft, and the
Mars-lander Viking spacecraft use servosystems. □ see
SERVOMECHANISM
SETA – see SEARCH FOR EXTRATERRESTRIAL ARTEFACTS
SETI – see SEARCH FOR EXTRATERRESTRIAL INTELLIGENCE
setting – see RISING
seven-segment display a device for displaying numbers and
some letters by making visible selected combinations of seven segments arranged in the form of a figure 8. The seven-segment display is
widely used in digital instruments (eg watches). There are two common types of seven-segment display, the liquid-crystal display and the
light-emitting diode display. □ see LIGHT-EMITTING DIODE, LIQUID
CRYSTAL DISPLAY
Seyfert galaxy a galaxy with a powerful energy source at its
centre. Over 150 Seyfert galaxies have been discovered. Most of them
are spiral galaxies and emit radiation at radio, infrared, optical, and,
particularly, X-ray wavelengths. They are probably less powerful
examples of quasars, and it is possible that all giant spiral galaxies,
including the Milky Way galaxy, spend some of their existance as a
Seyfert galaxy. □ see X-RAY ASTRONOMY, QUASAR
shield any material which is used to reduce or stop the passage of
radiation or particles through it. The material used depends on what it
is intended to absorb (eg gamma rays or heat) ⟨*a micrometeorite* ~⟩.
□ see DUST SHIELD, ¹GIOTTO, RADIATION HARDENING
shock wave a compressional wave formed whenever the speed of a
body in a fluid exceeds the speed at which it can be dispersed by the
fluid. The compressional wave banks up and, in the case of an aircraft
(eg the return of the Space Shuttle into the atmosphere), produces a
sonic boom.

shooting star the streak of light seen in a clear night sky caused by the rapid heating of bits of rock which enter the Earth's atmosphere at high speed from interplanetary space. On a clear moonless night, in the absence of meteor showers, about ten shooting stars an hour will be seen. □ see MICROMETEOROID, COMET, COSMIC DUST, ASTEROID, TEKTITE

short – see SHORT CIRCUIT

short-circuit *also* **short** an accidental connection between two parts of a circuit, usu the result of a path of low resistance to current flow. Circuits are generally protected against a short-circuit by a fuse which blows when excess current flows through it.

short waves radio waves which have wavelengths between about 120 metres and 20 metres, ie with frequencies between about 2.5MHz and 15MHz, respectively, and which are used mainly for amateur and long-range communications □ see LONG WAVES, MEDIUM WAVES, MICROWAVES, X-BAND, FREQUENCY, ELECTROMAGNETIC RADIATION

shutdown 1 the process of reducing rocket engine thrust to zero **2** the reduction of the power output of a nuclear reactor to zero

shutoff valve a valve that can stop the flow of a fluid ⟨*a* ~ *in a cryogenic propellant system*⟩ □ see BLEED, PNEUMATIC CONTROL ASSEMBLY

Shuttle cameras a series of small TV cameras positioned around the Space Shuttle that provide viewers on Earth with live pictures of operations inside the payload bay, and what astronauts are doing inside the cockpit. The cameras can be tilted, panned, and zoomed by the crew or from mission control at Houston. The cockpit camera is handheld and can be carried anywhere.

Shuttle glow a faintly glowing thin orange layer of light which covers the surfaces of the orbiting Space Shuttle. Although its origin is not fully understood, it appears to be due to chemical processes involving the ionization of oxygen atoms at high altitudes and oxygen molecules below 160km. It is important to know more about the Shuttle glow since it can confuse instruments in the cargo bay designed to look at faint stars and it could seriously affect the performance of the orbiting Hubble Space Telescope (HST). □ see HUBBLE SPACE TELESCOPE, SPACE SHUTTLE

Shuttle Imaging Radar (abbr **SIR**) an experimental radar camera carried aboard flights of the Space Shuttle (eg Mission 41G in October 1984), and designed to obtain high-resolution images of the Earth's surface. On Mission 41G, data obtained by SIRs at 46 megabits per second was to be transmitted to ground stations via the orbiting Tracking and Data Relay Satellite (TDRS). But there were problems with folding and unfolding the SIR antenna, which had to be stowed before reentry by two spacewalking astronauts, Leestma and Sullivan. Also, the TDRS satellite temporarily went out of control due to human error, at the control centre in White Sands, New Mexico. So SIR was not very successful on this flight! When it was first used on the second flight of the Space Shuttle in November 1981, it provided spectacular colour pictures of ancient watercourses below the featureless sands of the Sahara Desert. SIR's radar signals are at a frequency of 1.3 gigahertz (wavelength 230mm), and not only penetrate clouds but also see through several metres of dry sand. This ability was totally unexpected, and is likely to be of great value to geologists, archeologists, and planetologists. □ see RADAR, VENUS RADAR MAPPER, TRACKING AND DATA RELAY SATELLITE, SULLIVAN

Shuttle Pallet Satellite (abbr **SPAS**) a 1500kg reusable experiment platform that is designed to operate inside, or free-flying outside, the Space Shuttle's payload bay. SPAS is manufactured by the West German firm Messerschmitt-Bolkow-Blohm (MBB). SPAS-01 made its debut aboard flight STS-7 of the Space Shuttle in June 1983. In this flight, mission specialists Sally Ride and John Fabian used the 15m long arm of the Remote Manipulator System to park SPAS outside the payload bay, though some of the experiments were run with SPAS inside the bay. Six experiments were carried by SPAS-01 and these were activated by the mission specialists from inside the Space Shuttle. The experiments included MAUS-1 and MAUS-2 and a study of heat pipes. □ see EURECA, GETAWAY SPECIAL, MAUS, HEAT PIPE, SPACE SHUTTLE

Shuttle telephone a service which allows telephone subscribers to call in and hear live transmissions between the Space Shuttle astronauts and Mission Control, Houston

Shuttle toilet an arrangement for receiving and disposing of an astronaut's body waste. Though ten million pounds was spent on developing a zero gravity, 'high technology' toilet, astronauts have found it to be an unreliable device. Waste matter is dumped from a storage tank through a vent to space in the hull of the Shuttle. On mission 41B in February 1984, a lump of ice outside the hull blocked the vent and astronauts had to resort to using plastic bags so as not to fill up the tank before the blockage was removed. On two earlier flights of the Shuttle, astronauts had to free a jammed toilet mechanism using a crowbar! The use of bags, subsequently stored to be removed after every flight (the method used on earlier Apollo flights), may turn out to be the most suitable method for collecting and disposing of body waste. This 'low technology' solution (which women astronauts are less happy with) has reminded some observers that NASA once spent $1M dollars on developing a pen to use in weightless conditions when lead pencils are quite adequate for the task. □ see SPACE SHUTTLE, SPACE MEDICINE

sidereal of or expressed in relation to the stars ⟨~ *time*⟩

sidereal period the time taken by a planet or satellite to complete one orbit round a body, measured by reference to the background stars ⟨*the ~ of the Moon about the Earth is 27.322 days*⟩

sidereal year the time taken by the Earth to complete one orbit of the Sun by reference to a background star which is regarded as fixed in position ⟨*the ~ is equal to 365.256 sidereal days*⟩

Siding Spring Observatory an Australian observatory sited at an altitude of 1165m on Siding Spring Mountain in New South Wales. Its main telescopes are the 3.9m Anglo-Australian telescope, and the 1.2m UK Schmidt telescope, both of which are reflecting telescopes. □ see REFLECTING TELESCOPE, SCHMIDT TELESCOPE, HALE OBSERVATORIES

sievert (abbr Sv) a unit for measuring the effects of nuclear radiation on biological tissue. For X rays and gamma rays, 1Sv corresponds to the dissipation of 1 joule (J) of energy per kilogram (kg) of tissue. A single dose of about 10Sv on the whole human body is generally fatal. A typical chest X ray gives a dose of about 0.1mSv. The average annual dose from natural background radiation in the UK is about 2mSv. A return flight between London and Paris exposes the body to about one millionth of a Sievert (1μSv). Astronauts and cosmonauts journey well above the protective blanket of the Earth's atmosphere so they are exposed to higher levels of radiation from cosmic rays and the Earth's radiation belts. Radiation dose was formerly measured in units of rems. □ see REM, JOULE, NUCLEAR RADIATION, SPACE MEDICINE

signal any message transmitted from one place to another ⟨*a Morse code* ~⟩ □ see TELEMETRY, SEARCH FOR EXTRATERRESTRIAL INTELLIGENCE

signal-to-noise ratio (abbr **SNR**) the ratio of the strength of a signal to the strength of any background noise which might also be present. The higher the SNR, the more easy it is to understand the signal. □ see INTERFERENCE

silicon chip a small piece of silicon on which complex miniaturized circuits are made by photographic and chemical processes. The silicon chip is cheap to make in large quantities, can withstand rough treatment, is reliable in operation, needs very little electrical power, and it has therefore become the heart of computer, control, and communications systems. □ see MICROELECTRONICS, SUPERCOMPUTER

simulation the use of a model or make-believe of a real system, using a computer and one or more VDUs for displaying computer-generated or prerecorded pictures. Simulations are widely used for training astronauts (eg for learning to manoeuvre the Space Shuttle in orbit). □ see JOHNSON SPACE CENTRE, HYDROLABORATORY, CENTRIFUGE

singularity a point of zero radius at which the laws of physics break down. The concept of a singularity is nearly beyond comprehension, but there is nothing to stop a collapsing star becoming a singularity. This singularity is the centre of black hole. If the universe eventually stops expanding and collapses, it is likely to become a singularity. □ see BLACK HOLE, EVENT HORIZON, BIG BANG, PULSATING UNIVERSE

SIR – see SHUTTLE IMAGING RADAR

SI units [*Systeme International d'Unites*] an internationally agreed system of units for science and technology. Its seven basic units are the metre, kilogram, second, ampere, kelvin, candela, and mole. The system is coherent, ie the basic units can be combined to derive other units as, for example, the unit of energy, the joule, which can be expressed either as the watt × second or as the newton × metre. Multiples and submultiples of SI units are obtained using the prefixes mega-, micro-, etc. □ see JOULE, PASCAL, COULOMB, WATT

skin the outer covering of a body ⟨*the ~ of a propellant tank*⟩

skin temperature the temperature of the outer surface of a body ⟨*~ of the Space Shuttle on reentry*⟩ □ see THERMAL TILES

skirt the lower outer part of a rocket □ see NOZZLE

Skylab a space station based on the third stage of the Saturn 5 rocket and manned by US astronauts from May 1973 to early 1974 in a 435km high circular orbit inclined at 50° to the equator. Three crews of three astronauts each manned the space station for 28, 59, and 84 days. They travelled to and from Skylab aboard command and service modules left over from the Apollo programme. One of the two large solar panels which provided power, and a micrometeroid shield had been damaged when Skylab was launched, so astronauts fitted a reflective shield over its surface to prevent it overheating. Even though electrical power was reduced, this first space 'workshop' was used for a variety of scientific studies using instruments attached to the Apollo Telescope Mount (ATM): over 180 000 photographs of the Sun were taken, and 40 000 of the Earth. The experiments led to great advances in astronomy, and proved the value of making celestial and Earth observations from space. Skylab also provided the first opportunity (as the Soviet Salyut 6 was also doing) to study the physiological and psychological effects of long-term exposure to weightlessness. Skylab was expected to stay in orbit for 10 years, boosted by the Space Shuttle, when necessary, to maintain its orbit. However, by March 1978, Skylab's orbit was only 380km high, the Space Shuttle was not operational, and attempts to stabilize the orbit were unsuccessful. On 11 July 1979 Skylab burnt up in the atmosphere over the Indian Ocean. Brilliant flashes were seen from the city of Perth in Western Australia, and pieces of the space station were strewn over hundreds of miles of the Australian Outback, but neither injury nor damage to property was reported. □ see APOLLO, SATURN 5, SALYUT, US SPACE STATION, SPACE MEDICINE

slipstream a stream of air or liquid flowing round a body □ see STREAMLINE FLOW

SNR – see SIGNAL-TO-NOISE RATIO

soft having relatively low energy ⟨*~ X rays*⟩

soft landing the descent of a vehicle to the surface of a planet or one of its moons, without damaging any part of the vehicle (except possibly its landing gear) or its payload – compare HARD LANDING, □ SEE DROGUE PARACHUTE

SOHO – see SOLAR HELIOCENTRIC OBSERVATORY

solar of, related to, or caused by the Sun ⟨*~ radiation*⟩ – see SOLAR CELL, SOLAR SYSTEM, SOLAR WIND, SOLAR RADIATION, SOLAR MAX

solar cell a (transducer) device which converts solar energy or artificial light into electrical energy. It comprises a large-area semiconductor pn junction diode: when light falls on the junction, it causes a voltage across the junction which enables current to flow in a circuit connected to it. Thousands of solar cells are grouped together in paddle-like arrays as solar panels to provide the main source of electrical power in artificial satellites and interplanetary spacecraft. However, for spacecraft heading beyond the orbit of Mars, the solar radiation is too weak (about 4% of that at Earth), and alternative sources of power are used. □ see RADIOISOTOPE THERMOELECTRIC GENERATOR, POWERSTAT, OAST-1 SOLAR ARRAY

solar constant the amount of solar energy passing perpendicularly through one square metre at a given distance from the Sun. At the Earth's orbit, the solar constant is about 1.4 kilowatts per square metre ($1.4kWm^{-2}$). Clearly, designers of solar power systems for houses and solar panels for spacecraft need to know the value of the solar constant. □ see SOLAR MAX, EARTH RADIATION BUDGET SATELLITE, SOLAR CELL, SUN

solar flare an explosive release of energy from the Sun's upper chromosphere that sometimes ejects radiation and particles into interplanetary space. Solar flares vary in size but they all reach a maximum brightness within a few minutes and fade before an hour passes. Large solar flares are experienced on Earth: communication on short-wave radio is often subject to fading, especially in the daylight hemisphere of the Earth, and this is due to increased ionization of the D-layer of the ionosphere which restricts the passage of radio waves to the higher E-layer where they are normally reflected. Ultraviolet light and sometimes X rays, emitted during solar flares, are responsible for this ionization and hence the fading. These energetic radiations also produce attractive auroral displays in the Earth's atmosphere. □ see SUN, SOLAR WIND, SUNSPOTS, IONSPHERE, GEOMAGNETISM, AURORA, T TAURI STARS

solar flux – see INSOLATION

Solar Heliocentric Observatory (abbr **SOHO**) a unique spacecraft due for launch in 1992 and designed to make continuous observations of the Sun. SOHO is to be placed in a special (heliocentric) orbit round the Sun, inside that of the Earth, where the gravitational attractions by the Sun and the Earth are equal and opposite. The spacecraft will thus remain fixed on the Earth/Sun line about 1.6 million kilometres from the Earth. SOHO will make continuous observations of the solar surface, corona, and solar wind. One of its key functions will be to detect oscillations of the surface, which should give important clues about the structure of the Sun's interior. □ see EUROPEAN SPACE AGENCY, SUN

Solar Max a NASA spacecraft launched in February 1980 to investigate the Sun during the maximum portion of its 11-year cycle. The satellite returned useful data for nine months until a failure in its attitude control system made it lose the ability to point accurately at the Sun. However, in April 1984, Solar Max was captured and repaired by the crew of Shuttle Mission 41C (STS-13) which gave new life to the nine experiments on board. One of the experiments, the Activity Cavity Radiometer Irradiance Monitor (ACRIM), showed that during the period 1980 to the end of 1983 there was a net decrease of 0.08% in the total solar radiation emitted by the Sun. If this trend continues, it would result in a 2% drop in a century and produce a marked effect on the Earth's climate. A decrease of 0.5% in the solar radiation would lower the Earth's temperature by 1°C, and a drop of 13% would result in a global ice cover 1.6km (1 mile) thick! The successful capture and repair of Solar Max was marked by another space 'first': astronauts Nelson and van Hoften, who worked on the stricken satellite, set a new record for extravehicular activity (EVA) by remaining outside *Challenger* for 7 hours 18 minutes, just 19 minutes short of the time spent by Apollo astronauts on the Moon's surface in December 1972. □ see REMOTE MANIPULATOR SYSTEM, SUN, RECOVERY, SOLAR CONSTANT

solar paddle a solar panel □ see SOLAR CELL

solar panel – see SOLAR CELL

solar power satellite – see POWERSAT

solar radiation infrared, visible light, ultraviolet, radio, and other electromagnetic radiations emitted by the Sun □ see SOLAR WIND, ELECTROMAGNETIC RADIATION

solar sailing the use of solar radiation to propel interplanetary and interstellar spacecraft. When light is absorbed by or reflected from a surface, it exerts a force – not very much force – but if a spacecraft has a large enough 'sail' to catch the solar radiation, sufficient force would be generated to drive the spacecraft away from the Sun. The force arises because light and other electromagnetic radiations consist of photons which carry energy and momentum: the change in momentum as the photons are reflected or absorbed, imparts a force to the sail just like the force a tennis ball exerts on a racket. A Solar Sail Cargo vessel (SSCV) has been considered by NASA for ferrying supplies from Earth-orbit to Mars as part of a manned expedition to Mars: a suitable sail would have to be at least 2km square. A spacecraft fitted with a solar sail was also discussed for a rendezvous with Halley's comet in 1986, but it was never funded. During the Mariner 10 mission to Venus and Mercury in the early 1970s, Mariner was running out of propellant for its attitude control system, and so a plan was devised to use its solar panels as solar vanes to keep the cameras and radio aerials pointing in the right direction. Solar vanes for this purpose were also fitted on India's Insat satellite. However, the force exerted by solar radiation can be a nuisance: in space navigation it is essential to make allowances for it otherwise a spacecraft would end up thousands of kilometres off target. □ see SOLAR WIND, SOLAR FLARE, SUN, STAR, PHOTON, ATTITUDE CONTROL, INDIAN SPACE RESEARCH ORGANIZATION, ELECTRIC PROPULSION

Solar-Stellar Irradiance Comparison Experiment (abbr SOLSTICE) – see UPPER ATMOSPHERE RESEARCH SATELLITE

Solar System the Sun and all the bodies which orbit it. It has nine known planets: Mercury, Venus, Earth, Mars, Jupiter, Saturn, Uranus, Neptune, and Pluto. There are 35 known natural satellites of these planets. The asteroid belt between Mars and Jupiter contains countless minor planets. Comets exist in great numbers beyond the orbit of Pluto, and there are countless meteoroids and a few manmade spacecraft in heliocentric orbit. The Sun contains 99.86% of the mass of the Solar System and Jupiter contains most of the rest. If the distant comets are included, the Solar System can be considered to be a sphere with a radius of at least 10 000 astronomical units, with Pluto situated at a distance of about 50 astronomical units from the Sun. (The Earth is 1 astronomical unit from the Sun). The Sun and its entire family are moving in a nearly circular orbit round the centre of the Galaxy at an average speed of 250kms^{-1} so that the system completes an orbit every 220 million years. The Solar System is also moving in the direction of the constellation Hercules at a speed of 19.4kms^{-1}. Bar the comets, all the bodies of the Solar System rotate round the Sun in the same direction and in the same plane (the ecliptic), which supports the view that the Solar System condensed out of a contracting swirling cloud of interstellar dust about 4600 million years ago. Exploration of the Solar System using spacecraft such as Viking, Venera, Voyager, Galileo, Giotto, and others, is bent on finding out more about the Solar System's history. □ see SUN, MERCURY, VENUS, EARTH, MARS, JUPITER, SATURN, URANUS, NEPTUNE, PLUTO, COMET, ASTEROID, MICROMETEOROID, GALAXY, SPACECRAFT, BODE'S LAW

solar wind the flow of charged particles (mainly protons and electrons) outwards from the Sun through the Solar System. Near the Earth, these particles are moving at speeds between 200 and 900 kms^{-1}, though there are only about 8 million particles per cubic metre of space at this distance. The solar wind is a plasma and it is greatly distorted by the Earth's magnetic field to produce the magnetosphere. Interplanetary spacecraft, Pioneers 10 and 11 and Voyagers 1 and 2, have measured the solar wind throughout the Solar System. The satellite ISEE-3 (International Sun-Earth Explorer) was launched in 1978 to study the solar wind, solar flares, and sunspots, unperturbed by the influence of the Earth's magnetic field. □ see SOLAR SAILING, SUN, MAGNETOSPHERE, PLASMA, INTERNATIONAL COMETRY EXPLORER, SOLAR HELIOCENTRIC OBSERVATORY, SPACELAB

solid rocket booster (abbr **SRB**) two reusable rockets that burn a solid fuel and assist the Space Shuttle main engines (SSMEs) during the first two minutes of the ascent of the Space Shuttle from the launch pad. Each rocket is 35.3m long and 3.7m in diameter, and is packed with solid propellants consisting of 16% atomized aluminium powder (the fuel), 69.83% ammonium perchlorate (the oxidizer), 0.17% iron oxide powder (a catalyst to control the burning), 12% polybutadiene acrylic acid acrylonitryl (a binder), and 2% epoxy curing agent. At launch, the total mass of each booster is 589.67 tonnes, and together they provide a thrust of 24 575kN. These boosters have nozzles mounted on gimbals to control the direction of the rapidly moving gases (which leave the nozzle at about 9700kmh⁻¹ and help steer the entire Space Shuttle. After the SRBs have ceased firing, they are separated from the external tank, coast upwards, and then fall on a ballistic trajectory for almost four minutes. At an altitude of about 4700m, a barometric switch operates and the nose cap of each SRB is ejected. This allows a 16m diameter drogue parachute to be deployed. At about 2000m, a second signal from the barometric switch allows the three main parachutes to be pulled out by the drogue parachutes. Each parachute has a diameter of 41m and together they reduce the speed of the SRB to about 82kmh⁻¹ at splash-down. The SRBs are recovered for reuse on another flight, and are assembled and refurbished at the Kennedy Space Centre. In order that more payload can be carried into space in the payload bay of the Space Shuttle, a new lightweight SRB case has been designed. The new case is made of filaments of plastic reinforced with graphite fibres, wound into a cylinder; it weighs 15 tonnes less than the original metal case, but the latter will continue to be used on most missions where payload weight is not critical. □ see EXTERNAL TANK, THRUST VECTOR CONTROL, SPACE SHUTTLE, DROGUE PARACHUTE

solid-state camera – see CHARGE-COUPLED DEVICE

SOLSTICE [*Sol*ar-*St*ellar *I*rradiance *C*omparison *E*xperiment] – see UPPER ATMOSPHERE RESEARCH SATELLITE

sounding rocket a rocket designed to explore the Earth's atmosphere, or to make astronomical observations. Data obtained by instruments aboard a sounding rocket are either radioed back to Earth or are returned by parachute. Astronomers don't make as much use of sounding rockets as they used to since satellites enable measurements to be made over a longer period from a much higher vantage point. Nevertheless, some of the earlier sounding rockets did make some important discoveries about the Earth's upper atmosphere, and they detected the first X-ray source in the Galaxy. □ see CYGNUS X-1, RADIOSONDE, EARTH'S ATMOSPHERE

southbound node – see DESCENDING NODE

Soyuz a series of more than 30 Soviet manned spacecraft, the first of which, Soyuz 1, was launched in April 1967. Each Soyuz spacecraft is made up of three modules: an orbital module in which three cosmonauts can live and work, a propulsion and instrumentation module, and a reentry module which is occupied by the cosmonauts on the way up to orbit and is the only module to return to Earth. Since 1971, the Soyuz spacecraft have been used to ferry cosmonauts to and from the orbiting Salyut space station. For example, Soyuz T-11 was launched on 3 April 1984 and docked with Salyut 7 on 4 April. The spacecraft carried three cosmonauts (Malyshev, Strekalov, and Sharma – an Indian astronaut). The launch and docking with Salyut's rear unit was shown live on Soviet and Indian TV. These three cosmonauts (collective call-sign 'Jupiters') joined three other long-stay cosmonauts (Kizim, Solovyov, and Atkov – call-sign 'Mayaks') who had flown to

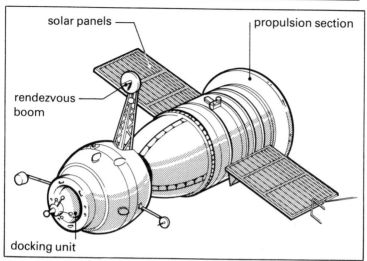

Soyuz

Salyut 7 on 8 February 1984. Salyut 7 now contained 6 astronauts, the most so far. The visiting Soyuz T-11 crew stayed until 11 April when they left using Soyuz 10, leaving the fresher Soyuz 11 for the long-stay crew. Soyuz-10 touched down at 10:50 in a ploughed field 56km east of Arkalyk. The Mayaks were visited by the three-man crew of Soyuz T-12 (call-sign 'Pamirs') on 18 July 1984. One of the Pamirs was a woman, Svetlana Savitskaya. The Pamirs returned to Earth on 26 July 1984 using the Soyuz T-12 spacecraft they had arrived in. The Salyut 7 cosmonauts had exceeded the old space endurance record of 211 days held by Berezovoi and Lebedev, when they returned to Earth on 2 October 1984 after a 237 day flight. Salyut 7 was not to be used again until cosmonauts Dzhanibekov and Savinykh returned to the space station aboard Soyuz T-13 on 6 June 1985. They began to make the station habitable again, which included recharging storage batteries and unfreezing blocked water pipes. On 23 June the unmanned Progress 24 supply ship delivered two tonnes of much needed consumables, followed by Progress 25 on 21 July which brought biological and astrophysical equipment, and fresh water for Salyut 7's tanks. Each supply ship was released to burn up in the Earth's atmosphere. On 17 September, the two cosmonauts were joined by Vasyutin, Grechko, and Volkov aboard Soyuz T-14. On 26 September Dzhanibekov and Grechko returned to Earth in Soyuz T-13 leaving Vasyutin, Savinykh and Volkov in the Salyut 7/Soyuz T-14 combination. □ see SALYUT, PROGRESS, SAVITSKAYA, SPACE SHUTTLE, SPACE STATION

¹space 1a *also* **outer space** the part of the universe lying beyond the Earth's atmosphere ⟨*a journey into* ~⟩ □ see SPACESHIP **1b** the volume in which all celestial bodies, including the Earth, exist ⟨*the universe of time and* ∼⟩ **2** an amount of room set apart or available ⟨*a* ∼ *for astronauts to work*⟩ □ see SPACE STATION **3a** an interval of time **3b** the time between pulses transmitted in a digital communications system

²**space** to place at intervals or arrange with gaps between ⟨*the radio aerials were* ∼d *at 1km intervals*⟩ □ see ARRAY

space adaptation syndrome – see SPACE SICKNESS

space age the present era of space exploration which began when the Soviet Union launched the first Earth-orbiting satellite, Sputnik, on October 4 1957 □ see SPACE SCIENCE AND TECHNOLOGY

space base – see SPACE STATION

space-based of or being in space ⟨∼ *astronomy with the Hubble Space Telescope*⟩

spacecraft *also* **interplanetary probe** a highly sophisticated device sent from Earth to make measurements in space between the planets of the Solar System and to observe the planets at close range. Many interplanetary probes have been sent on missions of this kind since the launch of Sputnik in 1957. All the planets, except the outer planets Neptune and Pluto, have so far been visited by interplanetary probes, some of which have entered the atmosphere of the planets to make observations on the spot. The Moon, of course, has been at the centre of investigation for more than 20 years. Even the Sun at the centre of the Solar System has been examined at close range from within the orbit of Mercury. In addition, special missions to asteroids and comets are underway. All these missions to the far reaches of the Solar System have been made possible by the remarkable advances in microelectronics, esp its use in the development of sophisticated communications, control, computing, and communications systems. □ see VOYAGER, VIKING, ¹GALILEO, ¹GIOTTO, PIONEER VENUS, INTERNATIONAL COMETRY EXPLORER, CASSINI

space greenhouse a type of Controlled Ecological Life Support System (CELSS) consisting of a self-contained Earth-orbiting garden for supplying astronauts with lettuce, tomatoes, carrots, potatoes, melons, strawberries, cabbage, etc. A feasibility study is being carried out by the Boeing Aerospace Company under a NASA contract into the design of a space greenhouse which may become necessary if astronauts are to remain in a space station for long periods of time. Nothing would go to waste in the process of growing food in this way for it would be a closed-cycle system: waste products (human, plant, and inorganic) would be recycled into a nutrient solution used to feed the plants. Although it is suspected that many plants will grow in zero gravity, artificial gravity could easily be supplied by growing the plants in rotating drums, and deep freezing the produce would be no problem. Solar panels extended from the garden would power artificial lighting for the drums, although sunlight would be in ready supply in cloudless skies. □ see SPACE STATION, WORLD SHIP, STARSHIP, MICROSPACESHIP

space insurance the business of guaranteeing repayment for loss or damage to satellites, spacecraft, and launch vehicles. Lloyds insured the two communications satellites, Palapa-B2 and Westar-6, for $75M and $105M, respectively. This was fortunate for the two countries, Indonesia and North America, they were intended to serve, since the Payload Assist Module (PAM) booster rockets on these satellites failed to ignite and the satellites were left in unusable elliptical orbits rather than in their intended circular geostationary orbits. However, Lloyds asked NASA to mount a rescue attempt during Mission 51A in November 1984. Astronauts, Joseph Allen and Dale Gardner, successfully captured and fixed the two satellites in the *Discovery*'s cargo bay. They were returned to their manufacturer, Hughes Aircraft Company, California, where they have been refurbished for sale at about $20M each (it is said). Lloyds paid NASA about $10M to mount the rescue mission. Despite this successful rescue mission, space insurance premiums have risen from 6% to 17% and some losses in 1985 have meant rates in the region of 25%. The rescue of satellites has proved the flexibility of the Space Shuttle in its operations above the Earth. □ see PAYLOAD ASSIST MODULE, SPACE SHUTTLE, SOLAR MAX

Spacelab a reusable self-contained laboratory, which is carried into space inside the Space Shuttle's payload bay, and in which scientists can work in a shirt-sleeve environment to carry out a range of experiments. Spacelab has been designed and built by the European Space Agency (ESA) as part of a cooperative programme between ESA and NASA. There are two main modules to a Spacelab: a pressurized module, comprising two 2.7m long by 4m diameter cylinders in which payload specialists live and monitor the experiments; and an unpressurized platform (pallet) where instruments can be directed to make observations in space. Of the 38 experiments aboard Spacelab 1, launched in November 1983, 26 were sponsored by ESA (8 were from the UK) and 12 by NASA. These experiments included observing X-ray emissions from the Sun and the energies of cosmic rays. Spacelab missions are now dedicated to specific areas, eg Spacelab 4 launched in January 1986 was dedicated to Life Sciences. On this mission, 24 experiments were planned, totalling 1400kg of a total payload of 8000kg. The areas covered reflect the concern about astronauts' well-being in the weightless conditions aboard orbiting space stations: cardiovascular, vestibular, renal/endocrine, haematologic, and immunological systems were areas covered. The payload specialists who carried out these experiments were not trained astronauts, but doctors with full-time jobs in health care and environmental protection. They are now entitled to call themselves astronauts. □ see SPACE SHUTTLE, PALLET, PAYLOAD SPECIALIST, SPACE SICKNESS, SPACE MEDICINE

spaceman (fem **spacewoman**) **1** one who travels outside the Earth's atmosphere □ see ASTRONAUT, COSMONAUT **2** a visitor to Earth from outer space □ see ENCOUNTER 2, UFO

217

space medicine techniques for ensuring the well-being of astronauts and cosmonauts. The environment aboard spacecraft and space stations poses a number of problems for people on long spaceflights; the main challenge to the human body is adapting to weightlessness. The organs concerned with vision, balance, and coordination (neuro-vestibular); the heart rate, blood pressure, and blood flow (cardiovascular); body fluids (homeostatic); and the musculoskeletal system are all profoundly affected by weightlessness (zero-g). The most obvious effect of zero-g is space motion sickness which seems to affect most space travellers to some degree; it develops within hours of launch and has seriously affected the success of some missions. Fortunately, it can be suppressed by drugs and the problem resolves itself after a few days. On the other hand, cardiovascular and homeostatic changes cause a feeling of stuffiness and fullness in the face as fluid moves from the lower limbs to the upper body; and there is an unexplained fall in the number of red blood cells. Calcium and nitrogen are lost from the body which causes weakness in the bones and wasting of muscles, and it is feared that irreversible changes in bone structure and metabolism will occur after about 8 months if protective measures are not adopted. Unfortunately, these problems do not seem to be eased by the use of thigh-cuffs, anti-g suits, and saline ingestion. The only way to counteract the effects of zero-g is to provide artificial gravity on board the spacecraft in the form of a centrifuge. To these problems of survival in space must be added the, as yet, unknown effects of sustained exposure to the higher radiation levels of space. For the US Space Station, it has been estimated that a crew member will receive an exposure of 15 rads per 90 day tour, which is higher than any acceptable radiation dose for Earth-bound exposure. In addition to the physiological problems, astronauts and cosmonauts have to adapt mentally to the rigours of spaceflight. Most space travellers have been the 'super-hero' type, but increasingly the crew of the Space Shuttle and Salyut are ordinary payload specialists who have a specific (scientific) task to perform once he/she has been delivered to space by the pilot and mission specialists. In choosing the crew, psychological factors such as general emotional stability, positive attitude towards self, and ability to live and work in close relationships with other crew members will be sought. And, of course, special attention has to be paid to the design of spacecraft and space stations to ensure that the best use is made of available space, that individual crew members have privacy, that food and drink is nutritious and palatable, and that waste disposal and personal hygiene present no problems. Not least, is the effect on astronauts and cosmonauts of the lack of normal signals (eg sunset and sunrise) which help to synchronize internal body clocks on Earth. The provision of regular work, rest, and feeding patterns will have to be adhered to if astronauts and cosmonauts are not to suffer from some severe form of jet lag when they arrive at their destination. □ see SKYLAB, SALYUT, SPACE SICKNESS, AEROEMBOLISM, CENTRIFUGE,

WEIGHTLESSNESS, VERTIGO, RAD, REM

spacenik someone who talks, writes, or otherwise gets involved enthusiastically in space activities

Space Plasma Lab – see PLASMA

space pollution – see POLLUTION

space radiator a device used on a spacecraft or space station to collect and dissipate waste heat to space. A conventional pumped liquid, loop cooling system (like a domestic refrigerator) is used on the Space Shuttle to dissipate about 25kW of waste heat. However, if a pipe is punctured by a micrometeorite, cooling fluid is lost and the system becomes useless. A different technique based on heat pipes is being developed for use on the US Space Station. Heat pipes transfer waste heat by the evaporation of a liquid moving along narrow tubes. Heat pipe space radiators should be able to handle the 300kW of waste heat which will need to be radiated into space from the Space Station in order to create comfortable living conditions for astronauts. Prototype heat pipe space radiators are being evaluated on flights of the Space Shuttle. □ see HEAT PIPE

space rescue – see RESCUE BALL

space science and technology a process that combines know-how and people in the exploration of space for the benefit of mankind □ see OPTICAL ASTRONOMY, RADIO ASTRONOMY, X-RAY ASTRONOMY, INFRARED ASTRONOMY, ASTRONAUT, COSMONAUT, MATERIALS PROCESSING, SPACE STATION, ARTIFICIAL SATELLITE, SOLAR SYSTEM

space settlement a large structure in space in which perhaps 1000 to 10 000 people could live and work in an Earthlike environment. Solar energy would provide power, and raw materials from the Moon or asteroids would be used for building satellites or for manufacturing materials and medicines for use on Earth. Large adjustable mirrors outside the settlement would direct sunlight into the interior to regulate the seasons and control the day-night cycle. Food would be grown in special agricultural stations at each end of the living area. Trees, birds, fish, and clouds would be a feature of the curved landscape inside the settlement. A cylindrical section 32km long and 6.4km in diameter, rotated once every 115 seconds would provide Earth-like gravity. Space settlements need to occupy stable positions in space so that there is no danger of them falling on the nearest planet or moon. Such points, known as Langrangian points, are located 60° ahead and 60° behind the Moon in its orbit about the Earth, and one of these might well be chosen as the site for a future space settlement. □ see STARSHIP, SPACE GREENHOUSE, CENTRIFUGAL FORCE, MOON BASE, LANGRANGIAN POINTS

spaceship a manned spacecraft, esp one making an interplanetary or interstellar journey (eg the fictional Starship Enterprise). It is interesting to note the analogy between the development of ships and spacecraft. Each originated with a small craft of limited capabilities (eg the US Mercury spacecraft), leading to the development of roomier craft (eg Soyuz and Skylab). The orbiting Soviet Soyuz spacecraft is serviced with the Progress 'cargoships', and the operation of the proposed US Space Station will require the services of ferries, orbital manoeuvring vehicles, fuel tankers, and container ships. □ see SPACE STATION, SOYUZ, SKYLAB, SPACE SHUTTLE, EXPEDITION TO MARS, WORLD SHIP, DAEDALUS

Space Shuttle a reusable Earth-orbiting spacecraft which is launched vertically with rocket assistance and which glides back to Earth to land like a normal aircraft. The assisting rockets are jettisoned after use and fall back into the sea for reuse. The underside of the Space shuttle is covered with heat-resisting tiles to protect the spacecraft from the heat generated by friction with the air during re-entry into the Earth's atmosphere. The Space Shuttle was developed to reduce the cost of manned flights into space, and to explore the possibilities of servicing and launching artificial satellites, of making extended observations of the Earth and space using on-board instruments, and of building space stations which could be assembled and manned using the Space Shuttle as a 'shuttle service'.

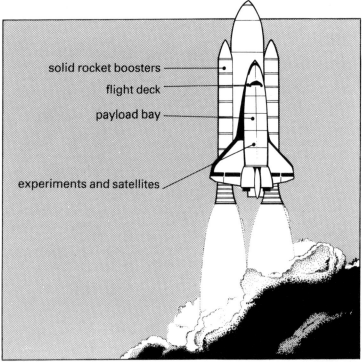

Space Shuttle

The part of the Space Shuttle that actually reaches Earth orbit is called the Orbiter. Four Orbiters have so far seen service: *Columbia,*

Challenger, Discovery, and *Atlantis,* the last making its maiden flight
on 3 October 1985. On 28 January 1986, *Challenger,* on Mission 51L,
exploded after take-off killing all 7 people on board. From October
1985, the Western Space and Missile centre on the Californian coast
became available for launches of the Space Shuttle, in addition to the
Kennedy Space Centre at Cape Canaveral. As from February 1984,
NASA no longer announces Space Shuttle missions by their STS
(Space Transportation System) numbers, but by Mission numbers and
letters: STS-10 (the tenth flight) became Mission 41B, the '4' indicating that it was a 1984 launch (the NASA year running from October),
the '1' meaning the flight was launched from Cape Canaveral, Florida
(a '2' indicating a launch from the Vandenberg Air Force Base,
California), and the letter 'B' that it was the second flight planned for
that (NASA) year. □ see STS-1 – STS-9, MISSION 41B – 61C, THERMAL
TILES, SOLID ROCKET BOOSTER, EXTERNAL TANK, MOBILE LAUNCH
PLATFORM, PAYLOAD BAY, SHUTTLE TOILET, PAYLOAD ASSIST
MODULE, SPACELAB, SPACE INSURANCE, US SPACE STATION, GETA-
WAY SPECIAL, SHUTTLE GLOW, HUBBLE SPACE TELESCOPE, MANNED
MANOEUVRING UNIT, REMOTE MANIPULATOR SYSTEM, SOLAR MAX

space sickness *also* **space adaptation syndrome** health problems
astronauts and cosmonauts experience in adjusting to weightlessness
in space. About 45% of astronauts have experienced nausea, vomiting, and general malaise. To speed up the process of finding
countermeasures for space sickness, NASA has recently set up a
Space Biomedical Adaptation Institute. The health of cosmonauts in
the Salyut space stations is very carefully monitored to see how long
they can remain in weightless conditions and still re-adapt to normal
gravity. □ see SPACE MEDICINE, SALYUT, SKYLAB

space station *also* **space base** a large (Earth-) orbiting structure in
which people can live and work for long periods. Skylab, Salyut, and
Spacelab are forerunners of larger space stations being planned for the
1990s. These smaller space stations are designed to test the feasibility
of using space stations for manufacturing (eg pharmaceutical products), and to assess and remedy the problems of living in weightless
conditions. However, a future space station will be rotated (or some
part of it will) so that an artificial gravity is produced to help its occupants combat the problems of weightlessness. In the long term, large
space stations will be used as assembly and launch platforms for manned spacecraft destined for a landing on Mars or for other
interplanetary journeys. Most of the studies being undertaken for
large space stations favour a modular building system so that the space
station can be expanded as the need arises. □ see US SPACE STATION,
SKYLAB, SOYUZ, SPACELAB, SPACE SICKNESS, SPACE GREENHOUSE,
SPACE SETTLEMENT, MATERIALS PROCESSING, MOON BASE, WORLD
SHIP, EXPEDITION TO MARS

spacesuit the clothing worn by an astronaut or cosmonaut when
working in the hostile conditions of space (ie when taking a spacewalk
outside an orbiting space station). The design of spacesuits has

evolved dramatically since the first astronauts and cosmonauts ventured outside their orbiting space capsules in the 1960s. Nowadays, when a crew member of Salyut or the Space Shuttle undertakes extravehicular activities, he or she wears complex life-support, control, and communications systems. No longer does an astronaut have to wear a spacesuit when boarding the Space Shuttle: the cabin is provided with an oxygen/nitrogen mixture just like the flightdeck of an aircraft. However, at launch, all the astronauts do wear helmets to provide a supply of oxygen if the cabin were to suddenly depressurize. □ see EXTRAVEHICULAR MOBILITY UNIT, EXTRAVEHICULAR ACTIVITY

space telescope – see HUBBLE SPACE TELESCOPE

spacetime a single concept which specifies an event (eg a supernova explosion) by four coordinates, three of which give position in space and one which gives time. The path of a body in spacetime records its history and is known as its world line. Einstein used the concept of spacetime in the special and general theories of relativity. In the special theory, spacetime is flat and can be described by three-dimensional Euclidean geometry. But in the general theory, where gravitational fields are present, spacetime becomes distorted: massive objects (eg black holes) curve spacetime. The universe may be curved by the matter within it to such an extent that it is closed. □ see RELATIVITY, BLACK HOLE, THREE-DIMENSIONAL, HYPERSPACE

space tourism space travel for holidays and tours. Tourism is big business and the attractions of a space holiday are obvious. But, at the moment, it is too expensive for the average holidaymaker to consider taking, a weekend break aboard the Space Shuttle. The cost of putting 1 tonne of payload into low Earth orbit using the Shuttle is about £2M per tonne compared with about £1000 per tonne for a flight across the Atlantic by Jumbo Jet. If sufficient numbers of people were prepared to pay, say, £5000 for a few days in space, the average space transportation cost would have to be lowered to about £30 000 per tonne. This would require the development of a 'spacebus', comprising a booster and an orbiter capable of delivering say, 30 tourists to an orbiting hotel. The booster would accelerate the orbiter to about Mach 4, and then return and land for reuse. The orbiter would continue into orbit and, under its own power, park at the orbiting hotel. And how would the tourists then spend their time? They would certainly amuse themselves in the microgravity gymnasium. And the panoramic views of the Earth, Sun (viewed through dark glasses!), Moon, and planets (seen through telescopes) would be unforgettable. Beyond these activities, the holiday might be an exercise in creative imagination! Or they might catch a taxi to 'Moon City', or make the much longer excursion to the Mars colony. □ see SKYLAB, SPACE SHUTTLE, MICROGRAVITY, EXPEDITION TO MARS

Space Transportation System (abbr **STS**) – see SPACE SHUTTLE, STS-1 – STS-9, MISSION 41B – MISSION 61C

space tug – see ORBITAL MANOEUVRING VEHICLE

spacewalk any activity (but hardly ever walking) an astronaut or cosmonaut does in space outside a spacecraft □ see SPACESUIT, EXTRAVEHICULAR MOBILITY UNIT

spacewalk radio the UHF radio carried by an astronaut when doing a spacewalk outside the Space Shuttle. These 4kg radios are about the size of a loaf of bread and provide telemetry signals on the astronaut's heart beat and other body functions, and deliver caution and warning signals when life support systems are running low. Since the radio is an astronaut's only link with the Shuttle, each radio has two transmitters and three receivers to provide a backup in the event of failure of one of the systems. □ see UHF, EXTRAVEHICULAR MOBILITY UNIT, RELIABILITY

span the distance from tip to tip of an aerospace vehicle's wings □ see AEROFOIL, AERODYNAMICS

Spartan a package of scientific instruments due to be carried into Earth orbit in the payload bay of the Space Shuttle to analyse ultraviolet radiation and X rays from the Sun and other celestial objects. Once in orbit, Spartan is released from the payload bay using the robot arm of the Space Shuttle's Remote Manipulator System to become a freeflying 'satellite' alongside the Orbiter. The instruments inside Spartan can be remotely operated by scientists within the Orbiter. Small thrusters rotate the package to point its instruments at different celestial objects. Once a sequence of measurements has been accomplished, Spartan is retrieved and brought back to Earth where its recorded data are analysed. Three Spartan payloads are being used: Spartan 1 examines X-ray emissions from clusters of galaxies and from the centre of our own Milky Way; Spartan 2 measures the characteristics of the Sun's solar wind using a 0.43m solar telescope; and Spartan 3 will carry out a survey of ultraviolet emissions from selected star fields in the search for young stars and white dwarfs, using a Schmidt electrographic camera which measures ultraviolet radiation in the 1230 to 1600 angstrom range of wavelengths. □ see ULTRAVIOLET ASTRONOMY, SOLAR WIND, WHITE DWARF, REMOTE MANIPULATOR SYSTEM, SCHMIDT TELESCOPE, SPACE SHUTTLE

SPAS – see SHUTTLE PALLET SATELLITE

special relativity a theory proposed by Albert Einstein in 1905 showing that if we take measurements in a system that is moving rapidly relative to us, then the everyday laws of physics do not hold. Special relativity agrees with Newton's Laws of Motion for low relative velocities, up to and far beyond the speed of spacecraft, for example. But for speeds close to the speed of light (ten thousand times the speed of the fastest spacecraft), special relativity predicts strange things. First, if we 'heard' the ticks of a clock in the speeding spacecraft, we would notice that the ticks got slower and slower as the spacecraft approached the speed of light: this is called time dilation and has been verified experimentally for an atomic clock on a satellite; time dilation also means that astronauts aboard the speeding spacecraft take much longer to reach old age than we who are left

223

behind, since all their body functions take place at a slower pace. Second, objects moving rapidly relative to us appear shorter in their direction of motion; this observation is known as the Lorentz contraction. Third, the mass of a rapidly moving body is not constant and increases as its speed increases. Time intervals appear longer by a factor $1/\beta$, distances appear shorter by a factor β, and mass appears to increase by a factor $1/\beta$, where $\beta = \sqrt{(1-v^2/c^2)}$, v is the relative speed of the two systems, and c is the speed of light. For example, if v=0.8c, ie the relative velocity if four-fifths the speed of light, $\beta=3/5$. Thus, time passes 60% slower, lengths are 60% shorter, and mass is 60% bigger relative to someone who measures these quantities on a spacecraft travelling at 80% the speed of light. One further consequence of special relativity is that the speed of light is the same for all observers, ie however fast the spacecraft moves away from or towards us, we would always measure the same speed for its radio signals; and this conclusion of special relativity has been proved experimentally. (One would expect that radio waves would get slower and slower as the spacecraft neared the speed of light just as, to a passerby, the exhaust vapour behind a passing car actually moves in the same direction as the car.) □ see GENERAL RELATIVITY, MASS-ENERGY EQUIVALENCE

specific impulse a measure of a rocket engine's performance, equal to the ratio of the thrust produced, F, to the rate at which matter is expelled by the rocket, ie specific impulse is equal to F/kgs^{-1}. □ see THRUST, MULTISTAGE ROCKET

spectral of or made by a spectrum ⟨*a spectrometer produces ~ lines*⟩ □ see ELECTROMAGNETIC RADIATION

spectrometer an instrument that produces a spectrum of a source of light or other electromagnetic radiation, and measures the wavelengths in the spectrum. Spectrometers are used in observatories on the ground, and in satellites and spacecraft, to determine the elements and molecules present in celestial objects. □ see SPECTRUM, EMISSION SPECTRUM, ABSORPTION SPECTRUM, PHOTOMETER, SPECTROPHOTOMETER

spectrophotometer an instrument that measures the intensity at different wavelengths of the radiation emitted from (celestial) objects □ see SPECTROMETER, SPECTRUM

spectrum (plural **spectra, spectrums**) the complete range of frequencies of sound, light, or other electromagnetic radiation ⟨*the visible ~*⟩ □ see ELECTROMAGNETIC RADIATION, EMISSION SPECTRUM, ABSORPTION SPECTRUM, SPECTROPHOTOMETER

speech recognition *also* **voice recognition** the process of getting a computer to recognize spoken words by comparing them with words stored in its memory as patterns of electrical signals. It is likely that astronauts will soon be using speech recognition equipment to direct the control of their spacecraft or the operation of instruments and robot arms, esp in the assembly and use of space stations. □ see SPEECH SYNTHESIS, MICROELECTRONICS

speech synthesis the generation of artificial speech using computers or purpose-designed integrated circuits. Speech synthesis is being increasingly used to make computers more user-friendly, and instrument readings (eg those in cars) can now be spoken. Indeed the first speech synthesizer is now operating in space aboard UOSAT-2 which is orbiting the Earth and transmitting the readings of its instruments in English. □ see UOSAT-2, SPEECH RECOGNITION, MICROELECTRONICS, SUPERCOMPUTER

¹speed 1 the rate at which a body or radiation moves ⟨*the ~ of a spacecraft*⟩ □ see SPEED OF LIGHT, VELOCITY **2** the rate at which a device operates ⟨*a high-speed computer*⟩ **3a** the sensitivity of a photographic emulsion ⟨*high-speed film*⟩ **3b** the light-gathering power of a lens or mirror ⟨*the ~ of a telephoto lens*⟩

²speed to move quickly ⟨*the spaceship ~ed on its way*⟩

speed of light (symbol **c**) the speed of light and other electromagnetic waves (eg radio) through a vacuum is equal to 299.7924 58 million metres per second (approximately $300 \times 10^6 \text{ms}^{-1}$), and it is less than this through air and other transparent substances. Even though the speed is so high, radio communication with interplanetary spacecraft takes time since the spacecraft are at such great distances from Earth. For example, when Voyager 2 crossed Saturn's rings in August 1981, command signals from Earth took 1 hour 26 minutes 35 seconds to reach the spacecraft. There are two experimental facts about the speed of light which are the basis for Einstein's Theory of Relativity: the speed of light cannot be exceeded, and it is the same for all observers regardless of their speed relative to the source of the radiation. □ see LIGHT-YEAR, RELATIVITY, ELECTROMAGNETIC RADIATION

spherical aberration a distortion of the image of a lens that has spherical surfaces, giving different foci for axial and marginal rays □ see ABERRATION 1, SCHMIDT TELESCOPE

spin the property of an elementary particle (eg an electron) or a nucleus that corresponds to intrinsic angular momentum: it can be thought of as rotation of the particle about its axis. According to the rules of quantum theory, spin can take only certain special values equal to a whole number or half a whole number multiplied by Planck's constant. Thus protons, neutrons, electrons, muons, and neutrinos have half a whole number of spin and photons or pions have zero or a whole number of spin. The quantization of spin and orbital angular momentum explains the fine detail (hyperfine structure) present in spectra. □ see ELEMENTARY PARTICLES, PLANCK'S CONSTANT, PAULI EXCLUSION PRINCIPLE, MESON, NEUTRINO, ANGULAR MOMENTUM, SPIN STABILIZATION

spin axis the line passing through the centre of mass, about which a body spins ⟨ *Uranus has a ~ inclined at almost 90° to its orbit*⟩ □ see SPIN STABILIZED

spin stabilized *of a satellite or spacecraft* provided with directional stability in space by being spun about an axis of symmetry. Most of the communications satellites launched from the Space Shuttle are 'spun-up' before release from the payload ⟨*Intelsat-6 is a ~ communications satellite*⟩ □ see COMMUNICATIONS SATELLITE, INTELSAT, BRITISH AEROSPACE, GYROSCOPE

spiral – see SPIRAL GALAXY

spiral arms regions of gas and dust in a spiral galaxy (eg our own Milky Way galaxy) where stars are born □ see SPIRAL GALAXY, MILKY WAY 1

spiral galaxy *also* **spiral** a flattened disc-shaped system of stars, rich in gas and dust, containing prominent spiral arms which wind outwards from a dense central nucleus. As in our own spiral galaxy, the Milky Way, star formation is still taking place in the spiral arms. Two-armed spirals are the most common spiral galaxy, though one arm and even three arms have been observed. □ see GALAXY

spoiler a plate, bar, or other device, that projects from the top of an aerofoil to break up the air flow, thus reducing lift and increasing drag. A spoiler is usu hinged or retractable so that it can be removed from the air flow when necessary. □ see AEROFOIL

Spot [*S*atellite *P*our L'*O*bservation de la *T*erre] a remote-sensing satellite built and scheduled for launch by France in 1985 to provide high-resolution photographs of the Earth's surface. Spot has two arrays of sensors for high-resolution images in the visible spectrum, which can distinguish objects on the Earth's surface only 10m apart, much better than the 35-metre resolution of Landsat's images. Moreover, Spot is able to look from side to side, unlike current Earth-observation satellites which can look only at the surface directly beneath them. Spot provides images for civilian users (eg oil and mining companies), and the images obtained are priced to compete with Landsat's prices. But they may not be affordable by the Third World which desperately needs information on the potential of their own mineral and other resources. □ see LANDSAT, SKYLAB, SALYUT

Sputnik 1 the first of the Earth's artificial satellites launched by Russia on October 4 1957. Sputnik 1 measured temperature and electron densities in the upper atmosphere before burning up as it entered the Earth's atmosphere on 4 January 1958. □ see EXPLORER 1, ARTIFICIAL SATELLITE, TELEMETRY, SPACE AGE

SRB – see SOLID ROCKET BOOSTER

stability 1 the property of a spacecraft, artificial satellite, rocket, or aerospace vehicle to maintain its attitude (eg to keep its aerial pointed Earthwards) **2** the property of a (rocket) fuel to retain its characteristics for long periods or in adverse conditions

stage 1 a self-contained section of a launch vehicle which may be separated from the rest of the vehicle after use □ see MULTISTAGE ROCKET **2** any part of a circuit or system which by itself, or together with similar parts, forms part of a larger system ⟨*a two-stage amplifier*⟩

staging the process of separating one or more rocket stages from a spent stage during the launch of a multistage rocket □ see MULTISTAGE ROCKET

star any of many large celestial objects that generate energy from nuclear fusion reactions in their interiors. Millions of stars make up a galaxy. A star is formed when an interstellar gas and dust cloud collapses under its own gravitational attraction. What initiates the collapse of a dust cloud is not clear: perhaps it is triggered by the compression wave generated by a nearby supernova. But, whatever the cause, as the dust cloud contracts, pressure builds up in it and the temperature rises in a hot core. This temperature is not high enough for nuclear reactions to take place and the star is called a protostar. If the mass of the protostar is below about one twentieth of the mass of the Sun, the central pressure and temperature are insufficient to trigger nuclear reactions. A core of a star above this mass limit eventually becomes hot enough (about 10^7K) to sustain nuclear fusion reactions which, for the greater part of its life, involves the conversion of hydrogen into helium. The more massive a star, the hotter it is and the faster it burns its hydrogen, but during this phase, the heat generated prevents further gravitational collapse. However, once the fuel runs out, gravity wins and the star collapses on itself and becomes a white dwarf, a red giant, or a supernova. Apart from the Sun, our nearest star, Proxima Centauri, is about 1.3 parsecs away. □ see SUN, PROTOSTAR, INTERSTELLAR MEDIUM, HERTZSPRUNG-RUSSELL DIAGRAM, SUPERNOVA, BLACK HOLE, WHITE DWARF, T TAURI STARS, POPULATIONS OF STARS, GALAXY

Star City a town about 40km northeast of Moscow, site of a space museum open to the public and a training centre for cosmonauts. Cosmonauts, engineers, scientists, and doctors live in Star City. Cosmonauts leave Star City for the Baikonur cosmodrome about 21 days before a flight. □ see COSMODROME

star dust the material between the stars, some of which is produced by exploding stars (supernovae), and some is the material from which new stars are born □ see INTERSTELLAR MEDIUM, SUPERNOVA, STAR, PROTOPLANET, INFRARED ASTRONOMY SATELLITE

STARLETT – see SATELLITE LASER-RANGING

starship a spaceship launched from Earth to explore the solar systems of other stars. If we do not destroy ourselves in warfare, and if we can afford to develop the enormous technical resources required, the commissioning of a starship to make the long journey to a nearby star is the logical outcome of projects like Apollo, Soyuz, Viking, Venera, Voyager, and the Space Shuttle. The unmanned spacecraft Pioneers 10 and 11 are already on their way to the stars, and Voyagers 1 and 2 will follow after they complete their exploration of the Solar System in the 1990s. A few problems must be solved before astronauts and cosmonauts embark on a starship (Enterprise?). For example: it may be technically difficult to talk with the crew over a few light-years of intervening space; the voyage would take a very long time, perhaps 50 years, using foreseeable rocket technology, so what do the crew do with time on their hands – would they have to be put in 'cold store' and be revived at their destination?; it would take a few years to get

an answer from the crew who are awake since radio waves travel at only the speed of light; all food and other resources would have to be grown or stored on board, and almost everything would have to be recycled; what sort of fuel would the starship use, and would it need to refuel? A starship would probably be built and equipped with materials carried to an assembly point in Earth orbit, on the Moon, or on Mars, by vehicles like the Space Shuttle – compare WORLD SHIP
□ see BRITISH INTERPLANETARY SOCIETY, DAEDALUS, ASTROBIOL-OGY, ELECTRIC PROPULSION, ANTIMATTER PROPULSION, HYPER-SPACE

Star Wars a programme using laser-equipped satellites to provide a protective shield for the security of the West in the event of attack by hostile warplanes and missiles. The laser defence system would replace existing antiballistic missiles (ABMs) which must be launched quickly if they are to meet hostile intercontinental ballistic missiles (ICBMs) which reach their targets in less than 30 minutes. (Sea-launched ballistic missiles take even less time). But lasers offer con-centrated power travelling at the speed of light. However, the laser beam has to lock on to its target long enough for damage to be done. The heat shield that protects ICBMs from the heat of re-entry into the Earth's atmosphere after their sub-orbital flight, would resist the effects of a laser. Laser-beam weapons are better used to disable ICBMs when they are at their most vulnerable, ie during the boost phase immediately after launch. A laser beam could penetrate the thin skin of a liquid-fuelled ICBM, weaken the rocket's structure and rup-ture its fuel tanks, or destroy vital components of the guidance system. The skin of solid-fuelled rockets (eg the Soviet SS-16s and US Minute-mans) are tougher. The obvious place for laser-beam weapons is on the 'high-ground' of space. Space-based 'battle stations' using chemi-cal lasers capable of delivering millions of watts of power, effective over thousands of kilometres, might be operational by the end of the century. The battle stations would use mirrors to focus the laser beam on fast-climbing ICBMs long enough to explode them. The mirrors would be segmented, each segment computer-controlled to compen-sate for the distortion caused by heating. The segments would be made of lightweight low-expansion materials such as a mixture of glass and graphite-fibre reinforcements, coated with vaporized silicone. The aiming-fire control system of battle station lasers would include sen-sors to detect the ICBM within 50 seconds of launch, precisely track it, select a point a few metres (depending on the ICBM's speed) in front of it, and fire the laser beam long enough to destroy it. One US plan envisions 18 battle stations, each assembled by components taken on three flights of the Space Shuttle, operating in 1750km high polar orbits. □ see INTERCONTINENTAL BALLISTIC MISSILE, LASER, WELLS

static 1 electrical interference on a communications channel caused by unusual atmospheric conditions, lightning, electrical machinery, etc ⟨*Voyager detected ~ from thunderstorms in Jupiter's atmosphere*⟩
□ see JUPITER **2** *also* **static electricity** the electrical effects caused by

charges at rest. It is essential to neutralize the build-up of static on spacecraft and satellites since it can interfere with the operation of on-board instruments, and produce a possibly dangerous electrical discharge, for example between the Space Shuttle and a satellite which is about to be captured for repair or refuelling. A system called the Flight Model Discharge System has been developed to monitor and neutralize electrical charge build-up on spacecraft.

station-keeping any manoeuvres (eg firing thruster rockets) that keep a satellite or spacecraft in a predetermined orbit

steady state the condition of a system in which important parameters (eg temperature and pressure) vary little ☐ see QUIESCENT SUN

Steady State a cosmological model which assumes that the density of the universe remains the same because new stars form as old stars die. This idea was first proposed by Hermann Bondi, Thomas Gold, and Fred Hoyle in 1948. They calculated that the average rate of creation of new material would have to be only about 10^{-10} nucleons (protons or neutrons) per cubic metre per year. Thus a Steady-State universe has no beginning and no end. However, the theory does not readily explain the microwave background radiation as does the more widely accepted Big Bang model of the universe. ☐ see BIG BANG, MICROWAVE BACKGROUND RADIATION, HOYLE, PULSATING UNIVERSE

stellar of or composed of (the) stars ☐ see INTERSTELLAR

stellar mass (symbol **m.**) the mass of a star expressed in terms of the Sun's mass, and ranging between about 0.05m. to about 60m.. The speed at which a star develops and its ultimate fate are determined by its mass: a star of small stellar mass has a long lifetime of perhaps thousands of millions of years; a massive star 'lives' for only a few million years before exploding as a supernova. The star becomes a white dwarf if its stellar mass is less than about 1.5m., and a black hole if its stellar mass is greater than about 3m.. ☐ see SUPERNOVA, BLACK HOLE, WHITE DWARF

stepping motor an electric motor whose shaft rotates in small steps (eg 48 steps per revolution). A stepping motor contains a number of coils which are fed in a particular sequence using digital signals. Stepping motors are widely used to make robot arms or other computer-controlled devices perform precise movements. ☐ see ROBOT ARM, SERVOSYSTEM, ROBOTICS

storable propellant a rocket propellant in which fuel and oxidizer can be stored for long periods in a rocket (eg an ICBM) awaiting launch. An example of a storable propellant is a blend of hydrazine and unsymmetrical dimethylhydrazine (UDMH) as a fuel and nitrogen tetroxide as oxidizer. ☐ see PROPELLANT

storage capacity 1 the amount of computer data which can be stored in a computer memory device ⟨a ~ of 5 megabytes⟩ ☐ see EXPERT SYSTEM, SUPERCOMPUTER **2** the amount of material (eg rocket fuel) that can be stored in a container ⟨the ~ of a fuel tank⟩

store 1 to put data into a computer memory **2** to keep materials in a container until needed

strain gauge a sensor attached to an object to detect how much distortion it undergoes when forces act on it. A strain gauge usu consists of a fine grid of metal foil, the resistance of which changes during tension or compression. The change of resistance is measured electronically and used to help engineers design structures (eg a vehicle to explore the surface of Mars) which have to withstand forces.
□ see SENSOR

stratosphere – see EARTH'S ATMOSPHERE

stream a continuous flow of something ⟨*a meteor* ~⟩

streamline flow the movement of air or liquid past a body, without turbulence. Aerospace vehicles are designed for streamline flow to reduce air resistance and instability – compare TURBULENT FLOW

strong interaction the electrical forces between neutrons and protons in the nuclei of atoms, and the strongest of the four natural forces between elementary particles □ see WEAK INTERACTION, GENERAL UNIFIED THEORY

STS [*Space Transportation System*] – see SPACE SHUTTLE, STS-1 – STS-9, MISSION 41B – MISSION 61A

STS-1 date: 12–14 April 1981; **launch**: Kennedy Space Centre; **spacecraft**: *Columbia*; **crew**: Commander John Young and Bob Crippen; **orbit**: 240km × 276km inclined at 40.3° to equator; **activities/findings**: first test flight of *Columbia*; maximum heat shield temperature on reentry 1510C; **problems**: 15 of 31 000 thermal tiles apparently lost through vibration at launch; **landing**: Edwards Air Force Base, California. □ see THERMAL TILES, YOUNG, CRIPPEN

STS-2 date: 12–14 November 1981; **launch**: Kennedy Space Centre; **spacecraft**: *Columbia*; **crew**: Commander Joe Engle and Dick Truly; **orbit**: 262km circular inclined at 38° to equator; **activities/findings**: second flight of *Columbia*; scientific experiments included high-resolution pictures of Earth using Shuttle Imaging Radar (SIR), air pollution, test of plant germination and growth in zero-g; first use of Remote Manipulator System (RMS), for moving objects about in the payload bay; **problems**: flight curtailed because of problems with fuel cells; **landing**: Edwards Air Force Base, California. □ see SHUTTLE IMAGING RADAR, REMOTE MANIPULATOR SYSTEM, FUEL CELL

STS-3 date: 22–30 March 1982; **launch**: Kennedy Space Centre; **spacecraft**: *Columbia*; **crew**: Commander Jack Lousma and Gordon Fullerton: **orbit**: 240km circular inclined at 38° to the equator; **activities/findings**: third test flight of *Columbia*; many scientific experiments included measurement of solar X rays; further evaluation of Remote Manipulator System (RMS); pharmaceutical experiment included a monodisperse latex reactor (MLR); **problems**: landing diverted to White Sands Missile Range, New Mexico, because of poor weather at Kennedy Space Centre. □ see REMOTE MANIPULATOR SYSTEM, MONODISPERSE LATEX REACTOR

STS-4 date: 27 June – 4 July 1982; **launch**: Kennedy Space Centre; **spacecraft**: *Columbia*; **crew**: Commander Ken Mattingly and Henry

Hartsfield; **orbit**: 240km circular inclined at 28.5° to the equator; **activities/findings**: fourth and final test flight of *Columbia*; pharmaceutical experiments included a monodisperse latex reactor (MLR); first Getaway Special (GAS) package included experiments to test effects of microgravity on growth of fruit flies, shrimps, and algae; Remote Manipulator System (RMS) lifted contamination monitor out of payload bay to test for chemical contamination by emissions from the payload bay; **landing**: Edwards Air Force Base with partial use of automatic landing system. □ see GETAWAY SPECIAL, MONODISPERSE LATEX REACOR, REMOTE MANIPULATOR SYSTEM

STS-5 date: 11–16 November 1982; **launch**: Kennedy Space Centre; **spacecraft**: *Columbia*; **crew**: Commander Vance Brand, Pilot Bob Overmyer, Mission Specialists Joe Allen and Bill Lenoir; **orbit**: 298km circular inclined at 28.5° to the equator; **activities/findings**: first operational flight using *Columbia*; the Mission Specialists launched two Hughes Aircraft Corporation communications satellites, SBS-3 (Satellite Business Systems) and the Canadian satellite Anik-C3, both launched by springs from a revolving turntable when the Shuttle was over the equator; Payload Assist Module (PAM) rocket motor boosted satellites into geostationary orbits; **problems**: planned spacewalk cancelled; **landing**: Edwards Air Force Base, California. □ see PAYLOAD ASSIST MODULE

STS-6 date: 4–6 April 1983; **launch**: Kennedy Space Centre; **spacecraft**: *Challenger*; **crew**: Commander Paul Weitz, Pilot Karel Bobko, Mission Specialists Story Musgrave and Don Peterson; **orbit**: 283km circular inclined at 28.5° to the equator; **activities/findings**: first flight of *Challenger*; first of three Tracking and Data Relay Satellites (the 2268kg TDRS-A) was deployed by springs from a tilt table in the payload bay; Musgrave and Peterson spent 3 hours 45 minutes on a spacewalk – first by US astronauts for nine years; materials processing experiments included a method of separating biological materials in an electrical field; further experiments with monodisperse latex reactor (MLR); three Getaway Special (GAS) payloads included experiments to test effect of microgravity on structure of snow crystals, seed germination, and growth of micro-organisms; **problems**: booster rocket failed to raise TDRS-A to geostationary orbit, but ground controllers raised it using on-board control thrusters; **landing**: Edwards Air Force Base, California.

STS-7 date: 18–24 June 1983; **launch**: Kennedy Space Centre; **spacecraft**: *Challenger*; **crew**: Commander Bob Crippen, Pilot Fred Hauk, Mission Specialists Sally Ride, John Fabian, and Norman Thagard; **orbit**: 298km circular inclined at 28.5° to the equator; **activities/findings**: Dr Thagard included in crew to study space sickness; Sally Ride was America's first woman astronaut; two communications satellites, TELSAT 7 for Canada and Palapa 3 for Indonesia, were released by springs and boosted to geostationary orbit by Payload Assist Modules (PAM); the Shuttle Pallet Satellite (SPAS) was deployed and tested using the Remote Manipulator System

(RMS); a number of Getaway Special (GAS) experiment packages were operated remotely in the payload bay; **problems**: intended to make first Kennedy Space Centre landing but poor weather forced switch to Edwards Air Force Base, California. □ see SPACE SICKNESS, REMOTE MANIPULATOR SYSTEM, RIDE, GETAWAY SPECIAL, SHUTTLE PALLET SATELLITE, OFFICE OF SPACE AND TERRESTRIAL APPLICATIONS

STS-8 date: 30 August – 5 September 1983; **launch**: Kennedy Space Centre; **spacecraft**: *Challenger*; **crew**: Commander Dick Truly, Pilot Dan Brandenstein, Mission Specialists Dale Gardner, Guy Bluford, and Bill Thornton; **orbit**: 294km × 301km inclined at 28.5° to equator; **activities/findings**: first night launch and landing, and *Challenger's* third flight; released INSAT-1B communications satellite. □ see INDIAN SPACE RESEARCH ORGANIZATION

STS-9 date: 28 November – 7 December 1983; **launch**: Kennedy Space Centre; **spacecraft**: *Columbia*; **crew**: Commander John Young, Pilot Brewster Shaw, Mission Specialists Robert Parker and Own Garriott, Payload Specialists Ulf Merbold (West Germany) and Byron Lichtenberg; **orbit**: 242km × 254km inclined at 57° to equator; **activities/findings**: double shift crew allowed 24 hour operations necessary to operate Spacelab which was being carried on its first mission; through video and sound satellite links, experimenters on the ground were part of the experiments on board; **problems**: landing delayed several hours because of computer fault. □ see SPACELAB, TRACKING AND DATA RELAY SATELLITE

subatomic 1 of the inside of an atom **2** of a particle (eg an electron) that is smaller than an atom □ see ELEMENTARY PARTICLES

subsonic 1 of or being a speed less than that of sound in air **2** moving at, or using airstreams moving at, a speed less than that of sound

subsystem one part of a bigger system. A gyroscope is a subsystem of a guidance and control system aboard some spacecraft. □ see SYSTEM

subtend to fix the angular extent of an object (eg the Sun), or the separation between two objects (eg a binary star) with respect to a fixed point ⟨*the angle* ~ed *at the eye by an object*⟩ □ see ARC SECOND

Suisei – see PLANET-A

Sullivan, Dr Kathryn D the first American woman to make a spacewalk on 11 October 1984. Sullivan was a mission specialist (MS) aboard the Space Shuttle *Challenger* (Mission 41G) which had been launched five days earlier. She spent about 3½ hours working in the payload bay with a fellow mission specialist, David Leestma. Their task was an exercise designed to test the feasibility of refuelling satellites from the Space Shuttle. Using a set-up called ORS (Orbiting Refuelling System) they successfully transferred hydrazine fuel from a tank to a simulated Landsat panel. (NASA has plans to refuel the Landsat 4 Earth resources satellite which has been in orbit since July 1982.) Mission 41G was also notable for carrying the largest crew (seven people including Commander Bob Crippen) ever to take off in a single spacecraft. □ see TERESHKOVA, RIDE, EXTRAVEHICULAR ACTIVITY, SHUTTLE IMAGING RADAR, EARTH RADIATION BUDGET

SATELLITE, MISSION 41G

Sun the central body of the Solar System, that is the nearest star to the Earth, and produces energy in its core by the conversion of hydrogen into helium. The Sun has a diameter of about 1.4×10^6km, and a mass of about 2×10^{30}kg. Most of the Sun's material is hydrogen and helium in the approximate ratio 3:1, with about 1% of heavier elements. The transformation of four hydrogen atoms into helium in the Sun's core is known as nuclear fusion. During this process, about 1/140 of the mass of the atoms is converted into energy as described by Einstein's mass-energy equation, $E = mc^2$. The Sun transforms about 5 million tonnes of its mass into energy every second, but this mass is negligible compared with the total mass of the Sun. During the Sun's lifetime of 10 thousand million years, it will use up less than 1% of its mass. As well as producing heat and light, the Sun emits protons, neutrinos, positrons, and gamma rays. Energy from the core is radiated outwards to within about 100 thousand kilometres from the surface where convection takes over to carry the energy to the surface. From the centre to the surface of the Sun, the temperature falls from about 15 million kelvin (K) to 6000K. The visible face of the Sun is known as the photosphere and has a mottled appearance on which sunspots appear from time to time. The movement of these sunspots shows that the Sun rotates faster at its equator than at its poles, at an average period of rotation of 27.2 days. Immediately above the photosphere is the lower region of the Sun's atmosphere, the chromosphere, which is a few thousand kilometres thick and in which the temperature rises to about 50 000K. The Sun's corona extends above this for millions of kilometres and eventually becomes the solar wind. The Sun belongs to a class of stars known as yellow dwarfs. Though the processes by which the Sun produces energy, and the dynamics of its atmosphere, have been studied extensively from Earth, a number of spacecraft are being sent to study it from close quarters. The health of the Sun is very important to life on Earth: it is believed that minute changes in the amount of heat radiated from the Sun cause major climatic changes on Earth. □ see SOLAR SYSTEM, SUNSPOTS, SOLAR WIND, SOLAR HELIOCENTRIC OBSERVATORY, ULYSSES, SOLAR MAX, STAR, POSITRON, NEUTRINO, PROTON, NUCLEAR FUSION

sunspots circular markings on the surface of the Sun, the larger of which can sometimes be seen with the naked eye. A sunspot has a darker central umbra surrounded by a lighter penumbra as compared with the bright photosphere. It is the centre of an intense localized magnetic field which suppresses the upward convection of hot gases from the interior. Thus the umbra is about 1000K, and the penumbra about 400K, cooler than the surrounding photosphere (6000K). Most sunspots cluster together in groups which may last for several weeks, and their movement across the Sun's disc indicates that the Sun rotates on its axis. The number of sunspots fluctuates over an average period of about 11 years, known as the sunspot cycle. □ see SUN,

SOLAR WIND, SOLAR FLARE, T TAURI STARS, [2]GALILEO

supercomputer a powerful computer system that is capable of performing calculations very rapidly. By 1987, NASA will have installed supercomputers at its Ames Research Centre, Silicon Valley, California, at a cost of about $120M. The high cost is justified by the fuel savings that will be made in reducing the drag on high performance aircraft. The computer facility will be used to examine the operation of rocket engines and to study the flow of air around models of aerospace vehicles, esp the Space Shuttle. One of these supercomputers, Cray-2, can perform calculations on data at the enormous speed of 250 million instructions per second (MIPS). Cray-2 will be involved in modelling what is likely to happen to the Galileo spacecraft when it flies into the hot dense atmosphere of Jupiter in 1988. And astrophysicists will use the system to model quasars and other violently active celestial objects. Weather forecasting and genetic engineering are other possible uses for the supercomputer. By the early 1990s, Cray-2 will be replaced by supercomputers capable of performing at 4000 MIPS, and having memory capacities of a billion words of 64 bits each. Universities, government laboratories, and aerospace companies will be able to remotely access the computer system from around the United States. □ see EXPERT SYSTEM, FIFTH GENERATION COMPUTER, [1]GALILEO, AERONAUTICS, UPPER ATMOSPHERE RESEARCH SATELLITE

superior planets the planets Mars, Jupiter, Saturn, Uranus, Neptune, and Pluto, which are further from the Sun than the Earth – compare INFERIOR PLANETS

supernova (plural **supernovae**) a very bright star which suddenly appears in the sky and which then fades away over a few weeks. A supernova marks the dramatic end of a few stars – those that are more than about 1.4 times as massive as the Sun. As stars grow old, they progressively produce energy by nuclear fusion reactions involving heavier and heavier elements. First to go is hydrogen, which is converted into helium; then helium burns to produce carbon, and so on. As each element gets used up, the star cools. With less heat being generated, the star's own inward-pulling gravity begins to take over from the outward-pushing radiation pressure, and its core starts to shrink under its own weight. At first, the contraction increases the temperature in the core which enables heavier elements to take part in fusion reactions. Then, when iron forms the core and is involved in fusion reactions, the dramatic end of the star is reached: the reactions absorb energy which rapidly reduces the radiation pressure, and the entire central region of the star falls in on its core. In a fraction of a second there is a giant implosion; but what happens next is not clear. One theory is that the infalling material bounces off the superdense core and triggers final fusion reactions. A flood of neutrinos is produced which blasts away the outer atmosphere of the star. Whatever the process, there is a titanic eruption – the supernova – and a shell of very luminous gas is expelled from the star at speeds of several thousand kilometres per second. The remainder of the star is crushed by the implosion forcing electrons to combine with protons to form neutrons. A superdense rapidly spinning neutron star is left with a diameter of between 20 and 30km, and a density a thousand million million times that of water. These spinning neutron stars are believed to be what are now called pulsars, and they mark the site of past supernovae. For tens of thousands of years, the surrounding space is filled with an expanding shell of glowing ionized gas (the supernova remnant). One such site is the Crab Nebula in the constellation Taurus, which is believed to be from the supernova recorded by Chinese astronomers in 1054AD. It was so bright it was visible during the day for almost a month. Since searches began in 1936, over 400 extragalactic supernovae have been discovered. In our own Milky Way galaxy two or three supernovae occur each century. Supernovae play an important part in the chemistry of galaxies for the exploding star is rich in heavy elements which are dispersed around the galaxy, and this material is incorporated in the debris from which new stars and planets are formed. The iron in our bodies, the uranium in reactors, and the carbon in trees were created by supernovae before the Solar System was formed. Without supernovae, Earth-like planets would not exist. □ see STAR, NUCLEAR FUSION, HYDROGEN BURNING, NEUTRINO, PULSAR, NEUTRON STAR, BLACK HOLE, NUCLEOSYNTHESIS, GALAXY, INTERSTELLAR MEDIUM, X-RAY ASTRONOMY, COSMOLOGY, UNIVERSE, ANTHROPIC PRINCIPLE, CARBON

supernova remnant the shell of gas and dust from a supernova explosion which is ejected at great speed into the surrounding space. Most supernova remnants are quite faint in visible light (the Crab Nebula is an exception), but they are fairly strong sources of X rays and radio waves. Perhaps the formation of stars in interstellar dust clouds is initiated by the compression produced by a nearby supernova remnant – the Solar System may have originated in this way. □ see CRAB NEBULA, SUPERNOVA, SYNCHROTRON RADIATION

supersonic 1 (using, produced by, or relating to waves or vibrations) having a frequency above the upper threshold of human hearing of about 20kHz □ see FREQUENCY **2** of, being, or using speeds from one to five times the speed of sound in air ⟨the Space Shuttle reenters the atmosphere at ~ speed⟩ **3** of supersonic aircraft or missiles ⟨the ~ age⟩ □ see HYPERSONIC, SUBSONIC

surge a sudden short-lived rise in the value of something ⟨a ~ of rocket power⟩ □ see PULSE

Surveyor a US programme which soft-landed seven unmanned Surveyor Spacecraft on the Moon between 1966 and 1968, to test the strength and properties of the lunar soil prior to the first Apollo manned landing of 1969. The spacecraft were soft-landed by retrorockets and equipped with TV cameras powered by solar cells. □ see LUNOKHOD

switch a device for making, breaking, or changing the connections to an electrical circuit ⟨to throw the emergency ~⟩

synchronous 1 happening or arising at exactly the same time **2** having the same period or phase □ see GEOSTATIONARY ORBIT,

MOON, IO, SYNCHROTRON RADIATION

synchronous rotation *also* **captured rotation** the orbital motion of a satellite about its primary (eg the Moon about the Earth) in which the satellite's period of rotation on its axis is equal to its orbital period. Thus the satellite always keeps one 'face' pointing towards its primary. As well as the Moon, the satellites of Mars, Jupiter, and Saturn are in synchronous rotation. □ see MOON, IO

synchrotron [*synchro*nous cyclo*tron*] *also* **particle accelerator** a device for accelerating fundamental particles (eg protons) to high speeds using radio frequency electric fields. The particles are confined to a circular path of constant radius by magnetic guiding and focussing fields that are increased synchronously to match the energy gained by the particles. □ see SYNCHROTRON RADIATION, ANTIMATTER PROPULSION

synchrotron emission – see SYNCHROTRON RADIATION

synchrotron radiation *also* **synchrotron emission** electromagnetic radiation produced by high-speed electrons spiralling in a magnetic field. This process is generally assumed to be responsible for the radio emission from supernovae remnants and extragalactic radio sources. If the electrons are travelling at high speed, the theory of relativity predicts that the radiation will be beamed forward along the axis of the spiral in the direction in which the electron moves. At lower speeds, the radiation is called cyclotron radiation, and is emitted in all directions from the spiralling electron. Synchrotron radiation is an example of nonthermal radiation since it is produced by the nonthermal acceleration of electrons in a magnetic field, rather than by the rapid movement of electrons at high temperatures. □ see SUPERNOVA REMNANT, RADIO GALAXY

system a group of related parts that exchange information and energy between themselves and with their environment, and that exists to serve a particular purpose ⟨*a satellite communications* ~⟩ □ see SUBSYSTEM, BLACK BOX, SYSTEMS ANALYSIS

systems analysis the study of an activity (eg growing vegetables in a space station), using computers and other means of modelling the activity, in order to clarify its goals and to discover the best ways of achieving them. Systems analysis has been of enormous value in complex engineering activities – over 20% of the technical and administrative sections of NASA's Jet Propulsion Laboratory (JPL) have the word 'system' in their title. The systems approach to studying problems emphasizes the value of the overview, ie 'the big picture', so avoiding the tendency to get lost in a welter of detail. Part of the popularity of the systems approach is thought to be due to the analogy of the system to a living organism, which is a collection of interrelated subsystems that survive and prosper: these are precisely the attributes which engineers design into their creations so they may, unwittingly, be guided by a set of patterns developed over 3500 million years of evolution. □ see SYSTEM

syzygy an arrangement of three celestial bodies in a straight line (eg the Sun, Moon, and Earth during a solar or lunar eclipse)

tachometer an electronic instrument for measuring speed of rotation. The speed is usu measured in units of revolutions per minute. Most tachometers use a photodiode to sense the reflection of light off a marker so that there is no direct contact with the rotating object. □ see ANEMOMETER, INSTRUMENTATION

take-off the ascent of a rocket, missile, or aircraft from a launch pad, at any angle ⟨*the ~ of the Harrier fighter*⟩ – compare LIFT-OFF

tandem launch the launch of two or more satellites by a single launch vehicle □ see LONG MARCH 2 AND 3, ACTIVE MAGNETOSPHERE PARTICLE EXPLORER

tank a container for storing a liquid, esp a rocket fuel ⟨*the Space Shuttle's ~s for cryogenic propellants*⟩ □ see EXTERNAL TANK

Tansei-5 – see PLANET-A

tape recorder a recording and playback device that uses magnetic tape to store data. Tape recorders were used on early spacecraft to store results of measurements, but nowadays data is stored in a semiconductor memory before being transmitted to Earth. □ see MEMORY, VENUS RADAR MAPPER

target any goal or objective towards which something is directed ⟨*Halley's comet, the ~ for the Giotto spacecraft*⟩ □ see PROJECT

Taurus A the source of radio waves in the constellation of Taurus, centred on the pulsar in the Crab nebula. The radio waves are produced by electrons spiralling in a magnetic field which permeates the nebula. At a frequency of 178MHz the strength of the radio energy is 1420 jansky. □ see CRAB NEBULA, SYNCROTRON RADIATION, RADIO ASTRONOMY, JANSKY

TDRS – see TRACKING AND DATA RELAY SATELLITE

Teal Ruby a prototype communications satellite which was placed in a 740km-high orbit inclined at 72.5° to the equator by the first Space Shuttle (Mission 62A) to be launched from the Vandenberg Air Base, California. It was designed to test the use of lasers in communications between ground stations and orbiting satellites. A small laser package aboard Teal Ruby tested the acquisition, pointing, and tracking accuracy of the White Sands ground station during its two 10 minute passes each day. □ see LASER, STAR WARS

technology transfer ideas and inventions in one area applied to the solution of problems and the creation of new products in an unrelated area. Microelectronics is an excellent example of how a product, the integrated circuit, which was stimulated by military interests, has been successfully applied in the areas of commerce, entertainment, leisure, education, scientific research, space exploration, etc. In a similar way, materials (eg Teflon) developed for the rigours of space technology have been transferred to domestic products. □ see OFFICE OF TECHNOLOGY ASSESSMENT

tektite

tektite a small rounded glassy object, 2 to 3cm across. Tektites are found strewn over large areas in certain parts of the world: 'australites' are found in Australia, 'moldavites' in Bohemia and Moravia, and 'indochinites' in Southeast Asia. Their varied shapes (eg pear-shaped, spheroidal, and disc-shaped), and surface-flow features, indicate that they were produced from molten material which cooled rapidly. Tektites contain more than 65% of silica, and it is thought they were formed from substances thrown from the Earth after the impact of giant meteorites. They may be connected with volcanic activity on the Earth or the Moon, and it is interesting that small tektite-like objects have been found on the Moon. ☐ see TUNGUSKA EVENT, COMET, GEOPHYSICS, METEORITE

telecommunications the use of electronic and other equipment to send information through wires, the air, and interplanetary space. Communications satellites are an increasingly important part of world-wide telecommunications systems. ☐ see COMMUNICATIONS SYSTEM, DIRECT BROADCAST SATELLITE, INFORMATION TECHNOLOGY, OPTICAL COMMUNICATIONS, TRANSPONDER

telemetry the transmission by radio of measurements made at a distance. Telemetry makes it possible to maintain two-way contact with crew in the Space Shuttle and with spacecraft such as Voyager. ☐ see SIGNAL, VOYAGER, SPACE SHUTTLE

telepresence the remote directing of a robot using sensors attached to the body of a human being, which transmit the movements of his or her hands, arms, and legs to the robot which then mimics the movements and performs the desired task. The idea is being considered to allow astronauts to operate robot arms in the payload bay of the Space Shuttle without going to the bother of putting on a spacesuit. However, before a human operator is given the sense of 'being there', TV cameras are needed that give three-dimensional images with a resolution equal to that of the eye. Telepresence also needs a handlike device which matches the touch and manipulative skill of the human hand. The development of artificial intelligence should make human control less and less important in the moment-to-moment action of robot arms. ☐ see ROBOT, REMOTE MANIPULATOR SYSTEM, ARTIFICIAL INTELLIGENCE, RADARSAT

telescope a device for collecting and producing an image of distant objects. On Earth, telescopes are used by optical and radio astronomers to gather and focus electromagnetic radiation passing through the two main windows in the Earth's atmosphere. However, above the Earth's atmosphere, telescopes can be designed to focus a much wider range of electromagnetic waves. The orbiting Hubble Space Telescope, for example, can image ultraviolet and infrared waves which would not pass through the atmosphere. A telescope has one or more large collectors to gather and focus radiation: a reflecting telescope has a paraboloidal collector (eg a radio dish), many optical telescopes use a glass lens and an X-ray telescope uses a mirror to focus and examine X rays. ☐ see WINDOW, HUBBLE SPACE TELESCOPE,

RADIO TELESCOPE, REFLECTING TELESCOPE, REFRACTING TELESCOPE, SCHMIDT TELESCOPE, X-RAY MULTI-MIRROR TELESCOPE

television camera – see VIDEO CAMERA

Television Infrared Orbital Satellite (abbr TIROS) – see NATIONAL OCEANIC AND ATMOSPHERIC ADMINISTRATION

temperature the degree of hotness or coldness of a body, measured on any of various scales, as in degrees celsius (C) or degrees kelvin (K). The temperature of celestial bodies is expressed in different ways depending on the property being measured (eg colour temperature or ionization temperature) ⟨*the colour ~s of stars*⟩. ☐ see ABSOLUTE TEMPERATURE SCALE, THERMOCOUPLE, THERMAL TILES

Tereshkova, Valentina (*b*1937) a Russian cosmonaut and the first woman in space. In 1962, she was launched in her own spacecraft and joined Valeri Bykovsky who was already in orbit. They remained in orbit for almost three days and Valentina was the only female spacefarer until Savitskaya flew in 1982. ☐ see SAVITSKAYA, RIDE

terminal guidance – see GUIDANCE

terminate to stop something happening ⟨*to ~ the countdown*⟩ ☐ see ABORT

terminator the dividing line between day and night on a celestial body such as the Moon, which corresponds to sunrise and sunset. Pictures taken of a body's terminator by interplanetary spacecraft often reveal interesting features. For example, in 1979, Voyager 1's cameras photographed a volcanic eruption on Jupiter's satellite, Io, as volcanic dust in Io's atmosphere reflected the slanting rays of the setting Sun. ☐ see ECLIPSE, OCCULTATION

terrestrial of or relating to the Earth ⟨*~ climate*⟩ ☐ see EXTRATERRESTRIAL

terrestrial-reference guidance – see GUIDANCE

test-bed a base, mount, or frame within or upon which a component, esp an engine, is secured for testing ☐ see HOT TEST

test flight a flight made to test the working of an aircraft or aerospace vehicle

tethered satellite a satellite to be deployed by the Space Shuttle in 1987, that will be kept in position by a tether so that it can be reeled back into the Shuttle's payload bay after use. The tether would be up to 100km long above or below the orbiting Shuttle and the difference between the gravitational forces acting on the Shuttle and the tethered satellite will keep the tether taut. This technique will enable the Earth's atmosphere to be sampled at heights which are too low for an orbiting satellite, including the Shuttle, because of the higher air resistance. The tethered satellite will also be able to sample cosmic dust, map the Earth's magnetic field, and find out how micrometeoroids react with the upper atmosphere, without the measurements being contaminated by exhaust gases and other Shuttle debris. Whereas some of the experiments will call for a synthetic tether (eg nylon and polytetrafluoroethylene), others would require an electrically conductive tether. The tethered satellite is a cooperative project between

NASA and Italy. □ see SPACE SHUTTLE, PAYLOAD BAY, COSMIC DUST, MICROMETEOROID

thematic mapper (abbr **TM**) – see LANDSAT

thermal of or caused by heat ⟨~ *insulation*⟩ □ see THERMAL TILES

thermal control system a system that regulates the temperature of a spacecraft. Sunlight in space is intense and would make the sunlit side of a spacecraft very hot, and the side in shadow would be very cool. Equipment on board also generates heat, so it is necessary to have some way of keeping the temperature inside a satellite or spacecraft within acceptable limits (usually in the range −10C to +40C). Heaters and thermal louvres are sometimes used, but controlling the temperature at the surface of the spacecraft is the preferred method: combinations of aluminized plastic film, polished metal surfaces, and paint are used for the surface finish, which is why space vehicles often look as though they have been wrapped in cooking foil! □ see ENVIRONMENTAL TESTING

thermal noise electrical interference on a communications channel caused by the agitation of molecules in components. Thermal noise is reduced by lowering the temperature of these components. For example, the electronic circuits in the Infrared Astronomy Satellite (IRAS) were cooled with liquid helium so as to reduce the 'hiss' which would have swamped the signals collected by its telescope.

thermal tiles heat resistant tiles which protect the Space Shuttle from the intense heat generated by its high speed reentry into the Earth's atmosphere. Each tile measures about 200mm square and is up to 100mm thick: more than 32 000 of them were laboriously glued to the surfaces of the first Shuttle Orbiter, *Columbia* in 1981. The tiles are custom-made and their size, thickness, and the material from which they are made, is determined by their exact location on the fuselage. They not only have to withstand heat, but also the vibration and deflection of the airframe during launch and other operations. White-coated tiles are designed to withstand temperatures in the range 400 to 670C, while black-coated tiles take over protection in the 670 to 1300C range. The Shuttle Orbiter *Discovery*, carries about 24 000 tiles, most of which are of this black variety. The Orbiters *Discovery* and *Atlantis* carry extra thermal insulation since they fly in polar orbits which cause greater heating on reentry. Areas not covered by tiles are insulated using a thermal blanket. Tiles are used in favour of an ablative surface (which is designed to melt or evaporate away when heated) since they are lighter. The basic raw material for the tiles is high-purity silica fibre which has been selected for its low thermal conductivity and high-temperature stability. A tile can be heated to red heat using a blow torch and then picked up immediately with bare hands, so good is it at shedding its heat. Though a good idea which works, the use of thermal tiles is not without its problems: the difficulty of making tiles stay put once glued in place has contributed to launch delays of the Space Shuttle since 1981. □ see SPACE SHUTTLE

thermistor a heat sensor made from a mixture of semiconductors and having an electrical resistance which varies with temperature. Thermistors are used for both temperature measurement (in thermometers) and control (in thermostats). □ see TRANSDUCER, THERMOCOUPLE

thermocouple a device made from a pair of dissimilar metals (eg copper and iron) which produces a voltage varying with temperature. The metals are joined together to make two junctions. Any difference of temperature between the junctions causes a voltage which varies according to the temperature difference. Thermocouples are used in electronic thermometers and have the advantage, compared with thermistors, that they can measure a higher temperature and that the heat-sensitive junction is much smaller.

thermodynamics a field of physics that deals with the theory of the conversion of heat into other forms of energy, and vice versa ⟨*~ is applied to the study of the performance of rocket engines*⟩ □ see SECOND LAW OF THERMODYNAMICS

thermometer a device for measuring temperature □ see THERMISTOR, THERMOCOUPLE, ABSOLUTE TEMPERATURE SCALE

thermonuclear of, using, or being changes to the nuclei of light atoms (eg hydrogen) at very high temperatures to release a large amount of energy ⟨*a hydrogen bomb is a ~ weapon*⟩ □ see CRUISE MISSILE, STAR WARS, NUCLEAR FUSION

thermosphere – see EARTH'S ATMOSPHERE

three-dimensional 1 having or described by three dimensions: length, breadth, and height **2** giving the illusion of depth ⟨*~ flight simulation*⟩ □ see TWO-DIMENSIONAL, FOUR-DIMENSIONAL

throttling any of a number of methods for altering the thrust of a rocket engine during flight. Adjusting the rate of fuel flow, changing the pressure in the combustion chamber, and varying the nozzle expansion are some of the methods used. □ see THRUSTERS, ORBITAL MANOEUVRING SYSTEM

thrust the force, measured in newtons, developed by a rocket engine. The force is produced by the high-speed ejection of exhaust gases from the engine's nozzle, and is equal to the rate of change of momentum of the ejected gases ⟨*the ~ developed by the Ariane rocket is equal to about 4 million newtons*⟩. □ see SPECIFIC IMPULSE, LINEAR MOMENTUM, BOOSTER ROCKET

thrusters *also* **verniers** small rockets on a spacecraft (eg the Space Shuttle) or a space station (eg Salyut) that make small corrections to the flightpath or trajectory □ see ORBITAL MANOEUVRING SYSTEM, ATTITUDE CONTROL, COURSE CORRECTION

thrust vector control (abbr **TVC**) the steering of a rocket by rotating its nozzle to direct the exhaust gases in different directions. The Space Shuttle is steered during ascent using TVC on its solid rocket booster. □ see SOLID ROCKET BOOSTER

time 1 the measurable period during which something exists, continues, or takes place ⟨*flight ~*⟩ **2** the point at which an event happens ⟨*Launch ~ was 1245 GMT*⟩ □ see COUNTDOWN, SPACETIME, TIME DILATION

time dilation the phenomenon whereby time slows down in a system which is moving swiftly relative to an observer. Time dilation is predicted by Einstein's special theory of relativity, and it has been verified experimentally. If two spaceships are moving relative to each other, it will appear to each that the other's clock has slowed down. Thus time passes more slowly on the other spaceship and its occupants age more slowly than the people on Earth. Time dilation is only significant at speeds near to that of light, so present-day astronauts do not show any obvious signs of staying young. They would have to move away from us at four-fifths the speed of light to grow old 60% slower. □ see SPECIAL RELATIVITY, GLOBAL POSITIONING SYSTEM, ATOMIC TIME

time lag the time taken for something to respond to, or deliver, a request for information. The time delay may be due to processes within a device (eg computer processing time), and/or to the time taken for the information (eg pictures from a spacecraft) to travel to the receiver.

TIROS [*T*elevision *I*nfrared *O*rbital *S*atellite] – see NATIONAL OCEANIC AND ATMOSPHERIC ADMINISTRATION

¹Titan the largest of Saturn's 15 known satellites and the second largest satellite in the Solar System. The Voyager 1 spacecraft flew by Saturn in November 1980 and showed that Titan has a reddish-brown atmosphere containing nitrogen and traces of methane. The lower levels of this atmosphere may be warmed by a weak greenhouse effect to about −100C. No clear surface details could be seen on Titan, but its low density (about 1.8 times that of water) suggests that it is made up mainly of ice and liquid water with a small solid core. In this respect, Titan may be similar to the two Galilean moons, Ganymede and Callisto, of Jupiter. □ see SATURN, VOYAGER, CASSINI

²Titan a powerful US cryogenic rocket which has launched a number of satellites and interplanetary spacecraft. An early version of Titan 3, in combination with a Centaur upper stage, was used to launch the Viking spacecraft to Mars in 1975. A variant of Titan 3 is currently being upgraded as a back-up launcher for national security satellites since the US Air Force is concerned about the reliability of the Space Shuttle. The company, Martin Marietta, has been awarded a US Air Force contract to build ten upgraded Titan 3s for launch at the rate of two per year beginning in 1988. Each will be capable of placing a Shuttle-equivalent 4500kg payload into geostationary orbit. □ see CRYOGENIC ENGINE, CENTAUR, SATURN 5, ARIANE, BIG BIRD

TM [*t*hematic *m*apper] – see LANDSAT

TNT equivalent the energy released in the detonation of a nuclear weapon in terms of the weight of TNT (trinitrotoluene) which would release the same amount of energy when exploded. It is usu expressed in kilotonnes or megatonnes □ see NUCLEAR WEAPONS, ELECTROMAGNETIC PULSE

tolerance the allowable variation within which some property of a product (eg its dimensions) are judged to be acceptable □ see RELIABILITY

tonne a metric unit of mass equal to 1000 kilograms ⟨*the Space Shuttle has a mass of about 2000 ~s at launch*⟩ □ see SI UNITS, KILOGRAM

torque a turning or twisting force that tends to produce rotation ⟨*the ~ applied to spin a satellite*⟩ □ see FORCE, PRECESSION, ANGULAR MOMENTUM

touchdown the (moment of) landing of an aerospace vehicle (eg the Space Shuttle) or of a spacecraft (eg Viking) on the surface of a planet or moon □ see APOLLO, ROLLOUT

track to follow the path of a rocket, spacecraft, or other object ⟨*to ~ a satellite using radar*⟩ □ see DEEP SPACE NETWORK, TRAJECTORY, COURSE CORRECTION

Tracking and Data Relay Satellite (abbr **TDRS**) one of three communications satellites in geostationary orbit which act as a communications link between spacecraft and data processing facility satellites on the ground. The TDRS system (TDRSS) was designed to replace NASA's 20-year old ground station tracking network, except for three Deep Space Network (DSN) stations equipped for communications with interplanetary spacecraft. Data from the TDRSS is transmitted to the 64m-diameter antenna at White Sands, New Mexico, at a rate of up to 50 kilobits per second and a frequency of 2.2875 GHz. Scientists operating experiments in Spacelab on board the Space Shuttle can use the TDRSS to send data, including text and graphics, to ground stations. The first satellite, TDRS-A, was deployed from the payload bay of Space Shuttle *Challenger* (STS-6), in April 1983, and TDRS-B and TDRS-C during Space Shuttle flights in 1985. □ see DEEP SPACE NETWORK, SPACELAB, HUBBLE SPACE TELESCOPE, GIGAHERTZ

trajectory the path that a spacecraft, rocket, or planet follows ⟨*Voyager's ~ through the Solar System*⟩ □ see COURSE CORRECTION, GUIDANCE, GRAVITY ASSIST

transceiver a combined radio transmitter and receiver in a single housing □ see SPACEWALK RADIO, TRANSPONDER

transducer a device for changing one form of energy into another. Transducers are used extensively in electronic instrumentation and control systems in space technology hardware. For example, a thermocouple is used in electronic thermometers since it changes heat energy into electricity; solar cells convert light energy into electricity to power spacecraft; microphones are used to change sound energy into electrical energy (eg for detecting the impact of micrometeorites on the hull of spacecraft). □ see SOLAR CELL, STRAIN GAUGE, THERMOCOUPLE

transfer to apply the skills and knowledge obtained in one field of science and technology to another unrelated field ⟨*space technology ~ed to commercial aircraft design*⟩ □ see TECHNOLOGY TRANSFER, OFFICE OF TECHNOLOGY ASSESSMENT

transfer orbit the path taken by a spacecraft in moving from one orbit to another (eg from Earth orbit to an orbit round Mars). The path that requires the least input of energy is an ellipse which just touches one point in the old orbit (the perigee of the transfer orbit), and one point in the new orbit (the apogee of the transfer orbit). When the spacecraft reaches apogee, it is given additional energy to inject it into the new orbit. The elliptical orbit that requires minimum energy to transfer a spacecraft between one body and another is known as a Hohmann transfer after the German engineer who first described it in 1925. However, the Hohmann transfer does take a long time so the flight time is reduced by giving the spacecraft more energy at perigee instead. The spacecraft would then fly by its destination were its speed not reduced to allow it to take up the required orbit.
□ see EXPEDITION TO MARS, GRAVITY ASSIST

transit 1 the passage of Venus or Mercury, the inferior planets, across the Sun's disc. Transits of Venus take place either in June or December and occur in pairs separated by 8 years. They are rare: the last was in 1882, the next is in June 2004. Transits of Mercury are more frequent and take place in May or November: the transit preceding that of May 1986 was in November 1973. **2** the passage of a satellite across the face of the planet it orbits. Some spacecraft (eg Voyager) see transits which are not visible from Earth. □ see ECLIPSE, OCCULTATION

translunar of or being space beyond the Moon □ see INTERPLANETARY

transmission the sending of information from one place to another

transmitter that part of a communications system which sends out messages ⟨the ~ on the Galileo spacecraft⟩

transonic 1 of or being a speed close to the speed of sound, ie between Mach 0.8 and 1.2 **2** capable of moving at transonic speed

transponder a combined transmitter and receiver that receives radio signals and automatically retransmits them, often at a different frequency. Transponders are used in communications satellites for relaying messages between Earth stations. □ see COMMUNICATIONS SATELLITE

trim to adjust an aerospace vehicle's controls to achieve stable flight

Trojan group a group of asteroids concentrated in the regions of two Lagrangian points in the orbit of Jupiter. Each point is located at the corner of an equilateral triangle, with the Sun and Jupiter at the other two points. The first Trojan, discovered in 1906, was called Achilles, and since then most have been named after heroes from both sides in the Trojan Wars. □ see ASTEROID, LAGRANGIAN POINTS

tropopause – see EARTH'S ATMOSPHERE

troposphere – see EARTH'S ATMOSPHERE

troubleshooting the location and repair, usu by skilled technicians, of faults in equipment □ see PAYLOAD SPECIALIST

Tsiolkovsky, Konstantin (1857–1935). Russian physicist who developed the theory of rocketry, esp the idea of the multi-stage liquid propellant rocket which was later to be used successfully by the Soviet Union and America. Tsiolkovsky wrote about space suits, satellites, and the colonization of the Solar System, and was the first to suggest the possibility of a space station. The message on the tombstone of his grave reads: "Mankind will not remain tied to Earth forever". Twenty-two years after his death the Soviet Union launched the first manmade satellite, Sputnik 1. □ see GODDARD, SPUTNIK, MULTISTAGE ROCKET

T Tauri stars very young stars, not more than 100 thousand years old, which emit strongly in the infrared and exhibit sudden changes of luminosity. Observations at radio wavelengths (eg using the Very Large Array radio telescope) show that T Tauri stars are ejecting considerable quantities of material at very high speeds. This material often forms jets, very like the jets seen from the most energetic quasars and radio galaxies. This activity is thought to be due to the violent adjustment necessary when a cloud of interstellar dust collapses to form a star. The compact object created is a 'mess' and the young protostar has to undergo a considerable redistribution of material within it, and of its angular momentum (spin), before it can properly call itself a star. The irregular light variation of T Tauri stars is thought to be caused by explosive releases of energy, analogous to colossal solar flares, originating in the lower regions of the stars' atmospheres (the chromosphere). □ see STAR, PROTOSTAR, VERY LARGE ARRAY, QUASAR, SUN, SOLAR FLARE, CEPHEID VARIABLE

tumbling the unwanted end-over-end motion of a spacecraft in orbit. A flashing light seen in the night sky is sometimes caused by a spent rocket in Earth orbit. NOAA-8 is an advanced US weather satellite which lost the use of its attitude control system in June 1984 and was tumbling out of control in orbit. □ see GROUND-BASED ELECTRO-OPTICAL DEEP SPACE SURVEILLANCE, NOAA-9

Tunguska Event a giant fireball in the sky followed by an explosion and devastation across a wide area that occurred in the early hours of 30 June 1908 in Central Siberia. Eyewitnesses told of great heat, noise, and strong winds, and of the forest blazing around them. A long flaming red object, many times larger than the Sun and wreathed by puffs of smoke and blue streamers, was seen rushing through the sky. An air wave was produced which was detected on barographs as it travelled twice round the world. Twenty five years afterwards, Soviet geologists surveyed the site of the event and found that an area of forest 80km in diameter had been laid flat by the explosion, but there was no impact crater. There are many theories of the cause of the devastation: it was a black hole, a lump of antimatter which was annihilated on contact with the Earth in a flash of gamma rays, an out of control alien spacecraft, and so on. Most likely, an icy fragment of a comet struck the Earth in 1908. It was probably 100m across, with a mass of perhaps 100 000 tonnes, and was travelling at 50kms^{-1} relative to the Earth before dissipating most of its energy in the atmosphere a mile or so above the Siberian forests. It's as well that these events are rare: an explosion of such magnitude near a city might well be mistaken for the explosion of a nuclear warhead and precipitate a nuclear war, though it would not produce any radioactive fallout or gamma radiation. □ see COMET, ANTIMATTER, BLACK HOLE, TEKTITE

turbulence irregular motion of a fluid ⟨*atmospheric ~ in thunderstorms*⟩

turbulent flow the flow of a fluid in which the velocity at any point changes randomly. Good design ensures that turbulent flow does not occur over an aerospace vehicle that has to move through a planet's atmosphere, for it wastes energy and makes the vehicle difficult to handle. – compare STREAMLINE FLOW

TVC – see THRUST VECTOR CONTROL

two-dimensional having two dimensions and therefore lacking perspective ⟨*a ~ image on a VDU*⟩ – compare THREE-DIMENSIONAL

Tyuratam – see COSMODROME

UARS – see UPPER ATMOSPHERE RESEARCH SATELLITE

UFO [*U*nidentified *F*lying *O*bject] an object seen in the sky by some-one who cannot explain its nature. In fact, most UFO sightings have been explained by known phenomena such as weather balloons, cer-tain types of clouds, high-flying aircraft, and the bright planet Venus. But inexplicable sightings have been made by experienced and level-headed people (eg airline pilots and astronauts). These and other puz-zling observations have given rise to speculation about little green men and alien visitors. However, at the moment there is no technical evi-dence that life forms exist elsewhere in the universe, though a number of astronomers are conducting a search for extraterrestrial life. □ see ASTROBIOLOGY, SEARCH FOR EXTRATERRESTRIAL INTELLIGENCE, ENCOUNTER 2

UHF [*u*ltra *h*igh *f*requency] (of or being radio waves having) a fre-quency in the range 300MHz to 3GHz □ see FREQUENCY, MEGAHERTZ, GIGAHERTZ, X-BAND

Uhuru ('Freedom' in Swahili) □ see SAS-1

UK Infrared Telescope (abbr **UKIRT**) a British telescope 4200m high on the mountain Mauna Kea, Hawaii. At this altitude, the tele-scope is above most of the atmospheric water vapour that strongly absorbs infrared from celestial objects. Seeing is also good for obser-vations in the visible spectrum. The telescope became operational in 1979 and is funded by the Science Research Council and run by the Royal Observatory, Edinburgh. Its mirror is 3.8m in diameter and infrared detectors at its focus are cryogenically cooled in order to reduce the thermal noise they generate. □ see SEEING, INFRARED ASTRONOMY, INFRARED ASTRONOMY SATELLITE, CRYOGENICS

UKIRT – see UK INFRARED TELESCOPE

UKS [*U*nited *K*ingdom *S*atellite] – see ACTIVE MAGNETOSPHERIC PARTICLE EXPLORER

UK Schmidt Telescope – see SCHMIDT TELESCOPE

ultra high frequency – see UHF

ultrasonic of or being waves and vibrations having frequencies above about 20kHz and which are therefore inaudible to humans

ultraviolet of or being electromagnetic radiation having wavelengths between light and X rays, ie between 400 nanometres and 2 nanometres ⟨*~ rays are present in sunlight and are responsible for a suntan*⟩ □ see NANOMETRE, ULTRAVIOLET ASTRONOMY

ultraviolet astronomy the study of celestial objects from their ultraviolet emissions in the wavelength band 25 to 350nm. Ultraviolet waves are strongly absorbed in the Earth's atmosphere so ultraviolet astronomy can only take place above an altitude of about 30km. At this altitude, observation is possible in the 250-350nm band. Ultravio-let emissions from celestial objects at shorter wavelengths are possible from satellites well above the Earth's atmosphere. □ see NANOMETRE, EXTREME ULTRAVIOLET EXPLORER, INTERNATIONAL ULTRAVIOLET EXPLORER, ROSAT, HUBBLE SPACE TELESCOPE, SPARTAN

ultraviolet spectrometer an optical instrument for analysing the intensity of ultraviolet radiation at various wavelengths. This is a common instrument sent aloft in artificial satellites and space probes for studying the radiation from planets, stars, and galaxies. □ see ULTRAVIOLET ASTRONOMY, EXTREME ULTRAVIOLET EXPLORER, INTERNATIONAL ULTRAVIOLET EXPLORER, SPECTROMETER

Ulysses (formerly known as the **International Solar Polar Mission**) a joint ESA/NASA project to study the Sun using a 370kg spacecraft built by Dornier Systems of West Germany and due to be launched during Mission 61F of the Space Shuttle. The project's relationship with Ulysses comes from a late medieval legend which tells of the last voyage of Ulysses beyond Gibraltar into 'an uninhabited world behind the Sun'. The unique feature of this solar mission is that the spacecraft will explore both poles of the Sun from a position far out of the plane of the ecliptic, the orbital plane of the Earth. Most space missions take place in a narrow band about ±7.5 degrees centred on the ecliptic. From its advantageous position, Ulysses will explore the properties of the solar wind, solar flare X rays, cosmic gamma-ray bursts, the heliospheric magnetic field, the interplanetary magnetic field, cosmic rays, and cosmic dust. In addition there will be a gravity-wave experiment. After launch, Ulysses will first head towards Jupiter in order to receive a gravity-assist from this planet that will fling it far out of the ecliptic and in a direction back towards the Sun. It will make its first observations above one solar pole in late 1989. Its path will then curve down to cross the ecliptic to begin its climb to the other pole which will be overflown in early 1991. Both the Galileo spacecraft, due for Jupiter, and Ulysses will share the same 'Jupiter launch window'. □ see EUROPEAN SPACE AGENCY, SOLAR HELIOCENTRIC OBSERVATORY, [1]GALILEO, ECLIPTIC

umbilical a line carrying electrical signals or fluids, between the ground or gantry and the upright rocket before launch □ see GANTRY

umbra 1 the darkest part of a shadow cast by an object illuminated by a broad, rather than a point, source of light. The lighter part of the shadow outside the umbra is known as the penumbra. An observer in the umbra cannot see any part of the source, but the source is only partly obscured when seen from the penumbra. □ see ECLIPSE **2** the darkest central portion of a sunspot which is surrounded by a lighter penumbra □ see SUNSPOTS

unidentified flying object – see UFO

United Kingdom Satellite (abbr **UKS**) – see ACTIVE MAGNETOSPHERIC PARTICLE EXPLORER

universal time – see GREENWICH MEAN TIME

universe all knowable things. With the advance of space science and technology, things (eg pulsars) once unknowable become knowable. The universe, once regarded as highly ordered and unchanging with Man at its centre, is now seen to be a complex and eventful collection of objects. And rather than enjoying some sort of privileged position in it, our importance to the scheme of things appears to be of no significance whatsoever. We live on a planet that orbits one of a hundred billion stars, in one of a countless number of galaxies stretching into space as far as our telescopes can see. And the forces that control events in the universe are so enormous that they could reduce our Solar System to gas dust in an instant. But perhaps Man does have a special place in the universe: certainly, as far as we know at the moment, we are the only intelligent lifeform in it that is trying to make sense of it all – compare COSMOS □ see COSMOLOGY, ASTROBIOLOGY, EARTH, SOLAR SYSTEM, SUN, STAR, GALAXY, ANTHROPIC PRINCIPLE, [1]GALILEO, COPERNICUS

University of Surrey Satellite – see UOSAT-2

unmanned *of a space craft* not carrying astronauts or cosmonauts ⟨*Voyager is an ~ spacecraft*⟩ – compare AUTOMATIC

UOSAT-2 [*University of Surrey Satellite*] a 50kg Earth-orbiting satellite that was launched 'pick-a-pack' with Landsat 5 in March 1985, and that can be used by amateur radio enthusiasts. Like UOSAT-1 launched in 1981, UOSAT-2 transmits information in English by means of a voice synthesizer. The readings of instruments and switches on board the satellite give data on solar radiation, temperature on board the satellite, and so on. The satellite carries a digital communications experiment which allows amateur radio users to access a random-access memory for messages left by other users. It also includes a camera using a charge-coupled device to obtain images of the Earth. □ see LANDSAT, SPEECH SYNTHESIS, TELEMETRY, COMMUNICATIONS SATELLITE, CHARGE-COUPLED DEVICE

uplink – see EARTH STATION

upper atmosphere the outer layers of the Earth's atmosphere, ie above about 30km, including part of the stratosphere, the mesosphere, the thermosphere, and the exosphere □ see EARTH'S ATMOSPHERE, UPPER ATMOSPHERE RESEARCH SATELLITE, SOUNDING ROCKET

Upper Atmosphere Research Satellite (abbr **UARS**) a new class of satellite to be placed in a 600km high orbit by the Space Shuttle in the Autumn of 1989, to measure temperature, pressure, atmospheric composition, and wind as functions of height, latitude, longitude, and time. Although the Earth's atmosphere has been studied extensively, in recent years there has been concern about its continuous existence in a form that sustains life. From being regarded as an essentially infinite and imperturbable medium, it is now regarded as a limited and finite resource. Indeed, since many Earth-like planets (eg Venus and Mars) have extremely hostile atmospheres, there is concern that our own may not be all that stable. Recent extreme changes in weather (eg droughts in Africa and severe winters in North America) may signal a significant change in climatic conditions. UARS will monitor the atmosphere above the troposphere (about 10km above the Earth's surface) which is particularly sensitive to changes caused by manmade pollution and variations in solar radiation. UARS will carry highly sensitive instruments to make a wide range of measurements. These include: ISAMS (Improved Stratospheric and Mesospheric Sounder), a sensitive spectrometer for measuring temperature and the distribution of minor constituents of the atmosphere; WINTERS (Wind and Temperature Remote Sensing) for measuring wnd speeds and temperature; and SOLSTICE (Solar-Stellar Irradiance Comparison Experiment) for measuring the amount of solar ultraviolet radiation reaching the atmosphere. When information begins to flow from UARS in late 1989, it will represent the state-of-the-art in computing and space technology: the enormous amount of data coming from the measurements will be analysed on a single central computer of great power and capacity with experimenters worldwide being connected directly by high-speed data lines through computers of their own. UARS is expected to provide new knowledge about our atmosphere and its response to changes which is certainly of great concern to everyone on planet Earth. □ see EARTH'S ATMOSPHERE, SPECTROMETER, ULTRAVIOLET SPECTROMETER, EARTH RADIATION BUDGET SATELLITE, SUPERCOMPUTER

upper stage the last stage (eg the third) to fire in a multistage rocket □ see MULTISTAGE ROCKET

upweight the weight at launch of a rocket's payload and all the supporting equipment it requires □ see PAYLOAD

Uranus the seventh planet of the Solar System, which has a diameter of about 52 000km, orbits the Sun at about 2900 million kilometres, and has 5 large moons – Miranda, Ariel, Umbriel, Titania, and Oberon, and 10 smaller ones discovered by Voyager 2 in 1986. Like Jupiter, Saturn, and Neptune, Uranus has an atmosphere composed mainly of hydrogen and helium, and a solid core believed to consist of rock and metal surrounded by liquid and gaseous hydrogen. Uranus is so remote from Earth that little is known about many of its features, such as what the dynamics of its atmosphere are, whether it has a magnetic field, or what its period of rotation is. However, in March 1977 experimenters aboard NASA's Kuiper Airborne Observatory (a C-141 aircraft equipped with a 0.91m infrared telescope) detected rings around Uranus. This exciting discovery was obtained by observing the variation in the light from a star occulted by material in the rings. The flight was made from Perth, Western Australia. Confirmation of the existence of these rings, and a harvest of other scientific data, was received when the spacecraft Voyager 2 swept to within 82 000km of Uranus on 24 January 1986 en route for Neptune. At this distance radio signals from the spacecraft took nearly three hours to reach the Earth. □ see NEPTUNE, JUPITER, SATURN, VOYAGER, SOLAR SYSTEM, INTERNATIONAL ULTRAVIOLET EXPLORER

US Space Station a proposal, endorsed by President Reagan in January 1984, that the United States should build a large manned structure in Earth orbit by the 1990s. The orbit would be 450km high,

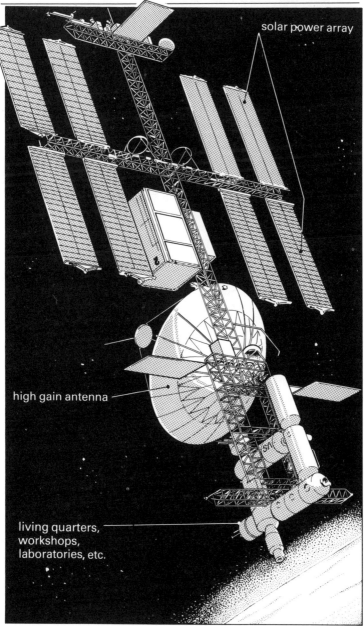

solar power array

high gain antenna

living quarters,
workshops,
laboratories, etc.

US Space Station concept

inclined at 28.5° to the equator, and be permanently manned by crews of six to eight astronauts who would be replaced every three months. The US Space Station is seen as a logical extension of Skylab and the Space Shuttle programmes, though there is strong opposition to the proposal on the grounds of its high cost (20 thousand million dollars by AD 2000), delays it would cause in the Space Shuttle programme, and the cautious response from potential users in industry. Critics argue that the Space Station is being approved on US domestic political grounds rather than on its merits. However, European, Japanese, and Canadian involvement is expected, especially as the modular design envisaged for the Space Station makes use of the Spacelab-type module which has been built in Europe for the Space Shuttle payload bay. Thus a modular structure whose component parts are limited by what the Space Shuttle can carry into space, is the most likely configuration for the Space Station. Spacelab modules carried into orbit by the Space Shuttle, will be attached to a framework and provide astronauts with comfortable living quarters: there are no plans to rotate any part of the Space Station to provide artificial gravity. There are many justifications for the US Space Station, which include research and development in life sciences, Earth sciences, telecommunications, pharmaceutical manufacturing, materials processing, and maintenance and repair. Both pharmaceutical manufacturing (the production of medicines) and maintenance and repair (of artificial satellites) are regarded as very useful roles for the US Space Station. In March 1985, NASA awarded parallel contracts to the aerospace companies Boeing and Martin-Marietta for the definition and preliminary design of the pressurized modules for the US Space Station which will be used as laboratories, living areas, and logistic transport. □ see SPACE STATION, SPACE SHUTTLE, SPACELAB, SKYLAB, SALYUT, SPACE SETTLEMENT, COLUMBUS, ORBITAL MANOEUVRING VEHICLE, SPACESHIP, SPACE MEDICINE

V-2 the rocket-powered bomb developed in Germany under the leadership of von Braun, which was first launched against London in 1944, but was too late to win the war for Hitler. The 'V' stands for 'vergeltung' meaning 'vengeance'. In all, 4300 V-2s were launched during the War, and of these, 1230 hit London. Von Braun's V-2s killed 2511 English people and seriously wounded 5869 others. □ see BRAUN

VAB – see VEHICLE ASSEMBLY BUILDING

vacuum a space from which all air and other substance has been removed, or in which nothing exists. Such a perfect vacuum cannot be made and does not occur, even in interplanetery and intergalactic space where cosmic dust, neutral or ionized atoms, and molecules may exist in small amounts □ see VACUUM CHAMBER, VACUUM PUMP

vacuum chamber a (large) hermetically sealed container or room from which air can be pumped to create a vacuum and in which devices and spacecraft designed to work in space can be tested. The largest (thermal) vacuum chamber in the US has a 15m diameter door and can test 15 tonne spacecraft in a chamber 4.6m wide and 10.7m high. It has eight independent thermal zones capable of producing temperatures in the range −190C to +120C, and was built to test Space Shuttle payloads. A space-type vacuum is created using pumps operating at cryogenic (super cold) temperatures which freeze gas molecules onto their surfaces so that they can be removed. □ see ENVIRONMENTAL TESTING

vacuum pump a pump for producing a vacuum by extracting gases from a chamber

Van Allen zones two regions in the upper atmosphere in which ionized particles spiral back and forth between the poles of the Earth's magnetic field. The lower region extends from 1000 to 5000km above the equator and contains protons and electrons captured from the solar wind. The upper region lies between 15 000 and 25 000km above the equator and contains mostly electrons. The regions were discovered by James Van Allen during analysis of results from the satellites Explorers 1, 3, and 4 launched from Cape Canaveral in 1959. □ see MAGNETOSPHERE, EXPLORER 1, ACTIVE MAGNETOSPHERIC PARTICLE EXPLORER, EARTH'S ATMOSPHERE

Vandenberg Air Force Base a launch pad on the California coast, USA, which has been developed for launching the Space Shuttle into polar orbits. The first Space Shuttle to be launched from this site was *Discovery*, which was placed into a 560km orbit inclined at 98.2° to the equator. Shuttles are prepared for launch by the reverse of the process used at Cape Canaveral: at the Cape, the Shuttle is assembled in a building then rolled out on the launch pad; at Vandenberg, the Shuttle is assembled on the pad and the buildings are moved back on rails. □ see CAPE CANAVERAL, SPACE SHUTTLE

Vanguard a series of three small Earth satellites launched in 1958 as part of the US programme for the International Geophysical Year (1957–58). Vanguard 1 was the first satellite to have a radio transmitter powered by solar cells. Vanguard tracked the Earth over a period of six years, and, from slight deviations in its orbit, showed that the Earth is slightly pear-shaped with the stem towards the north pole. It also enabled scientists to make the first long-term measurements of changes in atmospheric height and density. □ see EXPLORER 1, VAN ALLEN ZONES

vapour a gas which is capable of being made into a liquid or solid by pressure alone ⟨*water* ~ *in the atmosphere*⟩

vapour pressure the pressure exerted by a vapour that is in equilibrium with its liquid or solid form ⟨*the* ~ *exerted by liquid hydrogen*⟩

vapour trail – see CONDENSATION TRAIL

variable star a star that brightens and dims with time. Many thousands of variable stars have been recorded: some show changes of brightness over hours, others over years, and the range from maximum to minimum brightness is also very wide. Variable stars are classified as eclipsing, pulsating, or cataclysmic variables. Eclipsing variables are systems of two or more stars rotating about each other causing changes in brightness as they periodically eclipse each other. The star Algol, the second brightest star in the constellation Perseus, is an eclipsing binary system, and varies in magnitude from 2.2 to 3.47 over a period of 68.82 hours. Pulsating variables periodically brighten and dim as their outer layers expand and contract, cooling and heating up in the process. The fourth brightest star, Delta Cepheus, in the constellation Cepheus, is of this type, and similar stars are known as Cepheid variables. Cataclysmic variables include novae and flare stars (eg T Tauri stars). □ see T TAURI STARS, CEPHEID VARIABLE, NOVA, MAGNITUDE

VDT – see VDU

VDU [*visual display unit*] *also* **VDT** [*visual display terminal*] an input/output device having a screen and a keyboard to enable people to communicate with a computer. Most VDUs use a cathode-ray tube as the screen, although flat-screen displays, such as the liquid crystal display, are becoming more common. VDUs are widely used in all aspects of aerospace: they provide information to astronauts aboard the Space Shuttle and other manned spacecraft, and they are used in Earth stations as part of the global TV network and to receive pictures from interplanetary spacecraft. □ see LIGHT-EMITTING DIODE, LIQUID CRYSTAL DISPLAY

Vega either of two Soviet spacecraft launched on 15 and 21 December 1984 from the Tyuratam cosmodrome to study Venus's atmosphere and Halley's comet. These spacecraft arrived at Venus on 11 and 15 June 1985, and Halley's comet on 6 March 1986. The encounter with Venus was planned for two reasons: each spacecraft dropped an instrumented balloon into Venus's atmosphere and a lander to reach the surface of the planet; and each obtained a gravity-assist from Venus's gravitational field to speed it on its way to Halley's comet. For 46 hours, instruments attached to the 4m-diameter helium-filled balloons transmitted radio signals which were tracked by three 64m-diameter antennae of NASA's Deep Space Network, and other antennae worldwide. These antennae, and additional data from the Soviet internal network of radio telescopes, enabled the wind speed to be found in Venus's atmosphere to an accuracy of about 4kmh^{-1} at a distance of 108 million kilometres from Earth! These accurate measurements were made possible by a technique known as very long baseline interferometry (VLBI). The instruments also measured the frequency of lightning flashes and the temperature and pressure in the atmosphere. Vega-1 passed within 10 000km and Vega-2 5000km of Halley's comet. The spacecraft were travelling very fast (80kms^{-1}) at the encounter with the comet so they had some protection against damage by dust. France, the USA, and the USSR have cooperated in monitoring the progress of the Vega spacecraft and shared in the results obtained. The Vega spacecraft are called both 'Vega' and 'Veha', the latter strictly being their name in English, for it is made up of the first two letters of Venus and Halley. However, in Russian the 'h' sound is represented by the letter 'g'. Therefore 'Halley' becomes 'Galley', and 'Veha' becomes 'Vega' (not to be confused with the star, Alpha Lyrae, in the constellation Lyra). □ see VENUS, PLANET-A, [1]GIOTTO, HALLEY'S COMET, GRAVITY ASSIST, VENERA, PIONEER VENUS, PROTON

vehicle any aerospace machine (eg a rocket or spacecraft) which carries a payload through the atmosphere and/or space □ see LAUNCH VEHICLE

Vehicle Assembly Building (abbr **VAB**) the very large building at the Kennedy Space Centre in which the Space Shuttle is prepared for launch. The height of the VAB is 160m and it covers an area of 3.3 hectares and has doors 139m high. The VAB was originally used for the vertical assembly of the Saturn launch vehicles used in the Apollo, Skylab, and Apollo-Soyuz programmes. The VAB is also used for the refurbishment of the solid rocket boosters (SRBs) which are then mated, together with the external tank (ET), to the Space Shuttle before being transported by the mobile launch platform to the launch pad. □ see SPACE SHUTTLE, SOLID ROCKET BOOSTER, EXTERNAL TANK, MOBILE LAUNCH PLATFORM

velocity the rate at which something moves in a given direction ⟨*missile* ~⟩ □ see SPEED 1

Venera a successful series of Soviet spacecraft to cloud-covered Venus, beginning with Venera 1 launched in February 1961 and continuing to the present day. The early Venera failed to return data, but Veneras 4 and 8 ejected capsules into Venus's atmosphere and measured atmospheric temperature, pressure, and composition. Veneras 4, 5, and 6 were crushed by Venus's tremendous atmospheric pressure before they reached the surface. But Venera 7 returned data from the surface for 23 minutes on 15 December 1970, and Venera 8 for 50 minutes on 22 July 1972. Surface temperatures as high as 475C, pressures as high as 90 atmospheres, and an atmosphere of 98% carbon dioxide were recorded. Veneras 9 and 10 each comprised an orbiter module to relay signals from a lander. They both sent back the first pictures from the surface of Venus: an old eroded landscape was revealed. Veneras 11 and 12 were launched in 1978 and transmitted data from the surface. No lander survived for more than 110 minutes. Veneras 15 and 16 were launched in early June 1983. Both entered orbits of 1000 by 65 000km, period 24 hours, and began mapping the planet on 11 November 1983 using an on-board radar which was capable of 2km resolution. Venera 15 finished its radar mapping on 10 July 1984, but Venera 16 was still responding to signals from Earth a year later. On-board infrared cameras showed several hot spots on the surface of Venus, supporting the view that there are active volcanoes on the planet. □ see VENUS, PIONEER VENUS, VEGA

venturi a short tube having a smaller diameter in the middle than at the ends, that is used for measuring the rate of flow of gases and liquids (fluids) through a pipe. The venturi causes a reduction in pressure of the fluids which is a measure of the rate of flow.

vent valve a pressure release valve in a pipe line, which is operated externally – compare RELIEF VALVE

Venus the second planet of the Solar System, which has a diameter of 12 400km (about the same as the Earth's), orbits the Sun at about 67 million kilometres, and does not have any moons. Unlike Mars and Earth, Venus has an extremely dense atmosphere, 98% of which is carbon dioxide. It is completely surrounded by clouds which never allow astronomers a view of its surface. These very reflective clouds help to make Venus such a conspicuous 'evening star' or 'morning star' close to the Sun in the twilight sky. Radar observations from Earth showed that Venus has a retrograde motion on its axis so that its 'day' is about 117 days long. Observations from the spacecraft Mariner 10 showed that the atmospheric winds are very strong, reaching 350kmh[-1]. The atmosphere is extremely corrosive: Pioneer Venus which is in orbit round Venus, has shown that the top layer contains droplets of sulphuric acid. Lower down there are large particles of sulphur. The temperature on Venus increases from about 0C at the cloud tops to about 450C on the surface, where the atmospheric pressure is about 90 times that of the Earth's. It is no wonder that some of the Soviet Venera spacecraft which soft-landed on Venus transmitted for only a short period of time before being silenced forever. Continuous measurements from Pioneer Venus have shown that the amount of sulphur in the atmosphere of Venus increased by more than 50 times in 1978 and has been declining ever since. This suggests that a massive volcanic eruption has occurred on Venus. Measurements by the Venera and Pioneer Venus spacecraft indicate enormous volcanoes on two mountain ranges, Beta Regio and Atla

Regio: there is evidence of outpourings of young rocks (lava) in these areas. Venus is being intensively studied by the current Soviet Venera spacecraft, and the US Pioneer Venus. Two more Venera spacecraft are on their way to Venus, and in 1987 a US spacecraft called Venus Radar Mapper will be launched to map the surface from orbit. □ see VENERA, PIONEER VENUS, VENUS RADAR MAPPER, VEGA

Venus Radar Mapper (abbr **VRM**) a 1030kg US spacecraft to be launched by the Space Shuttle in April 1987, arriving at Venus in July 1988. There it will be placed in a near-polar elliptical orbit of 250km by 7800km, and will use on-board radar to map at least 70% of the surface of Venus, with a resolution of between 250m and 700m. The radar will be switched on for just over 36 minutes when VRM is at periapsis (closest distance to Venus) and the data representing images of Venus's surface recorded on two tape recorders. Then for about 111 minutes, the data collected by the radar will be transmitted to Earth at the rate of 268.8 thousand bits per second (kbs^{-1}). Since the same 4.7m diameter antenna on board VRM is used for data collection and for data transmission to Earth, the whole of the spacecraft will have to be rotated between each imaging session to get its antenna pointing towards the Earth. The mission is expected to end in April 1989 after mapping the entire surface of Venus. □ see VENUS, VENERA, PIONEER VENUS, POLAR ORBIT, DATA RATE

verniers – see THRUSTERS

vertigo a sensation of moving in one's surroundings (subjective vertigo) or of movement of the surroundings (objective vertigo) though no relative motion takes place. Inadequately trained astronauts and cosmonauts can experience space sickness through the effects of vertigo. □ see SPACE SICKNESS

very high frequency – see VHF

Very Large Array (abbr **VLA**) the world's largest radio telescope comprising 27 movable dish aerials located along three linear tracks in the form of a giant 'Y', each arm extending 19 to 21km over the New Mexico desert, USA. Each dish measures 25m in diameter and has a mass of 200 tonnes. Radio waves from, say, a quasar encounter these dishes with different phases and amplitudes due to the time delays between the arrivals of the waves at different aerials. For this reason, any two separated aerials act as an interferometer since the waves interfere with each other across the aerials. The array of 27 aerials instantaneously provide 351 independent interferometers of various lengths and orientations. As the Earth rotates, all these baselines connecting pairs of aerials intercept the radio waves over a curved path in space. By correctly adding together all the waves over a period of time, it is possible to produce an image of a radio source which shows as much detail as a single dish aerial equal in diameter to the array's longest baseline. This technique of mimicking a giant dish from smaller elements is known as aperture synthesis. About 75km of 60mm diameter waveguides entrenched in the desert, connect the aerials to the centre of the VLA where computers handle the prodigious amount of data collected. The first thing an astronomer sees as computers process the data, is an image on a VDU made up of 'contours' (like a contour map) in which (hopefully) different coloured peaks are seen indicating the presence of concentrations of radio energy from the cosmic object being studied. In this way, a wealth of information about the nature of the violent processes going on inside quasars and young (T Tauri) stars is being discovered. □ see ARRAY, RADIO ASTRONOMY, APERTURE SYNTHESIS, QUASAR, T TAURI STARS, QUASAT, ATOMIC TIME

very long base-line interferometry (abbr **VLBI**) – see QUASAT

veteran somebody who has been into space many times; an experienced astronaut ⟨ *Bob Crippen and John Young are* ~s *of the Space Age* ⟩ – compare ROOKIE □ SEE COMMANDER, YOUNG, CRIPPEN

VHF [*very high frequency*] (of or being radio waves having) a frequency in the range 30MHz to 300MHz. VHF is used mostly for high quality FM radio broadcasts and for television broadcasts. □ see FREQUENCY, UHF

vibrate to move to and fro; to shake, shudder, or quiver

vibration a (usu rapid) cyclic motion or shaking of a part or all of a body ⟨~ *of a launch vehicle at lift-off* ⟩ □ see RUMBLE, POGO

video of television and the processing and/or displaying of images on a VDU □ see VDU, VIDEO CAMERA

video camera *also* **television camera** a device that converts an optical image into electrical signals for recording and/or transmission. A lens system focusses an image of the scene to be recorded on a light-sensitive surface in the camera tube. In most cameras, but not in the charge-coupled device (CCD) camera, this image is scanned to convert it into a pattern of electrical (video) signals. Video cameras, and especially CCD cameras, are being used to record the image at the focus of ground-based telescopes and those aboard satellites and spacecraft. □ see CHARGE-COUPLED DEVICE, BETA PICTORIS, HUBBLE SPACE TELESCOPE, TELEPRESENCE

video signal the electrical signal produced in a television camera or received in a television receiver, which is related to the variation of light (and colour) in the image being televized. □ see VIDEO CAMERA

Viking either of two identical spacecraft launched from Earth in 1975 to search for life on Mars. Each spacecraft comprised an orbiter and a sterilized lander. After detaching from the orbiters, the 1000kg landers made a successful landing using a combination of parachute and retrorockets. Each lander carried two television cameras, a meteorological station, a seismometer, and a small laboratory with which to carry out analysis of soil samples picked up by a sampling arm. The TV cameras did not see any life forms, only red rocks and sand dunes under a pink and dusty sky. The seismometers picked up wind vibration and a few Martian earth tremors. The meteorological equipment measured temperatures between about −85C and −35C and mainly light winds gusting to 50km per hour. X-ray analysis of the soil samples revealed a high proportion of silicon and iron. Though most of the tests for biological compounds gave ambiguous results, one test did give a strong positive response. Whether lifeforms exist

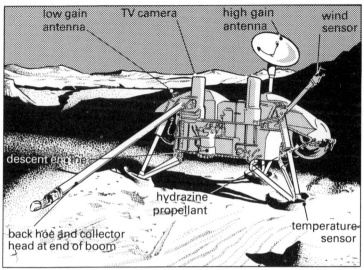

Viking lander

on Mars has yet to be proved and awaits further investigations, perhaps by astronauts or cosmonauts carrying out the experiments on the spot. □ see MARS, VOYAGER, CANALS, EXPEDITION TO MARS, PHOBOS AND DEIMOS

Virgo A an intense source of radio waves that is associated with the brightest galaxy, M87, in the Virgo cluster of galaxies. M87 is a giant elliptical galaxy which has a jet from which intense radio waves are generated by synchrotron radiation. The radiation has an intensity of 970 jansky. □ see RADIO ASTRONOMY, JANSKY, QUASAR

visual display terminal – see VDU

visual display unit – see VDU

VLA – see VERY LARGE ARRAY

VLBI [*very long baseline interferometry*] – see QUASAT

voice recognition – see SPEECH RECOGNITION

von Neumann machine one of the present generation of computers based on concepts introduced by John von Neumann in 1946. In a von Neumann machine instructions are stored and executed one after the other under the control of a central processor unit. □ see SUPERCOMPUTER, EXPERT SYSTEM, FLIGHT DATA SUBSYSTEM

Vostok a series of six one-man Soviet spacecraft that were launched in the early 1960s. Vostok 1 was launched on April 12 1961 and carried the first man, Yuri Gagarin, into space. □ see GAGARIN

Voyager either of two 825kg spacecraft launched towards Jupiter: Voyager 1 on 5 September 1977 and Voyager 2 on 20 August 1977.

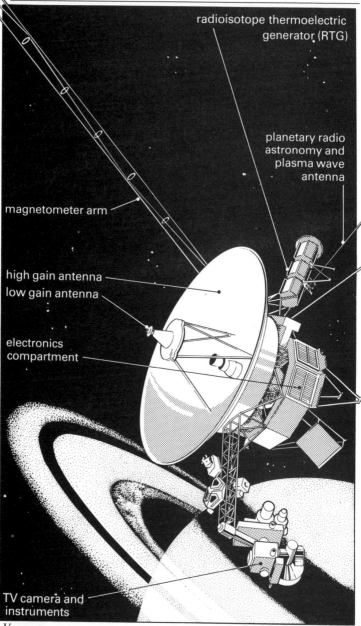

radioisotope thermoelectric
generator (RTG)

planetary radio
astronomy and
plasma wave
antenna

magnetometer arm

high gain antenna
low gain antenna

electronics
compartment

TV camera and
instruments

Voyager

Both Voyagers passed close to Jupiter's cloudtops and made important discoveries about its atmosphere, magnetosphere, and satellites. Using their TV cameras, the Voyagers sent back remarkable pictures including evidence that Jupiter, like Saturn, has a system of rings. Voyager 1 detected considerable volcanic activity on Io, one of Jupiter's satellites. During the flyby of Jupiter in 1979, Voyager 1 received a gravity-assist from Jupiter and was accelerated towards Saturn where it arrived in 1980. A considerable amount of new information about Saturn and its rings was obtained by both spacecraft. After Saturn, Voyager 1 left the ecliptic plane of the Solar System and will not meet another planet. Voyager 1 is travelling slightly faster than Voyager 2, about 20km per second, and has the distinction of being the fastest object made by man. By 1998 the rapidly moving Voyager 1 will have out-distanced Pioneer 10, now the most distant spacecraft, and will lead the fleet of two Voyagers and two Pioneers in their escape from the Solar System. Voyager 2 also received a gravity-assist from Jupiter and made a flyby of Saturn, but its gravity-assist targeted it towards Uranus (arrival January 1986) and Neptune where it will arrive in August 1989. Both spacecraft use their ultraviolet spectrometers to examine several celestial objects during their long journeys between the planets. During its epic journey, Voyager 2 has obtained about 32 000 images of Jupiter and Saturn with their rings and 13 major satellites. After leaving Saturn, Voyager 2 had 67kg of its original 104kg of propellant left to be used for attitude control and course correction. Its two radioisotope thermoelectric generators produce 422W of electrical power, which is 92W more than required. However, Voyager 2 is not as healthy as it was: it has only one working radio receiver to pick up signals from Earth, and the scan platform, on which the cameras and ultraviolet spectrometer are mounted, becomes stuck from time-to-time, especially after it has been moved rapidly. Engineers at NASA have extensively reprogrammed the computers on board Voyager 2 and neither problem is expected to seriously affect its future encounter with Neptune. To make sure that the scan platform did not become stuck when it was taking pictures close to Uranus, the computers were told to move the platform slowly. And the whole spacecraft was made to roll so that the cameras moved roughly the right way. When Voyager 2 reached Uranus in January 1986, it took over five hours to find out if a radio command sent from Earth had been received by the spacecraft. Should its one receiver have failed, so that commands could not be received from Earth, Voyager's computers already carried a simple program to tell the spacecraft what to do at Uranus. Both Voyager spacecraft carry a lot of information about life on Earth which might be of interest to intelligent beings elsewhere in the Galaxy. On a gold-plated record, with a cartridge, stylus, and instructions for playing it, is stored, in coded form, information about our genes and brains, and the sounds of human voices in 60 tongues, as well as the sounds of humpback whales. Photographs of people making things were included, as well as 90 minutes of music from all over the world. Though much of the information carried by the Voyagers is a compression of life on Earth in 1977 and may mean nothing to its recipients, the information about our genetic code might be meaningful to other intelligent beings who may be the product of a similar lengthy evolutionary development on their planet. □ see SATURN, URANUS, NEPTUNE, JET PROPULSION LABORATORY, FLIGHT DATA SUBSYSTEM, RADIOISOTOPE THERMOELECTRIC GENERATOR, GRAVITY ASSIST, DEEP SPACE NETWORK, ATTITUDE CONTROL, COURSE CORRECTION, DATA RATE, ULTRAVIOLET SPECTROMETER, SEARCH FOR EXTRATERRESTRIAL INTELLIGENCE, PHOBOS AND DEIMOS

VRM – see VENUS RADAR MAPPER

W the symbol for the unit of (electrical) power, the watt □ see WATT, JOULE, SI UNITS

warhead the part of a missile containing an explosive, chemical, or incendiary charge for damaging enemy targets – compare PAYLOAD □ see STAR WARS

watt (symbol **W**) the SI unit of power equal to 1 joule per second (Js^{-1}). The thermoelectric power generator on board the spacecraft Voyager 2 delivers 422W of electrical power. □ see SI UNITS, POWER 2, JOULE

wattage the amount of power, measured in watts, generated or absorbed by a device ⟨*the ~ of a solar panel might be 5kW*⟩

wave a regular variation in light, sound, voltage, pressure, or other property, which carries energy from one place to another ⟨*radio ~s from a pulsar*⟩ □ see ELECTROMAGNETIC RADIATION, SPECTRUM

waveband *also* **band** a range of radio or other frequencies ⟨*the ~ of the visible spectrum*⟩ □ see BANDWIDTH, SPECTRUM

waveform the shape of a signal (eg radar signals reflected from Venus) displayed on a cathode-ray tube or other recording device □ see PULSE

waveguide a hollow metal tube along which radio waves can travel without too much attenuation. High frequency radio waves (microwaves) can travel along a waveguide with little reduction in strength compared with coaxial cables; waveguides are therefore used as feeders in radio telescopes, to transfer signals to and from the aerial. In a more general sense, layers of plasma in the Earth's radiation belts act as waveguides. □ see RADIO ASTRONOMY, MICROWAVES, MASER, MAGNETOSPHERE, PLASMA

wavelength the distance between one point on a wave and the next corresponding point (eg from crest to crest). Wavelength is related to the frequency and speed of the wave by the simple equation: speed = frequency × wavelength. Thus a radio wave of wavelength 300 metres (medium waves) travelling through space at 300 million metres per second has a frequency of 1 million hertz (1MHz). □ see ELECTROMAGNETIC RADIATION, CARRIER WAVE

Waves in Space Plasma (abbr **WISP**) – see PLASMA

weak interaction one of the four general classes of forces between elementary particles. Weak interactions are much weaker than electromagnetic or strong interactions, though very much stronger than gravitational forces (the weakest of the four). Weak interactions are responsible for the relatively slow decay of particles like the neutron and muon, and for all reactions involving neutrinos. It is now generally believed that these four forces of nature can be explained by a general unified theory (GUT) of forces. □ see GRAVITY, GENERAL UNIFIED THEORY, RELATIVITY, MESON,

NEUTRINO, STRONG INTERACTION

weather satellite an Earth-orbiting satellite which makes measurements of atmospheric conditions so that more accurate weather forecasts can be made. The information obtained is transmitted to an Earth station as TV pictures and data. A weather station operates in polar orbit or geostationary orbit. In polar orbit, the satellite passes over the North and South poles enabling a complete coverage of the weather patterns to be obtained every 24 hours. In geostationary orbit, a continuous coverage of one large area is obtained. One important item of equipment aboard a weather satellite is a radiometer which scans the atmosphere to build up a picture of cloud cover and temperature at different levels in the atmosphere, including the temperature of the sea. By measuring the radiation emitted by the atmosphere in the infrared region, cloud pictures at night can be obtained. Information from weather satellites is available not only to forecasters but also to ships at sea, aircraft in flight, and to radio amateurs. □ see NATIONAL OCEANIC AND ATMOSPHERIC ADMINISTRATION, NOAA-9, REMOTE SENSING, COMMUNICATIONS SATELLITES, INFRARED, RADIOMETER, POLAR ORBIT

weight the force with which a planet or satellite attracts another (usu smaller) body to its centre of mass. The weight of a body of mass m, is equal to mg, where g is the local gravitational acceleration. □ see FORCE, ZERO-G, ARTIFICIAL GRAVITY, PAYLOAD, THRUST, NEWTON'S LAW OF GRAVITATION

weightlessness – see ZERO-G, MICROGRAVITY

Wells, H G (1866–1946) English writer whose books included science fiction, such as *The Time Machine* (1895), *War of the Worlds* (1898), and *The First Men in the Moon* (1901). These books anticipated discoveries and events of the 20th century. Thus time travel (though not yet a fact!) is embodied in Einstein's special theory of relativity; the 'heat rays', used by the Martian invaders who/which ravaged the English countryside in *War of the Worlds* are a fact, for they are the laser beam weapons being developed for the 'Star Wars' defence programme; and the first men have indeed travelled to the Moon. □ see RELATIVITY, LASER, APOLLO, CLARKE

Westar – see SPACE INSURANCE

White, Ed (1930–67) the first American astronaut to spacewalk from Gemini 4 in June 1965. Tragically, White and two fellow astronauts, Gus Grissom and Roger Chaffee, were burnt to death in the oxygen-rich atmosphere of their Apollo spacecraft just before they were to be launched from Cape Canaveral to make the first Apollo test flight. There has not been any other fatal accident to American astronauts since White lost his life. □ see GEMINI PROJECT, APOLLO, KOMAROV

white dwarf a star which has a mass less than 1.4 times that of the Sun and has undergone gravitational collapse. Stars more massive than this become very luminous supernovae. White dwarfs are very faint because, after collapsing, they have diameters less than 1% of the Sun's, but they are not always white and display a range of colours as they cool down, through white, yellow, and red before becoming cold black spheres. White dwarfs are the result of a star exhausting its nuclear fuel (mainly hydrogen) so that it produces much less heat. It contracts under its own gravity, electrons are stripped from their nuclei, and its density rises to more than 10 million kgm^{-3}. The light emitted by a white dwarf does not come from this central core but from a thin gaseous atmosphere which slowly leaks the heat of the star into space. There may be 10 thousand million white dwarfs in our Galaxy, the best known of which is Sirius B, companion to Sirius (Alpha Canis Major), which was discovered in 1844: it is smaller than the Earth yet has a mass equal to the Sun's. □ see STAR, SUPERNOVA, BLACK HOLE

white noise noise that sounds like a hiss and has no detectable frequency characteristic. On communications channels, white noise is caused by the random movement of electrons in wires and components, and special methods are used to reduce it (eg cooling the component). □ see THERMAL NOISE

Wide-Field Planetary Camera – see HUBBLE SPACE TELESCOPE

Wien's Displacement Law – see BLACK BODY

Wind and Temperature Remote Sensing (abbr WINTERS) – see UPPER ATMOSPHERE RESEARCH SATELLITE

window 1 any of a number of openings in the Earth's atmosphere through which light and radio waves from outer space can penetrate and reach the ground ⟨*the ~ used by radio astronomers*⟩ □ see RADIO ASTRONOMY, OPTICAL ASTRONOMY **2** an interval of time within which an event must occur if it is to have a successful outcome ⟨*the launch ~ for a flight of the Space Shuttle*⟩ **3** a small part of the display on a VDU selected for special effects or information □ see VDU

wind tunnel a tubular structure through which air flows at a known speed and which is used to determine the effects of air flowing past objects placed in it. Wind tunnels are an indispensable aid to designers of aircraft, missiles, and spacecraft which are designed to travel for all or part of their journey through a planet's atmosphere. The pattern of the air flow past the object is photographed, and instruments measure the pressure distribution round it, to determine the way the object will behave in practice (eg when entering the atmosphere of Venus). High speed wind tunnels are classified as supersonic or hypersonic. □ see VENTURI, SUPERSONIC, HYPERSONIC

WINTERS [*Wind and Temperature Remote Sensing* – see UPPER ATMOSPHERE RESEARCH SATELLITE

¹wire a single metal strand or collection of strands along which electricity flows easily □ see CABLE

²wire to make an electrical connection between two or more components, or between two circuits, using electrically conducting wire

wiring diagram a drawing showing the connections between components in an electrical circuit ⟨*the ~ of a radio*⟩ □ see SYSTEM

WISP [*Waves in Space Plasma*] – see PLASMA

Woomera Test Range an area in South Australia, used in the

1950s for the launch and recovery of the British Blue Streak, Black Night, and Black Arrow high-altitude research rockets. Facilities at the range were jointly operated by the Australian Weapons Research Establishment and the Royal Aircraft Establishment. ◻ see BLACK NIGHT

workstation a self-contained unit where computing, assembly, communications, or other tasks are performed. A workstation may be equipped with a VDU and keyboard linked to printers, modems, telephones, etc, as in an office ◻ see CONSOLE, EARTH STATION, VDU

world ship a hypothetical spaceship that journeys between the stars in a rather leisurely fashion, and has a self-contained bioregenerative life support system, ie a living system (an ecosystem) that is sealed off from its surroundings, but may exchange energy with the rest of the universe. The Earth is an example of such an ecosystem, alternatively called a Controlled Ecological Life Support System (CELSS). Small-scale ecosystems which have survived for years have been studied in laboratories, and may provide ideas for future world ships; at least the studies may help us to maintain a healthy 'Spaceship Earth' – compare STARSHIP ◻ see SPACE GREENHOUSE, DAEDALUS, MICROSPACESHIP

X-band a range of radio waves having frequencies from 5.2GHz, to 10.9GHz used esp for high-frequency radar and radio communications with orbiting satellites and interplanetary spacecraft □ see FREQUENCY, GIGAHERTZ, TELEMETRY, SHORT WAVES, MEDIUM WAVES, LONG WAVES, MICROWAVES, ELECTROMAGNETIC RADIATION

XMM – see X-RAY MULTI-MIRROR TELESCOPE

X-ray astronomy the study of celestial objects from their X-ray emissions in the wavelength band 10nm to 0.001nm. Since the Earth's atmosphere strongly absorbs X rays (fortunately), observations of X-ray sources has to take place at altitudes greater than 150km. Before the 1970s, detectors sent aloft in high altitude rockets and balloons detected about 25 sources of X rays within the Galaxy. The first of these sources, Scorpius X-1, was discovered by a sounding rocket in 1962, in the constellation Scorpius. Soon afterwards, a second source, Taurus X-1, was later identified with a supernova remnant in the Crab Nebula. During this period, one extragalactic source of X rays was identified and associated with the powerful radio galaxy Virgo A (M81). In the 1970s, artificial satellites in Earth orbit continued to find sources of X rays. The UK satellites Uhuru (also known as SAS-1) and Ariel V were launched from the San Marco platform off the Kenyan coast in 1970 and 1974, respectively, and produced the first detailed map of X-ray sources. These sources included X-ray binaries, clusters of galaxies, and Seyfert galaxies. In the second half of the 1970s, the satellites HEAO-1, SAS-3, and HEAO-2 (also known as Einstein) were launched. HEAD-2 carried a grazing incidence X-ray telescope which produced a resolution comparable with optical telescopes and provided accurate positions of many X-ray sources. Many of the sources emit 100 to 100 000 times more energy in X rays than the power of the Sun. Others generate bursts of X rays, or produce X rays periodically. Many X-ray sources are associated with neutron stars and black holes, which produce powerful gravitational fields. Gas from a nearby nebula, or a normal companion star, accelerates towards the more massive neutron star and becomes super-hot so that it emits X rays. In 1983, Exosat, the ESA's first X-ray satellite was launched to measure the intensity of X-ray sources at various wavelengths, using spectrometers. X-ray observations using telescopes are costly, since the mirrors require polishing to a very high surface accuracy. Therefore it is cheaper to make an array of telescopes which compromises angular resolution with increased sensitivity. The proposed X-ray Multi-Mirror (XMM) telescope will carry such a telescope comprising a cluster of mirrors each about 500mm in diameter. □ see NANOMETRE, X-RAYS, X-RAY MULTI-MIRROR TELESCOPE, ADVANCED X-RAY ASTROPHYSICS FACILITY, CRAB NEBULA, BLACK HOLE, NEUTRON STAR, WHITE

X-ray Multi-Mirror telescope

X-ray Multi-Mirror telescope (abbr **XMM** or **XMM telescope**) a sensitive X-ray telescope comprising an array of mirrors, which will be put into Earth orbit to look at the X-ray emissions from faint stars and the nuclei of distant galaxies □ see X-RAY ASTRONOMY, GRAZING INCIDENCE X-RAY TELESCOPE, EUROPEAN SPACE AGENCY, ADVANCED X-RAY ASTROPHYSICS FACILITY

X rays energetic and penetrating electromagnetic radiation having wavelengths lying between ultraviolet radiation and gamma rays. The energy of X rays ranges from about 100 electron volts (eV) to about 0.5MeV, which corresponds to a wavelength range of 12 nanometre (nm) to about 0.002nm. X rays are produced when electrons change speed, either by suddenly being stopped by matter, or by moving in a magnetic field. It is the latter process which gives rise to most of the X rays produced by celestial objects (eg quasars and pulsars) in space. □ see ELECTROMAGNETIC RADIATION, X-RAY ASTRONOMY, NANOMETRE

268

Yagi aerial – see DIPOLE AERIAL

yaw the side-to-side movement of the front end (and hence the rear end) of an aerospace vehicle – compare PITCH

Young, John (*b*1930) veteran US astronaut and commander of Apollo 16 who, with Charles Duke, landed the lunar module, *Orion*, in the Descartes Highlands of the Moon on 20 April 1972. He was also accompanied by Bob Crippen on the first Space Shuttle flight (STS-1) launched on 12 April 1981. □ see APOLLO, SPACE SHUTTLE, CRIPPEN

Z

zenith that point of the celestial sphere directly overhead – compare NADIR □ see CELESTIAL COORDINATES

¹zero having no value or size ⟨~ *gravity*⟩

²zero 1 to home-in on something **2** to set an instrument to a known starting condition

³zero the binary or decimal digit which represents a quantity of zero numerical value

zero-g *also* **weightlessness** the condition of having no weight. Contrary to popular belief, zero-g is not necessarily the condition of zero gravitational force, as would be the case at a great distance from any massive body. A parachutist experiences weightlessness when jumping from an aircraft before the parachute opens: like astronauts in an orbiting spacecraft, the parachutist is in free-fall and there are no forces opposing gravity to give the feeling of weight. Strictly, zero-g is impossible to achieve since all objects attract each other. So the term 'microgravity' better describes the small gravitational forces acting between objects in space stations. □ see MICROGRAVITY, FREE FALL, SKYLAB, SPACE STATION, SPACELAB, CENTRIFUGE

zodiacal light a diffuse glow seen in the west after dusk and in the east before dawn. The glow is most apparent in tropical latitudes on clear moonless nights. Spectroscopic analysis shows that the glow is due to sunlight scattered by interplanetary dust particles in a belt in the plane of the Solar System. □ see COSMIC DUST, SHUTTLE GLOW, SEEING

Zond a series of Soviet spacecraft launched in the 1960s to study Venus, Mars, the Moon, and interplanetary space. Zond 3 obtained the first high-quality pictures of the far side of the Moon in 1965. □ see VENERA, SOYUZ, SALYUT